LINKING UP THE ALPS

CRISTINA DEL BIAGGIO

LINKING UP THE ALPS

HOW NETWORKS OF LOCAL POLITICAL ACTORS BUILD THE PAN-ALPINE REGION

PETER LANG
Bern · Berlin · Bruxelles · Frankfurt am Main · New York · Oxford · Wien

Bibliographic information published by die Deutsche Nationalbibliothek
Die Deutsche Nationalbibliothek lists this publication in the Deutsche Nationalbibliografie; detailed bibliographic data is available on the Internet at ‹http://dnb.d-nb.de›.

British Library Cataloguing-in-Publication Data: A catalogue record for this book is available from The British Library, Great Britain.

Library of Congress Control Number: 2016934357

Published with the support of the Swiss National Science Foundation, the Geneva School of Social Sciences (former Faculty of Economics and Social Sciences) of the University of Geneva and the Foundation Schmidheiny.

Cover illustration: © DYNALP 2/Alliance dans les Alpes.

ISBN 978-3-0343-1630-9 pb. ISBN 978-3-0352-0338-7 eBook

This publication has been peer reviewed.

Hochfeldstrasse 32, CH-3012 Bern, Switzerland
info@peterlang.com, www.peterlang.com

Contents

La montagna può essere un insopportabile
incubatoio di faide, invidie e chiusure,
ma può essere anche il perfetto luogo-rifugio
di uomini straordinari, gente capace di opporsi
all'insensata monocultura del mondo contemporaneo.

[Paolo Rumiz, *La leggenda dei monti naviganti*, 2007]

While discovering the Alps for my PhD research, directing my eyes in the direction of the peaks, and exchanging with mayors of Alpine towns and protected areas rangers, my thoughts went to a book written by the Italian journalist and writer Paolo Rumiz. In the first part of his book *La leggenda dei monti naviganti (The legend of the sailing mountain)* he tells of his journey across the Alps. In the quote above, he perfectly renders the feelings I experienced towards the Alps. In his words, mountains can be, on the one hand, the place of an "unbearable hatchery of feuds, envies and closures"[1], and on the other, "the perfect place of refuge of extraordinary women and men, people capable of standing out against the absurd monoculture of the contemporary world"[2] (Rumiz, 2007).

If, as a young inhabitant of the Alps, I felt the weight of the mountain dwellers' closed-mindedness, I have discovered, by writing this book, another face of the people living in the mountains. It is with the people Rumiz describes as "extraordinary" that I had the fortune to work. Thanks to the time they devoted to me, I had the opportunity to explore questions that are specific to the Alpine context, but which also go well beyond it. The people I met could answer the main question I asked myself during my wandering in the Alps: "How do regions emerge in the contemporary world?" Moreover, they helped me to see the Alps as I had never seen them before. I am sincerely grateful to all of them.

1 Own translation from Italian.
2 Own translation from Italian.

1. Introduction

The natural and socio-cultural Alpine territory is being threatened by increasing motorized traffic, the ageing of its population, the national fragmentation of the politics on its territory, and concerns linked to climate change. The general issue that this book is to understand the main social and political responses proposed by local political actors to solve these important social and ecological challenges. Already in the 1950s, some particularly sensitive local politicians and managers of protected areas alerted national authorities to the social and ecological crises that were threatening the Alpine region.

Forty years later, one solution recommended was the constitution of networks of local political actors implementing at the local level the Convention on the Protection of the Alps, usually called Alpine Convention. Local political actors view this international treaty as an appropriate tool to respond to the challenges facing the Alps. By imposing the concept of sustainable development as the solution for assuring the survival of the Alps, the convention can be considered a model for other mountain regions facing similar problems. Signed by the eight Alpine countries (Austria, France, Germany, Italy, Liechtenstein, Monaco, Slovenia and Switzerland) and the European Union, it entered into force in March 1995. The aim of the convention, according to its second article, is to "pursue a comprehensive policy for the preservation and protection of the Alps by applying the principles of prevention, payment by the polluter (the 'polluter pays' principle) and cooperation".[3] Additionally, through this convention, the contracting parties hope to intensify trans-border cooperation inside the Alpine perimeter defined by the treaty.

To cover this wide scope, a considerable number of transnational networks of local political actors, connecting local representatives, researchers, managers of protected areas and ecological associations,

3 <www.alpconv.org/en/convention/framework/default.html>, consulted on the 01.12.2014.

were created. Their objective is to harmonize ecological balance, economic stability and social progress in mountain regions worldwide.[4] The networks have often been created on the initiative of the non-governmental organization *Commission Internationale pour la Protection des Alpes* (CIPRA), whose *raison d'être* is to "play a pioneering role in contributing to solving current problems and upcoming challenges aimed at bringing about a sustainable and ecological future for the Alpine region".[5]

Although the phenomenon of the constitution of networks of political actors is quite significant in terms of persons involved, money invested and projects undertaken, no in-depth scientific research has previously been conducted on this specific topic. This book endeavours to understand how and why local political actors, organized in pan-Alpine networks, chose to take mountain regions in general, and the Alps in particular, as the shared frame of reference for their involvement.

The discussion here will focus on two phenomena:

- Firstly, it explores if and how a pan-Alpine identity, detached from and/or combined with the more "traditional" national identities, is growing among and performed by local political actors on the basis of the Alpine Convention;
- Secondly, it analyses the contents and socio-political significance of the involvement of local political actors in the newly constituted pan-Alpine networks and region.

I will concentrate on three of the networks of local political actors formed in the Alps: Alliance in the Alps, Alparc and Alpine Pearls. These three networks have at least two characteristics in common: They are all willing to implement, albeit with different instruments, sustainable development at local level and they all recognize the Alpine Convention's map as the geographical limits of their actions. They are nevertheless different with regard to the kind of members they connect: Alparc ties together more than 900 large protected ar-

---

4 See the CIPRA webpage: <www.cipra.org/en/cipra/international/networks>, consulted on the 01.12.2014.

5 <www.cipra.org/en/cipra>, consulted on the 01.12.2014.

eas (national parks, regional and/or nature parks, nature reserves, biosphere reserves and natural world heritage sites); Alliance in the Alps currently links more than 300 communities (villages, cities and regions); and Alpine Pearls associates 27 "of the most gorgeous holiday regions in the Alps", according to the network's advertising material. Alparc and Alliance in the Alps are the oldest (Alliance in the Alps was established in 1994, Alparc in 1997) and most active networks. The third, Alpine Pearls, is independent from CIPRA (as is Alparc) and was established by 17 villages wishing to create innovative and sustainable, environmentally friendly tourism packages.

1.1 Thematic focuses

The territorial result of the commitment of local political actors for the pan-Alpine idea, notably via the constitution of networks of local political actors, is sufficiently distinctive to see in it a "nascent region", a region with a distinct territorial framework and *ad hoc* institutions created for working at that scale. In fact, in recent years, the Alps have been increasingly recognized as a region in their own right within Europe, namely through the signing of the Alpine Convention. According to Balsiger, it represents a case of "ecoregional institutionalization" (Balsiger, 2007: 4), whereby the Alpine region is becoming institutionalized on the basis of an area defined by the Alpine Convention.

This initiative has been accompanied by a mobilization of inhabitants, local representatives, researchers, managers of protected areas, and ecological associations, which I call "local political actors". This mobilization has often taken the form of networks sharing a concern for ensuring that the objectives of the Alpine Convention materialize at the local level. Most networks promote sustainable development and use a horizontal mode of functioning, i.e. by giving social and political local actors the opportunity to exchange experiences and information beyond the boundaries of language and culture. Together, they ensure the Alps also function as a reference area for collective action.

CIPRA has played a decisive role in linking these two types of initiatives. In the first instance, it performed lobbying work to promote and support the drafting and the signature of the convention. In the second, it initiated projects for the implementation of the convention involving the base; notably, it helped with the establishment of pan-Alpine networks of local political actors. The Alpine region is thus taking shape and becoming institutionalized under these two initiatives, and it constitutes one of the trans-border regions that are contributing to changing the map of Europe (McNeill, 2004: 89).

The research presented in this book responds to Cuypers and Randier's suggestion that the Alpine Convention should not be studied solely from a legal perspective, but also by considering projects such as the creation of networks of local political actors that seek to implement it (Cuypers & Randier, 2009: 30). In this regard, the research contributes to what Pickles declares to be one of the principal aims of current geography, i.e. "to resituate human beings as actors in the production and reproduction of social life and places" (Pickles, in: Pudup, 1988: 373). The focus will be on those "actors that produce places" such as regions. The intention is not merely to show *that* regions are constructed, but above all to highlight *how* social configurations of actors and mechanisms of regional construction are taking place in the Alps.

In order to understand these regional construction processes, I have adopted the concept of "regional assemblage", coined by Allen and Cochrane (Allen & Cochrane, 2007). Regional assemblages are made possible by tangled, multi-scalar and multi-actor interactions. The Alpine regional assemblage, comprising a variety of social and political actors that operate at different territorial scales, constitutes a perfect research laboratory that allows to go beyond the "classical" dichotomous conceptions in regional studies, which Harrison brought up in a recent article (Harrison, 2012). In fact, as Harrison underscores, regional researchers normally see regions as territorially embedded *or* as relational and unbounded entities. As Harrison suggests, it is time to recognize that these two elements combine in regional assemblages in today's new regional world (Harrison, 2012). The focus of this research is precisely to show how these two elements (regions as territories and regions as networks) coexist in contemporary regional structures.

This book also aims to react to the criticism expressed by Bulkeley about the ways scientists study governance in general, and environmental governance in particular. Bulkeley levels her criticism at approaches that do not leave analytical space for arrangements that are "hybrid, involving a mix of state and non-state actors, or multi-scalar, including actors drawn from different levels of governance simultaneously, and provide little insight into how different forms of hierarchical and network governance interact and intersect to produce particular governance forms, mechanisms and outcomes" (Bulkeley, 2005: 881). This research analyses the hybrid configuration of Alpine governance and offers elements to measure its social and political significance.

Last, but not least, this research contributes to the debate on regional identity. By distinguishing two facets of identity (objectivation and identification), it aims to overcome the obstacle that many academics highlighted in the study of identity in the social sciences, i.e. that identity is an "indefinable" (Bray 2002: 14) and "ambiguous" (Brubaker & Cooper 2000: 2) "catchword" (Häkli, 2001: 115), unable to help in the comprehension of social and political processes such as regional building. Brubaker and Cooper go as far as declaring that, "if identity is everywhere, it is nowhere" (Brubaker & Cooper, 2000: 1). That is why I propose to look at identity by splitting the concept into two processes. On the one hand, *objectivation*, i.e. the identification of an object (or a family of objects) by a subject (or a collectivity) who thinks that this object is a pertinent category through which to understand the world or part of it (Debarbieux & Rudaz, 2010: 11); on the other hand, *identification*, i.e. the sense and will, shared by a considerable number of individuals, of belonging to the same group (Debarbieux, 2006: 342). Using this approach, I hope to reveal how (multiple?) Alpine identities are constructed.

At a more general level, following Murphy's reasoning, if one accepts that one project of contemporary geography should consist of focusing attention on the complexity and individuality of regions in describing and analysing social processes (Murphy, 1991: 25), it is clear that my book fits into this task. In fact, my intention is to investigate how local political actors argue for the specificity and individuality of the Alpine region, by analysing the latter through the social processes taking place there. As Gilbert argues, the analyses proposed by new regional geography[6] should not imply an exhaustive synthesis but rather underline particular aspects of society. As she affirms: "Synthesis is restricted to the interaction of the particular social processes seen as playing a leading role in the creation of the region" (Gilbert, 1988: 219). This is the favoured approach in the research: promoters, taking the leading role in the pan-Alpine region, are identified. Furthermore, I present some behaviours, ways of saying and doing, as well as processes of objectivation of and identification with the region.

1.2 Axes of analysis

This book is based on research that focuses on three axes of analysis at the intersection of two sub-fields within human geography: cultural (mainly chapter 5) and political (mainly chapters 6 and 7) geographies. Using its broader definition, cultural geography can be defined as the study of the

> [...] ways that meanings and social understandings are constructed, contested and negotiated, and in exploring the diverse ways these fuse and splinter around intersecting notions of culture, place and space. [...] It is a manner of linking ideas and imaginations with the material world. It explores how social groups engage with their landscapes, how people construct and make sense of their places and spaces (Aktinson *et al.*, 2005: 3).

It explores, among other things, the "ways [in which] people enact identity and belonging" (Aktinson, *et al.*, 2005: 3). Following Blacksell, political geography can be defined as the study of how

6 For details on "new regional geography", see Section 2.6: "The advent of new regional geography".

> […] individuals and interest groups […] are continually vying with each other to promote their own interests thereby destabilising and changing the existing order and remaking the world in their own image. It is about the forces that go to shape the world we inhabit and how they play themselves out in the landscape across the world" (Blacksell, 2006: 1).

Because it simultaneously covers issues linked to identity building and processes of regional institutionalization, the present research can thus be considered as situated at the intersection of these two sub-fields.

The aim of the study is to expose the different modes in which the Alpine region is objectivized, identified and institutionalized through a process of social spatialization, defined by Shields as the "ongoing social construction of the spatial at the level of the social imaginary (collective mythologies, presuppositions) as well as interventions in the landscape (for example, the built environment)" (Shields: 1992: 31). This concept leads to a "rubric under which currently separated objects will be brought together to demonstrate their inter-connectedness and coordinated nature" (Shields: 1992: 31). This corresponds exactly to the process going on in the contemporary Alps: Local political actors bring together separated objects, i.e. the "national Alps", in order to construct a pan-Alpine project. They demonstrate their inter-connectedness via the construction of regional networks, which create new issues in the social imaginary (the creation of a pan-Alpine identity) and intervene in the Alpine landscape (by promoting sustainable development). As Shields points out, the term "social spatialization" makes it possible to name an object of study, in this case the pan-Alpine region, which encompasses the cultural logic of the spatial and allows us to analyse concrete actions, constructions, and institutional arrangements (Shields, 1992: 31).

1.3 Argument structure

The book is organized following the double process of social spatialization that Shields highlighted, focusing on the one hand on the questions linked to identity (what she calls "social imaginary") and on the other on issues concerning "concrete actions", i.e. the regional institutionalization's processes. The book is structured as eight chapters, including an introduction and a conclusion.

The first three chapters explain the theoretical and methodological framework of the research. Chapter 2 highlights the theoretical perspective in which this research is conceptualized. Chapter 3 presents the research methodology. Chapter 4 offers an overview of the three pan-Alpine institutions that will be considered in this study: the Alpine Convention, the Alpine networks of local political actors under examination and CIPRA.

The last three chapters are more strongly empirical and are built upon the material I collected during field work. Chapter 5 considers how the pan-Alpine region is identified and objectivized by local political actors. These two concepts explain the modalities of the rise of Alpine identities; they show the space of reference and involvement of the Alpine actors, i.e. the perimeter that motivates and gives meaning to the actions they undertake. Chapters 6 shows how local political actors, who identify with the Alpine region, organize in order to create an "institutional space", called "regional assemblage" (Allen, Cochrane, 2007), which enables the realization of projects. Chapter 7 analyses the political significance of these institutional arrangements in order to judge the potentiality of the establishment of the Alpine region. These analyses reveal the mechanisms underlying the construction of the "regional institutional architecture" (Balsiger, 2007: 5) and the regional assemblage (Allen, Cochrane, 2007) mechanisms taking place in the Alps, resulting in a complex system of governance at the mountain range scale.

2. Theoretical framework

The first part of this chapter has two aims. Firstly, it situates the research within the contemporary debate of regional building by underlining the complex relations that exist between social actors and the environmental context in which they act. The region, as analysed here, is affecting and being affected by those relations. It follows Painter's idea that regions should be understood "as both resources for, and the outcomes of, human action" (Painter, 2008: 343). Secondly, it intends to help the reader place the research within the wider spectrum and traditions of regional geography. The overarching goal is thus to justify the choice of the theoretical perspective I favoured in the research, i.e. the perspective of what is called new regional geography. In order to understand the implications of this choice, I will present an overview of the theoretical perspectives preceding the advent of new regional geography.

In the second part, I will discuss some of the key concepts structuring the research. I will begin by clarifying and laying out the content and application of the concept of identity and justifying the choice to divide it into two terms, namely *objectivation* and *identification*. On one level, I assume that the pan-Alpine identity of local political actors is based on the common reference to a particular, objectivized regional setting, namely the Alps. On another level, the identification with this regional setting can be a source of action and projects. Furthermore, I will debate the possibilities of transforming this regional setting, which local political actors and institutions identify as "natural", into a political one, through a process of *eco-regionalization*. Subsequently, I examine the institutions and actors that have allowed this "new space of action" to become more important. This favourable institutional framework allows the establishment of pan-Alpine projects, provides content and promotes the subsistence to the Alpine regionalization's political process.

2.1 The actors of the "new spaces of action"

In Europe, eco-regionalization's processes (i.e. processes of regional building based on natural elements such as mountain ranges or river basins) have opened up, through different mechanisms, the realm of possibilities for the creation of new geographical entities. The latter are not necessarily modelled on pre-existing territorial entities.

Despite a growing number of "new regions" being created in Europe, administrative regions still play a central political role. By joining the Assembly of European Regions in the late 1980s, they clearly proved their relevance and that they are important partners in every territorial rearrangement that takes place in Europe. The objectives of the assembly are to:

- Promote the principle of subsidiarity and regional democracy;
- Increase the regions' political influence within the European institutions;
- Support the regions in the process of European enlargement and globalization;
- Facilitate interregional cooperation across the wider Europe and beyond.[7]

The last objective, the facilitation of interregional cooperation, clearly suggests that administrative regions want to play an important role in every attempt to create new trans-boundary regions. They are even the promoters of increased regional cooperation. For them this is an opportunity to strengthen their power and authority, as Leitner *et al.* underline (Leitner *et al.*, 2008: 294), who also point out that external relations for cities and regions have traditionally been the domain of national authorities. Recently, however, new actors, such as local and regional authorities, have extended their scope of action (Leitner *et al.*, 2008: 294) and increased their cooperation. With the same proactive attitude, national states are also more likely to cooperate with each other, notably via the signing of international conventions and

7 <www.aer.eu/en/what-is-aer/aer.html>, consulted on the 01.12.2014.

treaties.[8] The financial incentives of the European Union stimulated this new enthusiasm for cooperation and gave rise to potential new spaces of action (such as the transnational regions) both inside and outside the European perimeter.

All these factors contributed to the emergence, since the 1990s, of a number of "new" regions in Europe, which Deas and Lord tried to list in a 2006 article. The result appears on a map, "a striking diagrammatic representation" as Deas and Lord call it (Deas & Lord, 2006: 1854), where 146 "unusual" regions, defined as "territorially contiguous new regional coalitions and institutions" (Deas & Lord, 2006: 1848), are identified. The Alpine Space is one of the 146 initiatives listed and mapped by Deas and Lord. These regions emerged by transcending territorially bounded entities and "in line with the European Commission's efforts to stimulate new interregional, intercity and transnational collaborative initiatives in economic development and planning" (Deas & Lord, 2006: 1849). Many scholars provide more in-depth analysis of some specific regions.[9]

Among the initiatives promoted by the European Union to encourage territorial integration, Interreg programmes (now European Territorial Co-operation) play an important role. They were created in the late 1980s to promote economic and social cohesion in Europe, stimulate cooperation and support weak regions. They are financed by the European Regional Development Fund (ERDF) and aim to support trans-boundary regional activities. A fourth phase of the programmes was launched in 2007, for the period 2007–2013.[10]

McNeill highlights that, thanks to these European initiatives, some trans-border regions, including the Alps, contribute to the "shift from a one-dimensional map of Europe as having fixed borders to one

8 A non-exhaustive list of international conventions, treaties, agreements, declarations, etc. in the field of the environment can be found here: <http://cms2.unige.ch/isdd/spip.php?rubrique1#mot_7>, consulted on the 01.12.2014.

9 See for example articles by Scott for the Baltic Region (Scott, 2003), Rocha Medeiros for the Iberian and the Swedish-Norway border region (Rocha Medeiros, 2010), Fall for the Carpathians (Fall, 2007; Fall, 2009), Virtanen for Karelia and Pomerania (Virtanen, 2004) and Häkli for the Pyrenees (Häkli, 2004)

10 The Interreg I phase took place in the period 1989–1993. It continued as Interreg II for the period 1994–1999. A third period, 2000–2006, took the name of Interreg III.

in which city-based, regional, national, and European scales of action are fluid" (McNeill, 2004: 89). Thus, the scales of action are changing, becoming more fluid and leaving room for new initiatives promoted by different actors, for whom those areas become a space of reference and involvement. This reinforces the idea that regions should not be considered as a fixed territory but as dynamic spaces that are "constantly evolving" (Pudup, 2004: 12908) under the influence of social relations. The commitment of devoted persons is a *conditio sine qua non* to create regions. In fact, regions are created thanks to the goodwill of persons and groups who identify with them and give concrete expression to their ambitions and images of that particular perimeter by implementing projects and institutions. In some ways, those people and groups "do not make places, but [...] are places" (Thrift, 1991: 462).

2.2 The crisis of nested territories

The blooming of those "unusual" regions (Deas & Lord, 2006) occurs while the world is facing "re-scaled territorialities" (Brenner, in Amin, 2002: 387), a "proliferation of scales and scalar complexity" (Amin, 2002: 387), where "nation-states become less and less satisfactory as units of analysis" (Hannerz, 1996: 48), because societies and nation-states do not inevitably inhabit the same space (Rumford, 2006: 163). These spatial reconfigurations take place, and this is not a coincidence, while nation-states seem to lose power to new actors (Agnew, 2001: 6; Häkli, 2004). In other words, this occurs as nested territories are facing a crisis, even though "state governance is still the major context for region (and identity) building" (Paasi, 2002b: 803).

2.2.1 *The creation of new regional entities*

The ongoing regionalization process in the Alps is a good illustration of what is happening in many other places, especially in Europe, with the creation of new regional entities. It shows that what is normally understood as "regions", i.e. a given scale between the state and the local (Scott, 1988), is currently being challenged. Therefore, attention should be paid to the multiple processes and scales of social reproduction (Paasi, 2002b: 806) that are not necessarily taking place in the customary spatial categories.

In this context, nationality is becoming less and less satisfactory as a category of identification. Since "the world is moving towards a 'forced' individualization" (Paasi, 2003: 475),

> [...] people's lives are increasingly being left as their own responsibility, so that people shape their lives and environments through personal identities rather than through categorizations such as nationality, class, occupation or home region (Paasi, 2003: 475).

This is true for ordinary people but also for local political actors, elites, who are shaping the destiny of the places they live in. So, for example, mayors, who have political authority over the smallest entity of the administrative hierarchy, are increasingly keen to spread their relations and actions far beyond the municipality's boundaries.

The emergence of new political actors weakens the pertinence of what Debarbieux calls the "magical triangle" (Debarbieux 2006: 345). The "magical triangle" assumes the superimposition between one culture, one territorial entity and one identity. Accordingly, the new and complex combination of political actors and spaces makes it impossible to analyse the new modes of governance with the ancient model of the "nested territorial politics" (Debarbieux 2006: 345–352).

It is interesting to note that, in the framework that Debarbieux exposes later in the same article, he recognizes that these new spatial configurations can be ephemeral, as in the case of event-places, or anchored more permanently in *territoires*.[11] The latter resemble the ter-

11 The concept of *territoire* is not used here as a specific notion designating the equivalent of the state, as used in geopolitics, and which denominates the "delimited space of governing" (Giraut, 2008: 59–60). The concept of *territoire*/territory has a different history and therefore different meanings in the French and Anglo-American geographical traditions. Without entering into the epistemological debate, I will use the French term "*territoire*" ("territory") and consider it with a common-use meaning. On the debate between French and Anglo-American meanings of this concept, see Del Biaggio (2015).

ritorial model that ordinary people are accustomed to, but with one difference: They are not regulated by state-institutions (Debarbieux, 2006: 350). Therefore, even though not everyone shares the view that the state-actor is undermined and less capable of regulating the spaces of people's lives, everybody seems to agree that collective identities are following new logics, are more autonomous and are less dependent on the nation-state authority (Debarbieux, 2006: 347).

From this point of view, it is possible to affirm, as Debarbieux does, that spatialities and temporalities are multiple and heterogeneous, and that, from time to time, they meet in a common project, which can take the form of regional building. Those processes also affect mountain chains, which, as Debarbieux and Rudaz point out, tend to become political objects and participate in the rise of transnational regions (Debarbieux, Rudaz, 2010: 323).

2.2.2 *The emergence of layered identities*

Worthy of note is that this leaves space for new forms of identities, collaborations and new actions and projects to arise. As Keating points out, due to more porous borders and spaces left open by nation-states, a renegotiation of identities and the emergence of layered identities are possible (Keating, 2003: 11). Paasi clearly states that we are now facing a "proliferation of discursively constituted and institutionally materialized and embedded spatial scales" (Paasi, 2009: 467).

In the European context, it is interesting to see that the European Union, as a result of its cross-border cooperation programmes, has given rise to new forms of regionalism, and thereby, to new forms of regional identities (Agnew, 2001: 107). The result of this process leads to the constitution of "re-scaled territorialities" (Brenner in: Amin, 2002: 387), a "proliferation of scales and scalar complexity" (Amin, 2002: 387) and an increased density of scales of reference and identification.

All those new collaborations and interplays of actors "amount to a more or less ordered assembly of institutional actors performing the 'region'" (Allen & Cochrane, 2007: 1171), where performing can mean creating an image of the region, which corresponds to what I shall call the "objectivation" process (see chapter 5). Moreover, however, it means to put this objectivation into action, notably by creating institutions to fix and govern the region, or simply by supporting some transnational concrete projects inside the newly defined regional perimeter. It is important to underline that those projects take a territorial form, since they are anchored in a specific territorial setting: for example, the Alps as defined by the Alpine Convention, or the Carpathians defined by the Carpathian Convention. Therefore, they contribute to the construction of new regional spaces and new institutional contexts designed to better reflect the scale of identification of the actors involved in "performing" it.

2.3 Regional assemblages: to name territorial entities escaping political borders

In order to introduce the concept favoured here, the one of "regional assemblage," I shall mention some critical points concerning two commonly used concepts, namely scale and level, to describe and understand regionalization's processes.

2.3.1 The limits of "scale"

Although the concept of "scale" seems to be quite useful and appropriate to capture those new spatial phenomena, it can be criticized for entailing too strong a geographical and topological image, as Allen affirmed in a conference held in Geneva in 2010 (Regional Environmental Governance workshop – REGov).[12] Allen considers this point

12 Allen, "Reflections on the regional question"; title of the Conference: *Regional Environmental Governance. Interdisciplinary perspectives, theoretical issues, comparative designs* (Geneva, 16.–18.06.2010). The record of the conference is available here: <https://cast.switch.ch/vod/clips/2ixcm5j794/flash.html>, consulted the 01.12.2014.

to be a handicap for comprehending processes, which by definition are much more scattered. Moreover, as Allen states, they entail a tension between the willingness to spread activities socially and politically beyond borders and the necessity to fix them in order to govern them. Geographers describe the processes of territorially fixing activities as "politics of scale". This is understood as "what people do when they try to fix a particular definition of the region to their advantage in order to convince people that the region is like this and not like that".[13]

The term "politics" has to be understood as a combination of different kinds of actors translating, connecting and convincing people that the region is the one they defined.[14] However, and this is where Allen and Cochrane's proposition becomes interesting, the actual result and the activities taking place in that perimeter escape the scale construction and, consequently, go beyond this geographical fixity. That is why it is necessary to find other concepts to grasp the phenomenon of regional constructions. Hence, I borrow from Allen and Cochrane the concept of "regional assemblage" (Allen & Cochrane, 2007), which seems the most appropriate to understand contemporary phenomena of regional building.

Geographers and political scientists are now using the concept of assemblage rather frequently. Quite emblematically, an issue of the journal *Area* was dedicated to this concept in June 2011. Assemblage is a particularly promising attempt to understand socio-spatial phenomena, which are spreading beyond the traditional administrative units. In the following paragraphs, I provide a few explanations as to why I decided to refer to regional institutionalization's processes in terms of assemblage. I will do so by pointing out some problematic aspects of other concepts coined for the same purpose (i.e. scale and level).

Politics of scale can be broadly defined as a way of "framing conceptions of reality" (Delaney and Leitner, 1997: 94–95). Brenner, who is not only a sociologist but an expert in phenomena linked to

13 Conference by Allen, "Reflections on the regional question" (REGov, Geneva, 17.06.2010).

14 Conference by Allen, "Reflections on the regional question" (REGov, Geneva, 17.06.2010).

urban and regional development, defines it as "the production, configuration or contestation of some aspects of socio-spatial organization within a relatively bounded geographical arena" (Brenner, 2001: 599), as the city or the region. Marston deduces from her empirical observations that the socio-spatial reorganization taking shape in this new "geographical arena" is not only discursive practices, but also the tangible outcome of the practices of everyday life as they articulate with and transform macro-level social structures (Marston, 2004: 173). Furthermore, I would add that this is not only the outcome of the practices of everyday life, but also the outcome of socio-political struggles, which discard and remake entrenched scalar configurations (Brenner, 2001: 592). Scale, in Brenner's perspective, consists of a spatial unit that has to be analysed in order to consider its incorporation and positionality in relation to a multitude of smaller and larger spatial units within a hierarchically configured geographical scaffolding (Brenner, 2001: 600). It is exactly this idea of incorporation and hierarchy, rooted in and inseparable from the concept of scale, which the advocates of the concept of assemblage are contesting.

2.3.2 The limits of "levels"

Other attempts to capture these widespread and heterogeneous phenomena are linked to theories on "multi-level governance". They suggest that governments are not the only ones that matter in terms of political decisions in new geographic arenas. Indeed, advocates of the idea of a restructuration of space in terms of multi-level governance highlight the fact that new social and political actors, such as non-governmental organizations and agencies are increasingly important. Thus, the concept of governance becomes distinct from that of government by defining the ways in which disparate but interdependent social agencies are coordinated to achieve specific social objectives. So, one can define the field of governance studies, as Jessop does, as concerned with the resolution of (para)political problems in and through specific configurations of governmental (hierarchical) and extra-governmental (non-hierarchical) institutions, organizations and practices (Jessop, 1995: 317). More succinctly, governance can be defined as "a continuum of governing, in which state and non-state actors play a variety of roles" (Bulkeley, 2005: 877). Thus, the concept of governance still entails a strong link with the concept of scale, as MacLeod and Goodwin highlight when stating that the new geography of governance studies "the presence of overlapping and competing authorities at different scales" (MacLeod and Goodwin, 1999: 505).

However, as Allen provocatively declared during REGov,

> [...] local political actors are not 'jumping scales' while getting in touch with the European Union; they are picking up the telephone and talking with somebody, and networking with somebody. Brussels is not 'up there' geographically, but institutionally and in terms of power.[15]

Or, in other terms:

> [The local] is not confined to moving through a set of nested scales from the local to the national to the international but can directly access other such local political actors whether in the same country or across borders (Sassen, 2003: 11).

In that sense, the array of actors involved inside a region are not acting "'up and down' the vertical scales of government" (Allen & Cochrane, 2007: 1171) but operate in parallel mobilizing strategies, with different degree of commitment, at one point in time in one particular area (Allen & Cochrane, 2007: 1171).

Similarly to "scale", the concept of "multi-level governance" focuses on levels, which implies a sort of institutional and topographic nested hierarchy (the famous Russian dolls model, in which the local is thought to be smaller than the regional and the national). Scale underlines the existence of a "form of vertical interdependence" (Allen & Cochrane, 2007: 1166), in which the local and the regional are not only thought to be smaller than the national, but less powerful as well. This way of thinking, therefore, hides the complex mosaic of new forms of governance and veils what Allen calls "the twist that physical distance is not a good indication about the proximity or social distance".[16]

---

15 Conference by Allen, "Reflections on the regional question" (REGov, Geneva, 17.06.2010).

16 Conference by Allen, "Reflections on the regional question" (REGov, Geneva, 17.06.2010).

I agree with Allen and Cochrane when they say that maintaining the term "level" in the concept of "multi-level governance" is problematic, because it entails the idea of an existing territorial hierarchy. Nevertheless, it helps to recognize the existence of a multitude of different actors acting in and for a region and to support the idea of existing "multi-party collaborative governance arrangements", as Karkkainen calls them (Karkkainen, 2004). These social and political actors share, recombine and coordinate the deployment of the varied resources and competencies of multiple actors (Karkkainen, 2004: 74). However, if this is the real phenomenon that one wants to capture, then Allen and Cochrane's proposition of "regional assemblages" is probably more suitable than the one of "multi-level governance", because it does not emphasize levels.

2.3.3 "Regional assemblage" for defining socio-spatial arrangements escaping traditional boundaries

The term "regional assemblage" was coined by Allen and Cochrane to describe and analyse the process that the two called "the reinvention of the South-East of England". As the two scholars suggest, "assemblage" is an attempt to "give a name to these territorial categories that escape political boundaries" and "undermine the idea of a region as a meaningful territorial entity" (Allen & Cochrane, 2007: 1163) through new arrangements that are taking shape from the networks of relations stretching across and beyond political administrative boundaries.

Without entering into the history of the concept, it is nonetheless interesting to note that "assemblage" (from the French "*agencement*") has different roots, as Devriendt, referring to the special issue of *Area*, sums up in his blog (Devriendt, 2011).[17] Indeed, "assemblage" seems to convey different traditions: Deleuze and Guattari refer to it, as do Callon, Latour, Foucault, Althusser and Harraway (Devriendt, 2011). What all the different uses of the concept have in common is an emphasis on the "redefinition of the socio-spatial in terms of the composition

17 <http://technogeo.hypotheses.org/370>, consulted on the 01.12.2014.

of diverse elements into some form of provisional socio-spatial formation" (Anderson & McFarlane, 2011: 124). What is interesting in our case is the use of assemblage in the analysis of regional and territorial constructions (Anderson & McFarlane, 2011: 124; Devriendt, 2011).

Allen explains that this tendency of contemporary social and political relations of spreading beyond borders is associated with the attempt by these same social and political actors to fix social and political processes geographically in order to govern them. In fact, as Allen adds, if everything is fluid, it is impossible to govern. That is why people define boundaries, i.e. in order to have a perimeter in which to give meaning to these relations.[18] The question remains open what to call those resulting socio-political and territorial constructs. The concept of "regional assemblage" seems the most suitable for this purpose.

Some key ideas related to the concept of regional assemblages are: first, underlining that actors acting *within* a particular region are not necessarily *of* the region and, second, distinguishing between the physical distance and the social distance that separate different actors from each other within the region.[19] In that sense, as Allen and Cochrane affirm, when

> [...] politicians and professionals mobilize around a particular spatial representation of the region and act upon it, [this] does not mean that the actual spatial forms are also contained within the boundaries of the reinvented region (Allen & Cochrane, 2007: 1167).

Instead, regions have to be considered a "complex, tangled mosaic of superimposed and interpenetrating nodes, levels, scales, and morphologies" (Brenner in McCann & Ward, 2010: 176–177). Thus, the term "regional assemblage" seems appropriate to define the contemporary regionalization processes.

In the case under investigation here, namely the Alpine region, the concept of regional assemblage is used to define a "natural mountain

---

18 Conference by Allen, "Reflections on the regional question" (REGov, Geneva, 17.06.2010).

19 Conference by Allen, "Reflections on the regional question" (REGov, Geneva, 17.06.2010).

region". In geography, the study of natural borders and natural regions has a long history and was conceived in different ways. A new way of analysing them appeared with the emergence of socio-constructivist approaches. In order to understand them, a small detour via the previous ideas is required. I will first explain how the realistic approaches of regional geography conceive natural regions, then describe the criticism towards them and, finally, give details of the "new regional geography" that questions the realistic approach of regions.

2.4 Natural regions for classical regional geographers

2.4.1 An analytical conception of "natural borders"

The geologist and mineralogist Palassou (1745–1830) declared that the Pyrenees have always been the "natural" border between France and Spain (in Nordman, 1998: 64). Some years later, the French geographer Gallois (1857–1941) affirmed that it is in "nature" that one has to seek for the principle of every geographical division (Gallois, 1908). Although not belonging to the same discipline, they both underline the importance that scientists should look to nature to define political and geographical borders. This idea actually dominated the French school of geography, formed at the turn of the 20th century at the instigation of Vidal de la Blache (1845–1918). This school occupied a "comfortable niche" (Pudup, 1988: 370) within geography at the beginning of the 20th century.

According to Palassou, a "natural border" exists, is visible, real and just waiting to be discovered by scientists and/or politicians, who will define it as a political boundary between two territories. Nature, from his perspective, defines borders and regions that human beings can identify. A contemporary geographer, Agnew, sums up this theoretical position by saying that regions invoke the idea of a homogeneous block of space that can be distinguished by its physical and cultural characteristics (Agnew, 1999: 92). The idea of territorial homogeneity, most of the time delineated by nature, constitutes the core of this paradigm as much as a belief in the existence of a natural divide between these homogeneous spaces.

Thus, the geographer's task is recognized as the analysing of and reporting on regional specificities and singularities all around the globe, which are studied analytically according to their natural contents. The aim is to give accounts of the "distinctive personality of regions" (Fall, 2005: 23), this regional "personality" being defined by the natural and cultural features. The individual and singular attributes of portions of space are highlighted and portrayed as if time has not left a trace on them. Regions are thus described by scholars as "little more than frozen scenes for human activity" (Pred, 1984: 279). Natural regions as defined by the supporters of this approach can be compared to landscape paintings, petrified in time and space.

2.4.2 *A normative conception of "natural borders"*

The regional frame is thus established by looking at natural features, considered as the only objective characteristics capable of defining a region. Furthermore, this natural framework is seen as something eternal and therefore the most appropriate tool to define regions. Not without facing criticism from others, some geographers propose the identification of those regional specificities as the basis for creating political delimitations, i.e. in a normative way. It is based on the assumption that natural borders are the "best choice", since they "[conform] to a pre-existing order" (Debarbieux & Rudaz, 2010: 59).

That is why the natural normative conception proposes to follow "natural borders" in order to designate political boundaries. Concerning the specific case of mountains, the idea emerged at the end of the 18th century that mountains can serve systematically as natural borders (Debarbieux & Rudaz, 2010: 51). The two contemporary geographers argue that such politics of natural borders related to mountains are functional and convenient (*commode*, in French), since they can be used to optimize the arrangement of national territories (Debarbieux & Rudaz, 2010: 53).

Furthermore, at the end of the 18th and beginning of the 19th century, there were attempts to use mountains (defined by their watersheds) as "handy" elements to delimit national boundaries (Debarbieux & Rudaz, 2010: 54). For example, the Pyrenean ridgeline is identified as the border that "naturally" separates France from Spain and a "model which should be imitated", says Turgot (in: Debarbieux & Rudaz, 2010: 57). In this case, the "natural border" is supposed to define the disjunction between the two national states clearly. Dion (1896–1981) refers, quite emblematically, to the "Pyrenean barrier" (in: Dornel, 2002). Ozouf-Marignier demonstrates that the individuals in charge of delimiting the French *départements* followed the same naturalist and essentialist reasoning (Ouzouf-Marignier, 1992).

This way of thinking is not uncontested. Already at the turn of the 19th century, the German geographer Ratzel contradicts the idea that mountains should be seen as timeless natural boundaries (Debarbieux & Rudaz, 2010: 65). Criticism of the concept of natural border quickly spread in the scientific world and now dominates academic debates. Two contemporary geographers, Gonon and Lasserre, criticize the concept of "natural borders" by formulating a simple but pertinent question: One can trace the border on a ridgeline, but why that particular ridgeline and not another one? In fact, a ridgeline's continuity between different mountain ranges, which could define a ceaseless border, does not necessarily exist. A river looks like a natural barrier. But, as the two scholars continue, on a bigger scale, where exactly will the line be drawn? Will it follow along the bank, along the thalweg or along the midline? (Gonon & Lasserre, 2003: 435). Starting from these questions, Gonon and Lasserre make us doubt the merits of the naturalist and realist approach of borders. Hence, the delimited regions are not as clearly identifiable as some authors suggest.

Another case illustrates this dilemma quite well. Albera refers to "physical geography's *clichés*" regarding the limits of the Mediterranean region (interview with Albera by Mens, in: Mens, 2008: 21). In fact, some scholars set the Mediterranean limits in the 19th century on the basis of olive cultivation. But, as Mens asks in his paper, why exactly was this criterion chosen to define the region? Why not the cultivation of tomatoes, for example, or of some other vegetable or fruit? And why not air temperature? These basic and somewhat banal questions suggest the fundamental idea that drawing a border is a matter of social choice and as such, as Gonon and Lasserre underscore, every border is per definition artificial; it is a construction of the mind that one has to physically trace in space (Gonon & Lasserre, 2008: 457). "Natural borders" are as artificial as highways or railway routes (Gonon & Lasserre, 2008: 457), and so are the "natural regions" that boundaries define.

2.5 Criticism of the realist approach of regional geography: the socio-constructivist proposition

2.5.1 The socio-constructivist criticisms of "natural regions"

Since the 1980s, there have been many critical voices with regard to the natural regional approach. Geographers are not the only ones criticizing the concept of "natural region". Sociologist Bourdieu writes an acerbic and clear commentary concerning this issue: For him, no criteria can establish "natural" classifications defining "natural" regions separated by "natural" borders. Reality is social, and the most natural classifications rely on features that cannot be considered natural.

Thus, regions are to a large extent a product of an arbitrary posture (Bourdieu, 1980: 66). More than 20 years later, Balsiger, following in Bourdieu's footsteps, writes that

> [...] although forests, coastal estuaries, or mountain ranges appear as undeniable physical realities, their delineation for the purpose of policy-making evolves around numerous decisions and categorization that are socially contested rather than biophysically given (Balsiger, 2006: 6).

Balsiger adds that "while regions may be defined according to various criteria, even the most taken-for-granted are the result of social interaction" (Balsiger, 2006: 6).

These criticisms are related to a more general theoretical debate that opposes the realist and naturalist paradigm to the one known as socio-constructivist. Socio-constructivists reject the naturalist and realistic theoretical framework, marked by a realistic, profoundly inductive and descriptive approach. Socio-constructivism aims to analyse the manner in which persons and institutions build reality, a reality that is constructed on the basis of their interpretations and representations of it. They base their conception on the epistemological perspective that knowledge is constructed and therefore cannot be discovered from the outside world. By definition, therefore, it goes counter to the naturalist approach.

Conversely, realism is an approach that consists in believing in the existence of objects that one wants to analyse, as per Ruby's definition (Ruby, 2003: 765). This existence is independent of the knowledge that a "knowing subject" develops. Realism interprets the act of knowing as a simple mechanism of copying or recording reality (Ruby, 2003: 765). Debarbieux borrows from Lalande the expression of "naïve realism," which conveys the belief in the power of scientific methods to reveal the "truth" and the conviction that objects as they appear to our eyes exist *per se* (Debarbieux, 2004a: 12).

In other words, constructivism poses as principle that cognition proceeds from an elaboration of statements building the object of knowledge, while realism understands concepts and objects as being present in a world constituted by realities that have always been there. In this latter context, the researcher that is perfectly neutral needs only to discover them by improving his or her methods of observation (Lussault, 2003: 200), insofar as to adopt a realist position signifies denying the idea of the existence of an activity of knowledge, capable of building objects (Ruby, 2003: 765–766).

2.5.2 Regions are social constructions

In the specific case of regional studies, socio-constructivists consider that regions are not "essential places out there", waiting to be discovered. They are not truly authentic but rather our (and others') constructions (Allen *et al.*, 1998: 2 and 34). If this approach is adopted, regions cannot be considered "mere 'containers' for social processes" (Pudup, 1988: 377) or "empirical givens" (Murphy, 1991: 24). Geographical objects in general, and regions in particular, cannot be regarded as set and assigned portions of space that "only await discovery and description by a concerned individual" (Fall, 2003: 87). As Debarbieux points out, the world is not made up of geographical objects that geography should discover; it comprises "things" that geography arranges in objects and in systems of objects with the purpose of building knowledge (Debarbieux, 2004a). This is true insofar as regions, deemed natural or based on other criteria, are socially constructed, culturally contested and politically charged, and not neutral and defined in an essentialist manner (MacLeod, 2001: 670).

It follows that socio-constructivist geographers' perspective contrasts with the one of their realist colleagues, who take the areal frameworks for their studies. The latter begin by asking *what* is going on in a particular region, with no interest, unlike the former, in *how* and *why* that region acquired significance in society. Therefore, with the increasing importance of the constructivism paradigm, and because of the essential role that observers play in the construction of their objects of reference and analysis, regions could no longer be considered "autonomous realms of existence" (Nigel Thrift in Murphy, 1991: 23).

As Gonon and Lasserre explain, human beings cannot grasp space in its totality. Because of their subjectivity, the observer takes in one or the other aspect of reality and then combines them in order to make sense of space (Gonon, Lasserre, 2003: 456). The result is that

> [...] objects are not simply given to analysis, but are themselves products, and must be conceptualized in such a way as to incorporate, not just their descriptive characteristics, but also the process of their production, the larger dynamic of which they are part (Massey, in: Pudup, 1988: 380).

Consequently, investigators move away from the notion of a naturalist division of the earth's crust, to which they prefer a social vision of spatial differentiation.

Considering that regions are socially constructed, which at present is a widely held assumption in the scientific community (Allen *et al.*, 1998; Balsiger 2006; Bourdieu 1980; Debarbieux 2004a; Debarbieux 2004b; Debarbieux & Rudaz 2010; Fall 2003; Gonon & Lasserre 2003; Häkli 2008; MacLeod 2001; Murphy 1991), and that they are the result of a complex process of production, reproduction and creation of spaces (Raagmaa, 2002: 56), they cannot be regarded as eternal. As they appear one day in the social history of a community, they are likely to be modified, or even to disappear, if there is a change in the referent identity in the society in which they were created. According to Murphy, commentators view regions as

> [...] constantly changing and socially significant historical and geographical formations, whose inhabitants interpret and respond to events and processes in particular ways as a consequence of the unique social and physical contexts in which they are situated (Murphy, 1991: 25).

The idea of regions "in formation" as a result of social actions (Gilbert, 1988: 216) is central to the new paradigm. This means that the essentialist vision of regions is abandoned and that the focus is on "contingency and becoming" (Painter, 2008: 343), depending on historical circumstances and geographical conditions (Painter, 2008: 354). There is a need to see regions constantly changing (Thrift, 1998: 40), being dynamic in time and space (Raagmaa, 2002: 57). Taylor, taking the Caribbean region as an example, suggests, albeit tersely, that regions "are created, reproduced, and finally disappear from the world scene". For Taylor, every region has a similar trajectory of rise and fall (Taylor, 1991: 186). Pudup reaches the same conclusion and declares that "the brave new regional world is an unstable one" (Pudup, 2004: 12908). So, questioning the essence of regions also signifies thinking about their birth, emergence, and disappearance (Paasi, 1986: 120). For Gilbert, contemporary regional geographers have to investigate regions made up of social relations, a web of complex interactions between social actors in a material environment both affecting and affected by these social relations (Gilbert, 1988: 215).

Therefore a region is conceived in this research as the result of how social and political actors decide to use its confines and features (in other words, its objectivation[20]) in their own projects and actions. The aim of this particular research is specifically to uncover how local political actors undertake this process of socio-political construction. In this specific research on the Alpine region, the focus is on two phenomena: firstly, on how local political actors conceive this "thing" called the Alps; and, secondly, on the actions that they have undertaken on this perimeter that they define as the Alps. Thus, Alpine

20 See chapter 5 for discussions on objectivation processes.

political actors, notably through the projects they implement, build their region according to their personal representation and vision of the territory in which they are involved. This region has, obviously, a materiality *per se*. It exists in reality, and it is conceived by local political actors as being a mountainous, natural region. The will to create out of it a region devoted to sustainable development highlights the intent of the political actors involved in the regional building process: their wish to build "another" region, shaped on criteria based on the respect of the natural and cultural Alpine environment.

2.5.3 Regions as constructs of the second degree

In order to avoid any misunderstanding, the status of materiality for both me as a researcher and for the Alpine local political actors should be clarified. The philosopher Schütz's conception of the social sciences is immensely helpful in this regard. One of the main points that Schütz underlines is that he considers the constructs of the social sciences as constructs of the second degree, i.e. constructs of the constructs built by the actors on the social scene, whose behaviour social scientists have to observe and explain in accordance with the procedural rules of their science (Schütz, 1954: 267). Schütz makes a distinction between the world of the people living in a "natural attitude", as he names it, and the posture of researchers, who have to analyse those natural attitudes. Natural attitudes, in the eyes of Schütz, have to be studied as objective, ideal and typical constructions, based on observations of the reality of their daily lives (Laoureux, 2008: 173; Schütz, 1954: 267). It is important to emphasize that "natural attitude" implies taking the world and the objects constituting it at face value, without questioning their real existence (Laoureux, 2008: 175).

This is the attitude of the pan-Alpine local political actors, who have taken the Alps and the perimeter of the Alpine Convention and are trying to implement inside this area the projects as they conceive them. Obviously, I have a different attitude: As a researcher, I decide how to *organize* and *conceptualize* the local political actors' descriptions of their social world and everyday life. Hence, I consider how local political actors give meaning and consistency to the reality they call "the Alps" or "mountain". This is not to deny their existence but rather to take an epistemological stance in order to uncover social behaviours. This means that there is not a naïve opposition between constructivism and realism and that this opposition is to be considered "unfortunate" (Agnew, 1999: 92), because there is always a "visceral tension between the idea that something is real and the idea that it is constructed" (Agnew, 1999: 92). The result of this tension, which is visible in the reality itself, is a mechanism that researchers are most skilled at uncovering.

2.6 The advent of new regional geography

2.6.1 Geography matters

With socio-constructivism gaining importance in geography during the 1980s, the interest for regional geography and the question of how regions are constructed garnered attention. Gilbert speaks of a "new" regional geography (Gilbert, 1988: 208). Some years later, some researchers, united in an enthusiastic chorus, announced that "regional geography is back" (Taylor, 1991: 184). Thrift, with the same fervour and not without a glimmer of hope, subsequently declared, in a lesson he gave in Leipzig, Germany: "There is an old Chinese curse, 'may you live in exciting times.' In the case of [the new regional] geography, now, I think the curse is actually a blessing" (Thrift, 1998: 46).

After years of almost complete silence due to the weight accorded to quantitative geography, regional geography has risen from its ashes in British and American universities. This has come to be thanks to the advent of the concept of "place", whose use notably became popular with the book by Tuan, *Space and Place: The Perspective of Experience* (Tuan, 1977). In other countries, the concept of "region" was not abandoned during the 20th century. In France, for example, Frémont's theory on the "*région vécue*" was one of the most important sources of inspiration for the geography of representations, which had significant success in France in the second half of the 20th century.

In English-speaking countries, it was in the 1980s, as Taylor points out, that the regional concept "fully recovered from its near banishment in geography's quantitative interlude" (Taylor, 1991: 184). The quantitative intermezzo having been set aside, the regional component was restarted on a new footing. In the 1980s, geographers realized that what had been considered as general structures were in reality geographically specific, context-dependent phenomena that had wrongly been treated as general by quantitativists (Sayer, 1989: 255). Massey and Allen summarize this idea in two words: "geography matters" (Massey & Allen, 1984). From this perspective, local singularities started to be taken into account again in analyses, and the search for universal laws was abandoned.

2.6.2 The approaches of new regional geography

It is in this context that "new" regional geography bloomed. The first mention of this concept appears in a text by Thrift, who, in 1983, put forward the expression "reconstructed regional geography" (Thrift, 1983: 38). Through Gilbert and Pudup, the concept is then modified into "new regional geography" (Gilbert, 1988; Pudup, 1988). This new approach is generally seen as the one capable of leading to a better understanding of the unique qualities of a region. In that sense, this revival of regional geography permits a better understanding of a particular place and its contribution to broader structures (Bradshaw, 1990: 316). The consideration of regional peculiarities, make researchers break with the idea of a theoretical geography. The latter aims to search for regularities, insofar as geographers, through these fundamental concepts of the new regional geography framework, rediscover the interest for the specific (Gilbert, 1988: 208), which gives new impulses to the analysis of regions.

Gilbert identifies three approaches of the new regional geography (Gilbert, 1988: 209–213). The last two are particularly interesting in this research:

- A *political economy approach*, grounded in Marxist theory, which considers regions as spatial organizations of social processes and is associated with the mode of production;

- A *cultural and humanistic approach*, concentrating on regional identity. Gilbert uses the concept of "cultural geography", while the term "humanistic geography" is employed by Bradshaw, who classes as "humanistic" those scholars that consider regions "sources of identification and meaning" (Bradshaw, 1990: 316). The region is thus defined as a specific set of cultural relations between groups and spaces. It is based on a belief that inhabitants share the same culture, which differs from that of other groups. Priority is given to human intentions that give meaning to places and regions;
- The *social interactionism approach*, which considers that regions constitute a means in the production and reproduction of social relations. The latter are structured in time and space. Regions are analysed as a physical frame in which social interactions take place. In this approach, Giddens' influence and that of his structuration model are evident, since regions are seen as an arena that allows (or restrains) social action (Bradshaw, 1990: 316). This school of thought is split into two traditions: the English-speaking, which puts the emphasis on "locality", and the French-speaking introducing the concept of *"territoire"*. Gilbert identifies Paasi, one of the scholars often quoted later on, as being from this second tradition.

2.6.3 The "common ground" of new regional geographers

Despite the enthusiasm of some geographers regarding this new paradigm, Paasi stresses that new regional geography cannot be considered a single approach but rather "an umbrella term for research reflecting how regions/places can be constituted by and constitutive of social life, relations and identity" (Paasi, 2002b: 802). Or, as Bradshaw affirms, "there is not a new regional geography, but *many*, each shedding light on the processes of regional differentiation" (Bradshaw, 1990: 316). Scholars agree that this approach has not yet been completely constituted. Nevertheless, authors who identify themselves with these new concepts share a certain number of ideas, among them the belief that regions are socially constructed and "constantly evolving" (Pudup, 2004: 12908).

Moreover, new regional geographers share the view that regions actively participate in a scaling process. As Thrift already wrote in 1998: "There is the attempt to see the region as a site of many intersecting processes operating at many different scales, rather than as a coherent and self-contained *whole*" (Thrift, 1998: 40). In order to understand the Alps today, one must take into account the important number of actors operating at different scales. Debarbieux points out that we are witnessing the emergence of complex governance at the Alpine scale, with a growing number of actors and levels of action (Debarbieux, 2008: 40). This complexity goes against the vision that regions are a coherent whole, an idea that is contested by scholars like Massey and Amin (Painter, 2008: 343).

The importance that new regional geographers attach to social interactions has often been criticized. It is true that, in this approach, natural elements have frequently been pushed aside by the weight given to social phenomena taking place within the region. However, we should not forget what Gilbert wrote in a founding article of the sub-discipline, in which she declares that regions have to be considered "medium[s] for social interaction synchronising people, nature, and social relations in time-space settings" (Gilbert, 1988: 212). Nature is therefore not left out. Thrift is even more explicit in an article he wrote in 1990, in which he describes four skirmishes, as he defines them, against the new regional geography. I will present one of them, which is of interest here, since it exposes the problem of the relation between people and nature: "The desire to take nature back is growing, whether as a desire to deconstruct ways of seeing like landscapes or as a means of recapturing a kind of authenticity" (Thrift, 1990: 274).

Nature is not put aside, quite the contrary. At most, it is incorporated into a more complex analytic system than the simple binary relation between nature and human beings. Gilbert is clear on that point:

> Traditionally concerned with the more straightforward relationships between people and natural environment, regional geography is now analysing the more complex links that relate people to nature through the action of society, seen either in political-economic and/or cultural terms (Gilbert, 1988: 213–214).

More generally, Harvey notes:

> One of the more interesting exercises to undertake in enquiring into the environmental-ecological debate, is to inspect arguments not for what they have to say about the environment, but for what it is they say about the 'community' and political-economic organisation (Harvey, 1993: 19).

In this research too, "natural issues" (like the identification with the mountain quality of the Alps) is studied in order to reveal social (i.e. linked with identity) and political (linked with governance) issues. In the next section, I clarify my position concerning the question of identity.

2.7 The role of identity

As previously discussed, one of the dimensions recognized as central to the new regional geography is "identity". The concept of identity has been criticized by a number of scholars for being "indefinable" (Bray 2002: 14) or at least "ambiguous" (Brubaker & Cooper 2000: 2). That is why it is important, at this point, to explain how this "identity catchword" (Häkli, 2001: 115) is employed in this book and to propose a solution for using it in a less ambiguous and indefinable manner.

2.7.1 Identity, a social construct

What emerges from the interpretation of identity is that identities are plural and constructed (Castells, 1999: 17). As the French geographer Di Méo summarizes, far from forming immutable traditions capable of withstanding time, identities and the cultures in which they are rooted appear as social *bricolages* constantly overhauled and directly influenced by circumstances at any given time (Di Méo, 2002: 175). Thus, what is happening in their world is constantly calling them into question (Di Méo, 2002: 178).

The idea of identity as a social construct emerged in the 1980s in the social sciences, notably under the influence of French sociologist Bourdieu. It incorporates the constructivist renewal of the concept of culture, until then considered ahistorical and less sociological (Avanza & Laferté, 2005: 136). This conception of identity refuses the essentialist vision and

> [...] rejects any category that sets forward essential or core features as the unique property of a community's members. From this perspective, every collective becomes a social artefact – an entity moulded, fabricated, and mobilized in accord with reigning cultural scripts and centres of power (Cerulo, 1997: 385–409).

From a constructivist point of view, "identities are seen as invented, created, reconstituted, or cobbled together rather than being biologically preordained or structurally or culturally determined" (Snow, 2004: 2215).

Considering identities as social constructs, as is generally admitted in social sciences according to Nebe (Nebe, 2003: 4), contrasts with the contemporary common usage of the concept, which regards identity as pre-existing and something that serves to indicate what in its own or in others' reality does not change (Avanza & Laferté, 2005: 135). However, to consider identities as something constructed does not mean that scholars dealing with this question do not think they are built on "something", "things" that can have a material consistency.

Snow synthesizes this idea as follows:

> [The conclusion that identities are constructed] should not be read as unequivocal refutations of the primordialist and structuralist arguments, since constructed identities are not fabricated whole cloth but typically knit together by drawing on threads of past and current cultural materials and traditions, structural arrangements, and even primordial attributes. These materials and attributes constitute the kinds of stuff from which collective identities [...] are fashioned, and thereby function [...] to constrain the construction process (Snow, 2004: 2215).

He goes on to affirm that

> [...] while collective identities are undeniably constructed, they rarely are constructed *carte blanche*; rather, they typically are forged not only with

> the materials suggested by the primordialist and structuralist perspectives, but with and through the experience of collective action itself (Snow, 2004: 2215–2216).

In the section dedicated to the "objectivation" and "identification" processes, I uncover the contents of the Alpine collective identity and affirm Snow's contention that it is not constructed *carte blanche.*

2.7.2 Going beyond identity

Brubaker and Cooper, in an article in which they invite us to go "beyond identity", remark that the concept of identity is too heterogeneous and that it would therefore be preferable to name the different meanings of the concepts with distinct terms. For Brubaker and Cooper, the concept of identity as commonly used by scholars entails heterogeneity because that single concept tends to signify too much (strong sense) or too little (weak sense), or nothing at all, because of its sheer ambiguity (Brubaker & Cooper, 2000: 1). The strong sense put "emphasis on sameness over time or across persons" (Brubaker & Cooper, 2000: 10), while those in favour of the weak sense expressly reject a fundamental or permanent sameness (Brubaker & Cooper, 2000: 10). Often, the meaning given to identity is not made explicit and therefore leads to confusion. That is why the two authors solved this problem by splitting it up into three concepts. Through the impetus of the constructivist paradigm and in order to diminish the impression of confusion that identity produces, Brubaker and Cooper propose three alternative concepts to grasp identity processes: "Identification and categorization", "self-understanding and social location" and "commonality, connectedness, groupness".

The first pair, identification and categorization, characterized by the *-tion* suffix, want to avoid the "reifying connotations" of the term identity and compels the investigation of agents (identifiers and categorizers) of the identification and categorization processes (Brubaker & Cooper, 2000: 14). This process occurs in everyday life and social relations, and it is therefore situational and contextual. The two scholars quote the modern state as one of the most important identification and categorization agents in recent history. The power of the state is built on important "material and symbolic resources to impose categories [...] and modes of social counting and accounting" (Brubaker & Cooper, 2000: 16).

The second pair, self-understanding and social location, designates a "'situated subjectivity': one's sense of who one is, of one's social location, and of how (given the first two) one is prepared to act" (Brubaker & Cooper, 2000: 17). Given this definition, these concepts can only capture one's own representations of the self but not other people's understandings. That is why it is important that they be combined with the other concepts in order to cover the multiple facets of "identity".

The trio of commonality (the sharing of some common attribute), connectedness (the relational ties that link people) and groupness (the sense of belonging to a distinctive, bounded, solidarity group) designates the "emotionally laden sense of belonging to a distinctive, bounded group, involving both a felt solidarity or oneness with fellow group members and a felt difference from or even antipathy to specified outsiders" (Brubaker & Cooper, 2000: 19).

Although these concepts represent an interesting way to capture the complexity of the multi-faceted concept of identity, the distinctions made by Brubaker and Cooper are criticized by Avanza and Laferté for being barely employable and understandable (Avanza & Laferté, 2005: 140). The terms favoured by Brubaker and Cooper are indeed not very intuitive and are detached from common language (Avanza & Laferté, 2005: 140). Despite their criticism, Avanza and Laferté share the fundamental idea that the means given to the concept of identity are, generally, too ambiguous and scattered to satisfy scientific analysis (Avanza & Laferté, 2005: 136). In other words, if identity is everywhere, it is nowhere (Nebe, 2003: 3).

For that reason Avanza and Laferté suggest three alternative concepts to those proposed by Brubaker and Cooper: identification, social image and belonging (Avanza & Laferté, 2005; also see Del Biaggio, 2010), which they define as follows:

- *Identification* can be qualified as a social action where identity attribution is external and practised on individuals within social

institutions, following a codified technique. Although Avanza and Laferté identify the state as the most emblematic actor of the process of identification, the authors do not recognize the bureaucratic apparatus as the only one capable of classifying persons and societies;
- The *social image* relates to the social production of discourses and symbols (representations) that designate *territoires* and groups;
- *Belonging* defines the participation of people to the collective "thing" and to the group.

As the definition of the six groups of concepts shows (three by Brubaker and Cooper and three by Avanza and Laferté), the distinctness between the two theories cannot be considered only as a result of the vocabulary, but also of the meanings and conceptions the authors ascribe to "identity". This is true, even though in their paper Avanza and Laferté mainly reproach Brubaker and Cooper for the low comprehensibility of their terms and not so much the inappropriateness of their theory.

However, the two propositions have in common that they both offer an interpretation of the concept of identity that should not be read as an attempt to fragment it into distinct and unrelated terms, but rather as a valuable attempt to give names to different significations of a multifaceted and "indefinable" (Bray, 2002: 14) concept. This helps to clear up the linkages between the different meanings in a sort of crossed-relational game. With the same spirit and motivations with which Brubaker and Cooper on the one hand and Avanza and Laferté on the other proposed to split up the term identity, I propose to combine two already existing terms in order to analyse the identity process taking place in the Alps: objectivation and identification. Before giving a definition of those concepts, however, it is important to explain why I partly reject the previous propositions and suggest other concepts.

2.7.3 Grasping identity: objectivation and identification

The decision to favour those two particular concepts has been dictated by the impossibility of placing the different pan-Alpine social and political actors and their actions in the categories proposed by the aforementioned four scholars. Avanza and Laferté assert, for example, that "identification" defines an *external* identity attribution practised on individuals. A question can be raised about the meaning of and the importance accorded to the idea of an "external" attribution of identity. In the Alpine case, if it is true that, for the most part, scientists and politicians undertook the identification process of the Alpine region with a "scaling down" approach (Debarbieux, 2008: 43), it is also true that it was accompanied by an *internal* process of identification. Some Alpine non-governmental organizations such as the *Commission Internationale pour la Protection des Alpes* (CIPRA), the International Scientific Committee on Research in the Alps (ISCAR) and local political actors connected through pan-Alpine networks carried out this process.

In any case, the distinction between "internal" and "external" is not so apparent. For example, pan-Alpine networks can be considered *internal* Alpine agents, since they came into being as a result of the initiative of local political actors living and working in the Alps. Yet, together with scientists and politicians, they play an equally important role in the attribution of an Alpine identity. Additionally, it is reductionist to consider them only as *internal* actors, i.e. as monolithically structured institutions imposing their identification's attributes on people. In fact, even within the networked institutions, there is a fragmentation between, for example, leaders and "simple" members, who do not identify themselves and others in the same way with the Alps. Some members can consider the leaders as *external* to their reality (because they do not share the same beliefs and ideas), although geographically speaking they all belong to the same territory, the Alps. The same questions can be directed at the concept of belonging. In fact, Avanza and Laferté define it as the "participation of people to the collective 'thing'". But who are these "people", really? Here again, "people" cannot be considered a monolithic structure, and one will find different ways of belonging depending on which group one focuses on: local politicians, representatives of environmental NGOs, scientists or the "broader population"?

Thus, in order to show "who is doing what" in the Alps, I judged it necessary to search for alternatives, whilst maintaining the persuasive idea to split up the concept of identity. The validity of this operation is well justified by Avanza and Laferté. The two assert that opening the concept of identity into different terms offers the advantage of better identifying the multiple social institutions at work. It helps to better understand the complex interdependencies between individuals, institutions and social groups, which collectively produce social "cuttings" that are constantly renegotiated and only rarely correspond to the initial intentions of the committed actors (Avanza & Laferté, 2005: 148).

Objectivation

These "cuttings" are the result of an objectivation process. In fact, in a socio-constructive perspective, as discussed above, there are no objects *a priori*. The world, as highlighted by Debarbieux, is "made of 'things' that the observer composes into objects and systems of objects within a process of knowledge building" (Debarbieux, 2004: 397). Of course, as Paasi emphasizes, this classification is a matter of choices, since the objectivizing subject chooses some elements from the ones available to constitute an identity narrative (Paasi, 2009: 469). Debarbieux and Rudaz call this mechanism "objectivation". It consists in the identification of an object (or a family of objects) by a subject (or a collectivity) who thinks that this object is a pertinent category for understanding the world or a part of it (Debarbieux & Rudaz, 2010: 11). Behind this definition, there is the idea of a separation between the subject, who objectivizes, and the object, objectivized, as Rudaz underlines (Rudaz, 2005: 20). In that sense, it can be considered a constantly evolving process of social construction of materiality.

The objectivation process is different and unique for each human being, who tries to make order, i.e. organize the world into coherent and relevant units (Rudaz, 2005: 21; Gervais-Lambony, 2004: 470), among the multitude of objects constituting reality. It can also be a collective process, in which society builds a territorial frame in order to form a suitable structure to undertake actions and projects in a particular geographical area. This structure makes sense, because people thinking about it see coherence between different features that make it distinct from other geographical areas.

Paasi conceptualizes this process as the creation of an "identity of regions", i.e. the features of nature, culture and inhabitants that distinguish a region from another (Paasi, 2009: 469). In his theory of regional institutionalization, Paasi distinguishes four stages, one of them being the emergence of a territorial shape, which results from the definition of the "identity of regions". In fact, the territorial shape is defined as the development of the social practices through which regions achieve their boundaries and are identified as distinct units in the spatial structure of society (Paasi, 1986: 124). In this definition, two aspects are highlighted: first, a process of objectivation and delimitation of a region takes place through the territorial shaping; and second, a process of definition of the contents of that delimitation appears, following the social practices taking place on it. There is therefore, in Paasi's definition, a dialogue between regional construction and regional practices. As Debarbieux points out, the relation should not be understood as unidirectional (only practices defining the region) but as dialectic, too. In fact, the opposite causal and intentional relation must also be taken into account (regions describe the perimeter where practices can be undertaken): "Since they are intentional and pertinent, geographical objects are always related to practices and projects" (Debarbieux, 2004: 402).

Identification

A common point in the writings devoted to identity reviewed above is the recognition of the capacity of individuals and groups to feel part of a collectivity and to put some distance between them and other groups; it is what I call identification. Identification processes allow human beings to make a distinction between "us" and "them", between "me" and "him or her", between "what we have in common" and "what distinguishes us from each other". Indeed, it allows for the creation of a collective identity, defined by Debarbieux as the sense and will shared by several individuals of belonging to the same group (Debarbieux, 2006: 342). The "same group" can be understood in a social sense (as having the same ethnic origins, same religion) but equally in a geographical sense: the feeling of being part of the same group, *because* this particular group shares the same (or what is believed to be the same) space. Additionally, as Petite underlines, a "group" can form not only through a shared feeling of belonging but also through the actions that its members undertake together (Petite, 2009: 13).

Following the same order of ideas, Jenkins establishes a strong link between identification processes and social action (Jenkins, 2000: 8). This conception has been notably developed, in the mid-1980s, by French sociologist Touraine, who chose an evocative title for his book: *Le retour de l'acteur*, which has been translated as *Return of the Actor* (Touraine, 1984). Touraine underscores the fact that "national feeling" among the population is weaker today, notably because the state is no longer an entity capable of creating a sense of integration among social actors (Touraine, 1984: 82). As a consequence, people do not just react to a situation, but they are keener to produce a situation in order to find new ways of integration (Touraine, 1984: 69). Touraine ventures the hypothesis that collectivities have an increased capacity to actively produce social reality. This thesis ties up with what Castells conceptualizes as the "project identity", which occurs

> [...] when social actors, on the basis of whatever cultural materials are available to them, build a new identity that redefines their position in society and, by doing so, seek the transformation of overall social structures (Castells, 2004 [1997]: 8).

In these processes, as pointed out by Langlois, evidence of adhesion to the group cannot only remain abstract. "Identity work" (Snow, 2004: 2216), a concept that Snow coined, precisely underlines the necessity of identities to be materially and symbolically expressed by collectivities. As Langlois indicates, identity must be transposed into collective symbols in order to become a reference for the members of the group (Langlois, 2004: 7164). Those symbols are not, as previously underlined, "fabricated whole cloth" (Snow, 2004: 2215) but put together in a sort of *bricolage* from different referents, which can have a materiality (such as mountains) or not (such as primordial attributes, like ethnicity or religion). In that sense, people generate, invoke and maintain symbolic resources in order to bind and distinguish the collectivity, internally as well as externally (Snow, 2004: 2216).

2.7.4 Territorialized identities

The identity references described above can be geographical, if one admits that *territoires* and regions are constructed. Thus, the process of territorial construction presupposes that, despite their differences, individuals and groups share, on the one hand, social practices and, on the other, manners of thinking and representing the space in which they live. Escobar formulates this idea as follows:

> Links to places are certainly neither natural nor fixed, and while boundaries do not exist in a 'real' sense, their construction is an important aspect of the active material and cultural production of place by groups of people that, while heterogeneous and diverse, do share what Virilio calls the *hic et nunc* (here and now) of social practice (Escobar, 2001: 152).

In a way, Escobar seems to identify a link between identity and space (the *hic*).

This link is clarified by Di Méo, for whom identity results from a tripartite construction, in which human beings, society and geographical space interact with each other (Di Méo, 2002: 178). Thus, the process linking geographical spaces, which can take the form of regions, and identities, has to be considered as dialectic. On the one hand, regions are constructs built by social practices and discourses, which calls to mind what I labelled the "objectivation process of identity". On the other hand, regions are the spatial framework inside which social action is possible. This spatial framework is thus not only objectivized by actors but also recognized as a pertinent framework in which they identify and act (the "identification process of identity").

The reference to the common spatial framework is, consequently, not only the condition of the action ("I feel part of this space, therefore I act") but is also its purpose ("I act to feel part of this space").

Those processes underline a strong link of co-construction between identity and places. Murphy's words are enlightening in that sense:

> People's lives are infused with an elaborate network of interrelated and overlapping regional constructs with functional and symbolic significance. [...] They shape activities, ideas and decisions, which in turn shape conceptual and functional compartmentalizations of space (Murphy, 1991: 32).

Thus, one cannot look at identity only as an individual or social category. Rather, it should also be acknowledged as a spatial category (Paasi 2001: 10). As Gervais-Lambony underscores, identity discourses almost always entail a spatial dimension. Since identity is a discourse that allows belief in an "orderly world", it consists in a narrative of the past (time) and of the space (*territoire*) (Gervais-Lambony, 2004: 470). In that sense, as French geographer Di Méo argues, the *territoire* forms the visible and readable figure of social identity, because it can be mapped, and its limits can be drawn. It gives identity a material consistency made up of signs and symbols that are inserted in objects, things, landscapes and places (Di Méo 2002: 175). *Territoire* is conceptualized by Di Méo as a portion of space that shows an economic, ideological and political (therefore social) appropriation of space by social groups. These groups have a specific representation of themselves, of their history and of their singularity (Di Méo, 1998: 38). It is therefore the tangible object of an immaterial social reality (Di Méo 2002: 178). Di Méo, in the same article, adds that identity leans on cut and delimited *territoires* for a precise purpose and uses *territoires* as an efficient bond of social groups.

Since *territoires* do not have to correspond to nation states, identities can spread beyond national borders. In fact, in the contemporary context, researchers should not only analyse how social identities are invented and institutionalized inside national states but should equally observe the affirmation of chosen and shared identities, which grow within collective projects (Debarbieux, 2006). Therefore, collectives of people involved in social mobilisations can undertake actions and projects that are more compatible with their feelings of belonging but have been neglected by the established institutions.

These mobilizations can take form in the constitution of new *territoires*, supported by *ad hoc* institutions created for that purpose.

The significance of social relations is therefore crucial to an attempt to define what a *territoire* is. For this reason, collective identity cannot be considered as existing in individuals; it develops, grows and changes within social interactions (Pohl, 2004: 12918). Identity is not created forever; it can change, under the effect of new projects, new ideas, and new territories to invest.

Thus, the concept of collective identity finds an interesting niche in the study of regional geography, since "belonging to a region may elicit a sense of identity that challenges the hegemonic identity narrative" (Paasi, 2003: 477). One of the most emblematic narratives that regional identities can challenge is doubtlessly the national one. In the contemporary world, new narratives can supplant or overlap the state-centred ones, considered as hegemonic until a few decades ago. In this context, there is room for the creation of "regions as meaningful socio-political spaces" (Paasi, 2002a: 138), independent of other meaningful socio-spatial spaces as nation states can be. This occurs because regions and regional identities are social facts that can generate action as long as people believe in them (Paasi 2002: 139).

2.8 Eco-regionalization: the bioregionalist paradigm

2.8.1 Thinking the globe in natural subdivisions

As previously discussed, regional geographers are attributing great importance to social interactions taking place in regional assemblages. The introduction of social components in regional studies constitutes not only the strength of the approach but also its Achilles' heel. Regional geography is sometimes criticized for according too much weight to social relations (see for example Holmén, 1995). New regional geographers are accused of pushing nature aside. In this section, I describe how nature is present in this research; in doing so, I demonstrate that new regional geography is not necessarily unnatural. I explain how regional objectivation and identification lead and give substance to regional institutionalizations' constructs. In the case of the Alps, these correspond to an *eco*-regionalization process, because the mountain range is becoming a potential political arena.

Local political actors use the identified natural perimeter as the framework for regional and transnational actions and projects. This process raises questions about the match between natural and political (institutional) areas, an issue that has occupied a number of scientists for centuries. In the 20th century, scientists and environmental activists tried to think in terms of "natural regions" to determine the merits of this entity to protect nature. Their identification is the fruit of scholarly expertise and a political and ethical approach (Debarbieux & Rudaz, 2010: 307). Udvardy can be counted as one of them. The biologist tried to divide the Earth into eight biogeographical regions according to unifying features of geography, fauna and flora, and his theory was taken up by the World Conservation Union (Fall, 2001: 628). Other environmental organizations utilize similar concepts to designate "areas of commune characteristics" and primarily make use of natural criteria to define them. The World Wildlife Fund for Nature (WWF), in its Ecoregional Conservation Plan, identified 825 terrestrial ecoregions across the globe and a set of 426 freshwater ecoregions.[21]

The WWF identified the Alpine region as one of the "Global 200", or "the most biologically distinct ecoregions of the planet".[22] Ecoregion is defined as a

> [...] large area of land or water containing a geographically distinct assemblage of natural communities that (a) share a large majority of their species and ecological dynamics; (b) share similar environmental conditions, and; (c) interact ecologically in ways that are critical for their long-term persistence.

Conservation International counted 35 "biodiversity hotspots".[23] Out of a concern for nature conservation, these organizations view the globe in subdivisions other than the ones drawn from criteria centred

21 The present and subsequent definitions by the WWF are taken from: <www.worldwildlife.org/science/ecoregions/item1847.html>, consulted on the 05.12.2014.

22 <www.worldwildlife.org/science/ecoregions/item1847.html>, consulted on the 05.12.2014.

23 <www.conservation.org/How/Pages/Hotspots.aspx>, consulted on the 05.12.2014.

on nations. These organizations call for a subdivision into natural regions, that is, according to them, better adapted for the purpose of biological conservation. Since the 1970s, the question of the regionalization of collective action on the basis of natural criteria has been advanced, in another form, by the bioregionalist current of thought. Some bioregionalists promote an administrative reorganisation, or, in extreme cases, the creation of new political entities, based on a naturalist division.

2.8.2 Describing bioregionalism

Bioregionalists found fertile ground within the paradigm of "deep ecology", a movement the Norwegian philosopher Naess founded in 1973 that recognizes "the inherent value of all living beings and the use of this view in shaping environmental policies".[24] From this definition, one can deduce the strong political component intrinsic to the ecologist movement, as the aforementioned quotation clearly points to its aim, namely to "shape environmental policies".

Thus, bioregionalism consists, in the eyes of its supporters, in more than just a type of ecosystem management, because it requires fundamental changes in beliefs, attitudes and values regarding the interaction of humans with their natural environment (Diffenderer & Birch, 1997: 3). The Canadian biologist Van Newkirk articulated the concept of bioregion in 1971 (Hamm & Rasche, 2002: 13). Bioregionalism, as the advocates of bioregions call their field of investigation, thrived into the 1970s amongst the alternative scene of the U.S. West Coast, notably thanks to the activist Berg, founder of the Planet Drum Foundation,[25] and through the ecologist Dasmann. Some more recent texts, such as *Dwellers in the land* (Sale, 1985) and *Boundaries of home: mapping for local empowerment* (Aberley, 1993), also contributed to the spreading of this ideology.

The founding text of bioregionalist thought is identified with the publication of Berg and Dasmann's article "Reinhabiting California"

24 <www.deepecology.org/deepecology.htm>, consulted on the 05.12.2014.
25 <www.planetdrum.org>, consulted on the 05.12.2014.

in the journal *The Ecologist* (1977). In this article, the authors highlight that a bioregion is simultaneously a "geographical terrain" and a "terrain of consciousness", and that these two entities should correspond (Berg & Dasmann, 1977: 399). Bioregionalists therefore propose a restructuring of everyday life, in which everyone should/could live in harmony with nature (Frenkel, 1994: 290). In order to pursue this aim, bioregionalists "reject existing political units such as modern counties, states and nations, and replace them with bioregions" (Frenkel, 1994: 291). However, bioregions should not correspond to political boundaries but to natural conditions (Hamm & Rasche, 2002: 3–4).

It is possible to recognize what was discussed above as the desire to designate geographical and political areas following naturalistic zoning. In other words, as Debarbieux suggests, bioregionalism consists in a philosophy that tries to promote natural regions as the new frameworks of collective identification and practices (Debarbieux, 2006: 350). In that sense, bioregionalists wish for a new territorial grid and propose, in the words of Fall, a "fascinating mixture, bounded by both biophysical and societal discourse" (Fall, 2005: 25–26).

But, what is a "bioregion"? Berg expresses the concept as follows:

> A bioregion is defined in terms of the unique overall pattern of natural characteristics that are found in a specific place. The main features are generally found throughout a continuous geographic terrain and include a particular climate, local aspects of seasons, landforms, watersheds, soils, and native plants and animals. People are also counted as an integral aspect of a place's life, as can be seen in the ecologically adaptive cultures of early inhabitants, and in the activities of present-day reinhabitants who attempt to harmonize in a sustainable way with the place where they live.[26]

So, bioregionalism

> [...] offers an alternative paradigm based upon principles, including: division of the earth into nested scales of 'natural regions'; development of localized and self-sufficient economies; adoption of a decentralized structure of governance that promotes autonomy, subsidiarity and diversity; integration of urban, rural and wild environments (Aberley, 2006: 29).

---

26 The definition is taken from the Diggers' Internet site, a group to which Peter Berg belonged (Alexander, 1999: 2): <www.diggers.org/freecitynews/_disc1/00000017.htm>, consulted on the 05.12.2014.

2.8.3 *Adjusting the political order to the natural one*

Bioregionalists thus invoke the adjustment of the political order to the natural one. In fact, they presume, as Harvey suggests in a critical article about this movement, that bioregions are produced by nature or by history rather than created through a variety of intersecting processes operating at quite different temporal and spatial scales; bioregions are thought of as "things" rather than unstable products of shifting processes (Harvey, 1993: 46). Harvey disapproves of the bioregionalists' static vision of regions, which calls to mind one of the main characteristics of the definition of region from the field of "classical regional geography" and criticized by new regional geographers. Moreover, one of the most important characteristics of bioregionalism is a belief in the pertinence of natural regions, in contrast with the administrative and political ones (Alexander, 1999), which are considered arbitrary. This implies political consequences, which Berg and Dasmann already clearly identified in their article: bioregions, as watersheds or mountain ranges, should form a "separate state" (Berg & Dasmann, 1977). "Watershed governments", for example, are seen as appropriate entities to maintain local-life places (Berg & Dasmann, 1977: 401). Attention should no longer be focused on nation states but on bioregions. As Olsen reports in a critical article: "Bioregionalists suggest that the earth can be divided into distinctive and discrete ecosystems, each system or region exhibiting a unique pattern of geographical characteristics and life forms" (Olsen, 2000: 73).

Most contemporary territorial boundaries are thought to be artificial, and therefore not natural. Bioregionalists argue that human communities, if they are to be more ecologically harmonious, should reflect 'natural' (i.e. bioregional) boundaries (Olsen, 2000: 73). Or, if one decides to implement the bioregionalists' precepts, the already existing political boundaries should be questioned, and the creation of new regions should be encouraged. This aspect is nevertheless only seldom openly suggested (Fall, 2003: 89), even in North America, where the bioregionalist ideology spread first (Berg, 1977; Sale, 1985).

In fact, although bioregionalism found an important echo in the United States as a scholarly as well as political idea, it has had little impact on European scholars and politicians. Only in Great Britain, and to a lesser extent in the German-speaking countries (Hamm & Rasche, 2002), has it gained support. The concept of "bioregion" does not evoke much for the great majority of European geographers and scholars, nor does it stir politicians or citizens. Bioregionalism, as a practice and a philosophy, seems not to have found many followers in Europe.

Simonis states that "no real bioregion may be found in Europe" (Simonis, 1997: 12), if one goes by the bioregionalist indicators that Diffenderer and Birch proposed while analysing two bioregions: the Adirondacks and the Sierra Nevada. In fact, the implementation of European bioregions is difficult to undertake, essentially because of the "historical and structural conditions" present in Europe (Simonis, 1997: 14). So, even if there is a growing determination by different actors to display a proactive attitude in constituting new regions, the weight of anchored "structural conditions" is too much to be able to make a *tabula rasa* of the existing administrative grid or to create new territorial structures, as the bioregional precepts would impose.

Although Simonis, in his article on European bioregionalism, finds no examples of "real" bioregions in Europe, he does offer a single example that, in his eyes, could represent a "bioregion to come" (Simonis, 1997: 12). This is the Alpine region. Simonis puts forward some aspects of the Alpine context that can be considered bioregional. After taking a negative stance on the problems facing the Alps (transportation, demographic changes, intensive tourism), he supports the idea that the Alpine Convention

> [...] provides a general framework for integrating ecosystem management and economic development of the region. The Convention contains guidelines both for maintaining the ecological functions and strengthening the economic basis of the region; it calls for permanent information flows and improved regional cooperation (Simonis, 1997: 13).

He sees these as steps in the direction of a constitution of an Alpine bioregion.

Simonis, however, regrets that the "necessary institutional changes" have not been made following the signing of the convention and therefore face implementation difficulties. According to Simonis, socio-political mechanisms are the main factors behind the Alpine bioregion not having been achieved. Nevertheless, in its biophysical configuration, it could constitute a bioregion:

> Though the Alpine region, with regard to its biophysical structure, its resource inventory and natural carrying capacity, may qualify as a major bioregion in Europe, the concomitant change (and harmonization) of those beliefs, attitudes and values that affect humans' interaction with the natural environment [...] seems still to be missing (Simonis, 1997: 14).

Following Simonis' point of view, one can wonder whether some Alpine actors, maybe just a minority group, support the same idea that bioregionalists promote, i.e. the use of a natural perimeter for political purposes in the name of, in the Alps, sustainable development. In other words, I will examine whether the Alpine region, for the Alpine actors under investigation, is "just" an analytical concept or can also be considered a normative concept.

2.9 Networks, key actors in the Alpine regional assemblage

In the Alps, one of the most interesting phenomena that have taken place since the 1990s is the establishment of networks of local political actors. These illustrate quite well the double process that Allen and Cochrane have in mind when they speak about the concept of regional assemblage. In fact, local political actors, to support their activities, built *ad hoc* institutions that transcend existing administrative national borders. At the same time, those activities and institutions are fixing new regional borders. The new collaborations are considered to be the best solution for environmental problems that social and political actors think the entire Alpine region is facing and/or will face in the future.

Those networks, not only in the Alps but also in other regional settings, have often been created to satisfy short-term projects and have been institutionalized at the end of the project's life cycle. They show a "move towards new forms of rule" (Keating, 2003: 2) that do not yet have a proper name. As Keating says, for the moment, these "new forms of rule" include concepts such as governance, networks and neo-institutionalism (Keating, 2003: 2). Above, I discussed the concept of "governance", and in the next section I will discuss the ongoing debate around the term "institutionalization". At that point, I will open the discussion around the concept of "network" in order to explore the role that pan-Alpine networks of local political actors play within the Alpine regional assemblage framework. In the Alps, institutional networks of local political actors link institutions of equivalent levels (for example municipalities, or protected areas) and act at a given common scale, i.e. inside the Alpine Convention's limits.

The first question to answer is: What is a network? One of the most common definitions of network in social sciences come from Leitner *et al.*:

> Networks make connections where none existed before, connecting previously socially or geographically distant actors together in ways that may potentially create new shared understandings and collaborations" (Leitner *et al.*, 2002: 288).

This is, indeed, a first explanation: Networks are structures capable of establishing links, where none were previously.

Additionally, an interesting way to understand the reality of networks, and one that seems appropriate for the Alpine context, is captured in the following definition by Balsiger: "[Networks are] ties between actors on the basis of their common relationship to 'events'". His use of the term "events" allows Balsiger to include "joint participation in social gatherings or professional conferences, or co-membership in organizations or coalitions" (Balsiger, 2006: 12). These events allow the involved actors to share common interests and are likely to last over time.

To push the involvement of actors in a longer-term scenario, we can affirm, as Vertovec does, that "some networks are marked by patterns of communication or exchange of resources and information along with participation in socio-cultural and political activities" (Vertovec, 2001: 573). In this sense, networks seem to play an important role for the re-negotiation of collective identities as "many people's transnational networks are grounded upon the perception that they share some form of common identity" (Vertovec, 2001: 573). As was discussed in the previous chapters, these identities are not fixed in time and space, but they evolve depending on the needs of the actors, the projects they support and their interpersonal interactions, or, as Balsiger suggests, through the exchanges taking place during common events. This first definition is useful for the Alpine case, since one of the main goals of pan-Alpine networks is indeed the exchange of knowledge and know-how in the field of sustainable development.

A second question worth raising concerns the links between networks and territories. If, from a sociological point of view, networks can have an impact on identities, they can also have an effect on geography. French geographer Fourny analyses this issue through her research on the Alpine Town of the Year network. Fourny affirms, while using the material collected for her research on the Alpine Town of the Year network, that the link between identity and action observed by looking at the rhetoric of the network is shifting the status of the *territoire* of reference. The Alpine Town of the Year network refers to the Alps in order to justify its common action and its role in the management of that space. The members defend their public action as an answer to a specific problem. In so doing, Fourny says, the Alpine towns, connected via a network, contribute to politically building the Alpine *territoire*, to create a public space, an object that will be collectively debated (Fourny, 1999: 179–180).

As a result of the networks' activity, a process of redefinition of the *territoires* of actions, and so a redefinition of borders, is also taking place in parallel with the re-negotiation of collective identities. This process can be seen in the EU's cooperation projects, such as Interreg. In fact, as Bray admits, they "have helped to redefine borders as complex zones in which multiple identities can be expressed and negotiated" (Bray, in Keating, 2003: 12) and where identities are performed into actions and projects.

Leitner conceptualizes as follows how new types of horizontal cooperation, no longer linked to old territorial units, are drawing new geographies: "Transnational networks represent new modes of coordination and governance, a new politics of horizontal relations that also have a distinct spatiality" (Leitner, in Marston *et al.*, 2005: 417). Or, as Bulkeley underlines, networks activities are not without their boundaries, in "the ways in which they operate and the ways in which they are framed, configured and crystallised" (Bulkeley, 2005: 888). So, "networks span space rather than covering it, transgressing the boundaries that separate and define these political entities" (Leitner, in Marston *et al.*, 2005: 417). If this is true, Allen would suggest that actors, in order to govern these "transgressing entities", also somehow need to fix those new spaces of actions. This creates a tension between the necessity to spread activities beyond given boundaries and the need to fix these same activities in order to govern them.

This conception of territories and networks recalls the double observation, made by Bulkeley, that on the one hand scalar boundaries are fluid and contested, and, on the other, networks are simultaneously bounded, contrary to what is commonly thought (Bulkeley, 2005: 888). This recognition, Bulkeley hopes, "may provide the basis for further constructive dialogue" between the two concepts (Bulkeley, 2005: 888).

The two notions, networks and territories, seem to coexist in regional assemblages, or, as Bulkeley calls them, in "new networked arenas" (Bulkeley, 2005: 897). These confirm that "geographical scales and networks of spatial connectivity can be seen as mutually constitutive rather than mutually exclusive aspects of social spatiality" (Bulkeley, 2005: 888). Indeed, Bulkeley emphasizes that "networks, scales and territories are not alternatives, but are intimately connected in both a politics of scale, and in creating new arenas of political authority and legitimacy" (Bulkeley, 2005: 896). Bulkeley takes the example of how climate change is governed, but other examples could be used to illustrate this link between networks and territories.

These "new spatialities" defined by networks, which are no longer contained in a topographically defined space, give rise, indeed, to new modes of governance. Perkmann explores the explicit link between transnational cooperation, which can take the form of networks of local political actors, and new forms of governance.[27] He starts his ar-

27 Although I prefer the concepts of "assemblage" and "functional space" for the reasons I explained above, I will also use in this section the terms that I contested (such as "scale", "governance", "level"), because the scholars I am quoting here use them.

ticle by diminishing the impact of networking on governance, affirming that there is a tendency to "overemphasize cross-border regions as emerging territorial units equipped with self-governing capacities" (Perkmann, 1999: 660). For Perkmann, the only positive impact of this kind of cooperation is the financial one. Although he is critical of the long-term effects of cross-border cooperation, he nonetheless underscores a point that should not be underestimated and moderates his initial point of view: "Cross-border cooperation governance is in fact helping to create new opportunities for actors that might change their strategic landscapes both in border areas as well as on a European level" (Perkmann, 1999: 661).

So, it is possible to defend the argument that social and political actors promoting cross-border cooperation are able to create opportunities for the establishment of new territorial arrangements where action can take place. Häkli summarizes this idea when he affirms that cross-border regionalization can challenge the state through new scalar constructs (Häkli, 2008: 475). He continues by saying that

> [...] emerging transnational political spaces can be conceptualized as scales that help actors to skirt the traditional state-centred patterns of networking. Transnational scales, then, are produced and reproduced in processes that set alternative perimeters to networks of cooperation between actors who seek strategic advantage from this cooperation (Häkli, 2008: 475).

Although Perkmann, contrary to Häkli, does not see how trans-border regions can challenge the domination of nation states, he concludes his article by saying that "cross-border cooperation is symptomatic of the newly emerging European policy characterized by the operation of cross-level networks encouraged by structural funds" (Perkmann, 1999: 665). Networking practices that have been implemented in the past few decades in Europe allow us to think of the European space as a place with new regional assemblages, and within which political action can be undertaken, in a way that is not limited by the traditional historical spatial grid. In that sense, "cross-border cooperation has fostered the development of governance disconnected from politics rooted in national territories" (Häkli, 2004: 65).

The scholars I have cited above help us understand how links between networks of local political actors and trans-border cooperation, on the one side, and between networks and the re-negotiation of collective identities on the other can be conceptualized. Yet, a question can be raised about the capability of those same networks to create new spatial modes of functioning – in other words, the capability of those initiatives, based on networks of local political actors, to institutionalize as regions.

2.10 Regional institutionalization

Paasi's theory on regional institutionalization seems particularly interesting for the purpose of understanding regionalization processes. Paasi intends regionalization as a

> [...] socio-spatial process in which a territorial unit emerges as part of the spatial structure of the society concerned, becomes established and identified in various spheres of social action and consciousness, and may eventually vanish or deinstitutionalise in regional transformation (Paasi, 1991: 243).

The Finnish geographer emphasizes that "the areal extent of the idea of region set out here is not confined to any specific regional level" (Paasi, 1986: 121), which means that the new regional arrangements do not necessarily have to match the existing administrative boundaries. The scholar highlights later in the article that a region, as he understands it, can also be a part of a city, a county, a province or a national state. Thus, and although Paasi rarely suggests it openly, his theory can be applied to regions, that do not (yet?) officially exist.

If one considers regional institutionalization such as it is presented above, the true interest of this theory is to highlight the process of construction of a "human and social category" (Paasi, 1986: 119), a process that needs not be associated, in a strict sense, to political and administrative regions. In an article devoted to Upper Silesia, a region that spreads over three nations (Poland, Czech Republic and Germany), geographer Bialasiewicz illustrates how this social and historical process takes place and how it is evolving. In order to capture the reality of Upper Silesia, Bialasiewicz defines the process of regionalization

> [...] not simply as the administrative apportionment of segments of space but, rather, as the dynamic process of constituting regions as particular 'imagined communities' in space, legitimating distinct territorial divides and allocating governance (Bialasiewicz, 2002: 112).

This definition adds still more arguments in favour of a larger definition of how a "region" must be thought.

But how does Paasi define the process of regional institutionalization? If we follow his theory, this involves four not necessarily chronological steps that I itemize with the help of an article he wrote in 1991 (Paasi, 1991: 244–246):

- The development of a "territorial shape", which refers to the "localisation of social practices (for example, economy, politics, and administration) through which the regional transformation takes place and a region achieves its boundaries and is identified as a distinct unit in the spatial structure";
- The formation of a "symbolic shape", which establishes "specific structures of the territorial symbols for a region. Symbolic orderings of space and time provide a framework for experience through which we learn who or what we are in society (Harvey, 1989). The increasing number and use of territorial symbols is crucial for creating the symbolic significance of a region". Paasi quotes Martin, who explains the function of symbolic systems as follows: "One major function of symbolic systems in all societies and groups is to legitimate, reinforce and celebrate the status quo – both the existence of the social *per se* and the particular disposition of roles, values, identities, privileges and so on in the place and time";
- The emergence of institutions, which serve "to strengthen the significance and role of territorial symbols and signs";
- The establishment of a region, which refers to "any continuation of the institutionalisation process after the region has an established, albeit not necessarily administrative, status in the spatial structure and social consciousness of the society".

MacLeod considers the theory of regional institutionalization to be valuable. Although he only refers to administrative regional units in his 1998 article, his words seem equally applicable to other, not administrative, regions:

> In the contemporary period of socioeconomic and political turbulence across global space, the theory [of the institutionalization of regions] may provide some valuable insights as to how particular socially cohesive territories emerge, institutionalize, and come to be established as specific sites for economic and political governance (MacLeod, 1998: 840).

This process and these stages confirm that regions are not "given" but definitely built through the determination of social and political actors, who, by means of different processes (the designation of a territorial area charged in symbols and established through institutions) are making this territorial entity exist in the social consciousness of the group that is defining it and defined by it.

In that sense, regions are not only social facts, but they entail a strong normative element:

> "Region building" always includes normative components because institutional structures are structures of rules, power and trust, in which boundaries, symbols and institutions merge through material practice (Paasi, 2002b: 805).

Therefore, an institutionalized region also signifies a "space of political governance" (MacLeod, 1998: 840). This occurs in a European context in which, according to some scholars, the construction and development of new regional assemblages is promoted and supported, notably via the Interreg programmes. In this context, networks of local political actors play an essential role, since they become, via their social and political actions, promoters of the creation of new regions, as we will demonstrate by illustrating the role of pan-Alpine networks in the establishment of the pan-Alpine region.

2.11 Hypothesis

With the help of concepts such as regional assemblage and institutionalization, the framework has been established to understand current processes of regionalization. The emergence of new spaces of action allows a great range of people and groups of people to actively promote initiatives amongst them. Since the 1990s, in the Alpine context, these have mainly taken the form of networks of local political actors, the main subjects under investigation in this research. In the following sections I explain how I analysed these networks, highlighting my research focuses and the axes of analysis.

For the members of the three networks under analysis here (Alparc, Alliance in the Alps and Alpine Pearls), the mountain range constitutes, thanks to different ways used for *objectivizing* it, the common reference for local political actors organized in pan-Alpine networks. They define themselves as belonging to the Alps and refer to them whilst undertaking pan-Alpine projects. The distinctive characteristics of the Alpine territory (its landscape and its topographic specificities), objectivized by local political actors, seem to be a powerful engine for collective action, which, in turn, helps build a common identity.

Working together towards protecting the mountain *territoire*, local political actors share experiences and know-how through their networks; they also share common representations of Alpine nature, landscapes, topography and society. Because of this shared identity, they act to construct the Alps as they imagine them: a *territoire* shaped by sustainable development. Thus, the "social glue" that unites them to work on common projects is their adherence to the concept of sustainable development. Local political actors mobilize with the conviction that their commitment can produce a change in the ways people think of and live in this mountain area. By doing so, they are the reference for this source of social change. The awareness of this capability is associated with a desire to act on the space in which they live and on their own society: The *identification* with the Alpine region can therefore be considered an engine for social and political action.

So, the double feature of identity, objectivation and identification, helps explain the process of regional assemblage at the Alpine range scale, based on a "model of multi-actor cooperation" (Balsiger, 2007: 33; Debarbieux, 2008b: 40). Alpine actors create pan-Alpine institutions because they have a shared belief of what the Alps are (and especially of what they should be) and because they feel part of it. The role of some particularly active individuals (called leaders) is decisive, and they are to be considered the real "engines" of the regionalization process. The institutions built by pan-Alpine actors cover the perimeter of an area thought to correspond to what is believed to be the natural border of the Alps.

Of course, the two processes, the creation of an Alpine identity on the one hand and the establishment of institutions on the other, are not chronological. They clearly have to be considered as mechanisms of co-construction. It is because somebody believes in a thing called "Alpine region" that he or she organizes institutions to support it, but, at the same time, regional institutions contribute by giving meaning to the Alpine region. This book examines how the three processes (objectivation, identification and institutionalization) combine in the Alps and measures the social and political consequences of this convergence.

I argue that, although there is clear involvement of a number of social and political actors for the Alpine cause, and although *ad hoc* institutions have been created in order to support this action, the Alps cannot yet be considered a real institutionalized region, as defined by Paasi (Paasi, 1986), since pan-Alpine regional processes are in all likelihood in the hands of only a few particularly active persons and, as observed in other regions as well, without much concern from ordinary people (see for example Häkli, 1998; Strüver, 2004).

3. Methodology

This chapter begins with a presentation of the research design conceived for the study. I explain why I chose to use an interpretative method based on a single case embedded research design. The case under investigation, namely three pan-Alpine networks of local political actors, comprises three principal units of analysis: the Alliance in the Alps, Alparc and Alpine Pearls networks. In the second part of the chapter, I introduce the three methods used to collect primary data (questionnaires, interviews and first-hand observations) and the strategy adopted to analyse them. In the third part, I propose some reflections on my position in this research.

3.1 Research design

Robert Yin suggests four basic types of research designs for case studies (Yin, 2003: 40). In this research I apply the single-case embedded research design. Yin visualizes this type of design as follows:

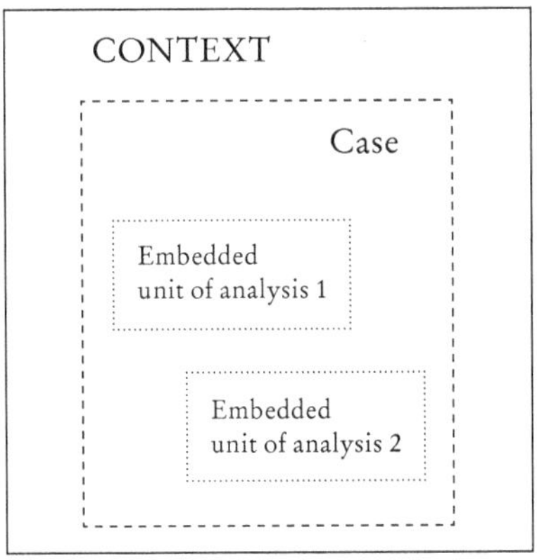

Fig. 1: Research design, based on Yin, 2003.

In the context of the Alpine regionalization process, the analysis will focus on a case study, i.e. networks of local political actors. The case study, highlighting how social configurations of actors and mechanisms of regional construction are taking place in the Alps, is composed of three main units of analysis (namely the Alliance in the Alps, Alparc and Alpine Pearls networks). The three chosen networks will be analysed as emblematic examples of a more generalized phenomenon: the constitution of a number of networks involving local social and political actors on the Alpine range scale.

Among the characteristics of case study research listed by Yin (Yin, 2003), some are helpful to understand why this research strategy has been applied in the current study. Yin stresses that the case study research is particularly suited to describe a contemporary phenomenon and to generate theory (analytical generalization). In other words, it is the best research strategy for describing, understanding and explaining a phenomenon (Hamel, 1993: 39). Hamel summarizes the design as follows: "The detail and depth of the description rendered by the case study permit an understanding of the empirical foundations of the theory" (Hamel, 1993: 33).

That is why the case study research is best suited to a thorough understanding of a phenomenon, given the complex characteristics and issues observed in the field. A case study generates concrete, practical and context-dependent knowledge (Flyvbjerg, 2001: 70), which is considered one of the major strengths of this kind of approach, as the case study research aims to examine complex issues by focusing on one or a few units of analysis to provide detailed and contextual analysis.[28]

To sum up, the strategic aim of case study research is to "provide an analysis of the context and processes which illuminates the theoretical issues being studied" (Hartley, in Kohlbacher, 2006). The possibility of going deep with a case is often criticized for being unproductive. Responding to this oft-heard criticism, Flyvbjerg asks where, by contrast, a singular focus on deduction and general principles (frequently seen by the scientific community as the ideal way of

28 Personal notes of the course held by Michael Gibbert, *Case studies: design, methods and reporting*, Swiss Summer School 2009, Lugano.

researching), leads. Flyvbjerg has a clear answer: "This leads to a dead end" (Flyvbjerg, 2001: 71).

Case study research, thus, is not interested in deduction and general principles, but aims towards analytical generalization (as opposed to the statistical generalization). Yin highlights that, in this type of generalization, a previously developed theory is used as a template with which to compare the empirical results of the case study (Yin, 2003: 32–33). In sum, analytical generalization compares theory to findings. While this perspective is favoured, it is not essential to obtain the same results in every organization, since generalization can be developed by observing how each case potentially represents values of some generic variables or processes (Becker, 1990: 239–240).

Taking a study on prison cultures as an example, Becker shows that although different results were produced by analysing different prisons, these differences did not invalidate the generalization of that theory on prison culture. In fact, variations in the results mean that "generalizations are not about how all prisons are just the same, but about a process, the same no matter where it occurs, in which variations in conditions create variations in results" (Becker, 1990: 240). He concludes with a clear statement: "That is actually a classier form of generalization anyway" (Becker, 1990: 240).

For the purpose of analytical generalization, the number of cases (sample) is not important. What does matter is matching cases with theory and not, as regards statistical generalization, being able to generalize from sample to population.[29] Hamel, quoting Bourdieu, convincingly shows why analytic generalization based on a single case study is a valid tool for enhancing knowledge:

> In Pierre Bourdieu's recently published discussions with his student, Loïc Wacquant, he appropriately noted that "there was no need for Galileo to constantly repeat the slope experiment to construct the falling body model. A well-constructed single case is no longer singular" (Bourdieu, 1992: 57). (Hamel, 1993: 35).

As Charles Ragin explains, it does not make any sense to criticize single-case studies for being inferior to multiple-case studies, since

29 Personal notes of the course held by Michael Gibbert, *Case studies: design, methods and reporting*, Swiss Summer School 2009, Lugano.

the former "are multiple in most research efforts because ideas and evidence may be linked in many different ways" (Ragin, in Flyvbjerg, 2001: 82). These links are created by researchers in constant contact with their field. For these reasons, case studies provide good material to "learn something", as Eysenck declares: "Sometimes we simply have to keep our eyes open and look carefully at individual cases – not in the hope of proving anything, but rather in the hope of learning something!" (Eysenck, 1976: 9).

The importance of analysing the phenomenon under investigation in the context in which it occurs is one of the main characteristics of the case study research strategy. What Mintzberg already stated for management studies can easily be applied to the general fields of the social sciences. He defends the idea that measuring social facts means first of all getting out into the field. While, as he argues, psychologists can analyse the phenomenon they are studying inside their laboratory, the social scientists cannot. Since social phenomena are characterized by their inherent complexity and dynamic nature, it is impossible to simulate their functioning in a laboratory (Mintzberg, 1979: 586).

Going into the field is a *sine qua non* condition in order to study social facts. Speaking about processes of scalar structuration, Brenner stresses that there is only one way to find out whether a given social process generates sociologically or politically significant outcomes, namely through context-specific inquiries (Brenner, 2001: 606). That is why a large part of this research is devoted to interviews and personal observations from the field. According to Flyvbjerg, this closeness of the case study to real-life situations is important for two reasons: First, it allows for a nuanced view of reality; second, case studies are well suited to produce a concrete and context-dependent knowledge, which is what social sciences should aim for, as they cannot produce predictive theories (Flyvbjerg, 2001: 72).

3.1.1 Units of analysis: three networks under investigation

In the Alps, there is a growing number of networks of local political actors. One can mention the network of municipalities, "Alliance in the Alps"; the "Alpine Town of the Year" network; "Alparc", the network of protected areas, the "Network Enterprise Alps" (NENA); the scientific network of Alpine research (ISCAR) and the *Club Arc Alpin*. These are members of the *Commission Internationale pour la Protection des Alpes* (CIPRA), as shown in fig. 2: Pan-Alpine networks, © CIPRA.

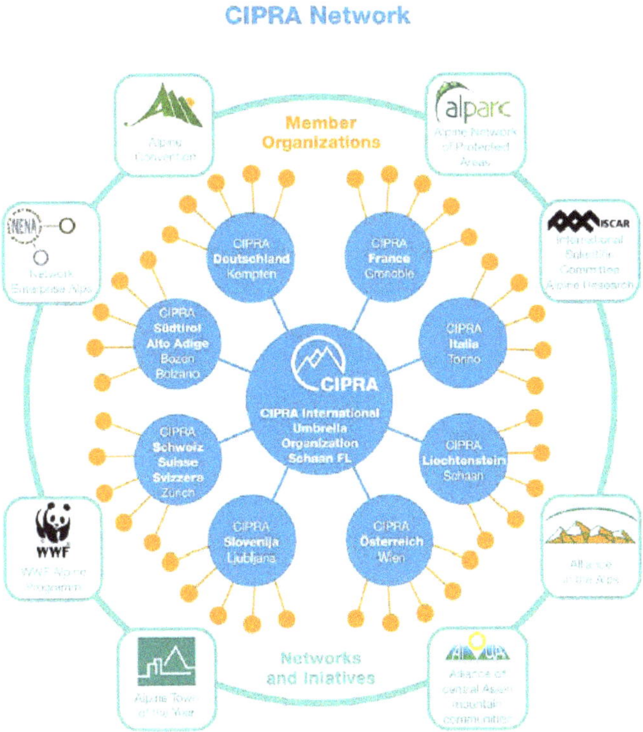

Fig. 2: Pan-Alpine networks, © CIPRA.

Other networks, independent from CIPRA, were created, for example Best of the Alps, which links "classic Alpine resorts, all bearing a resounding name"[30] with Alpine Pearls, a network of Alpine resorts promoting an "environmentally sustainable, culturally diverse, qualitatively high and adventure-packed holiday experience".[31] All of

30 <www.bestofthealps.com/050dav_01_en.htm>, consulted on the 05.12.2014.
31 <www.alpine-pearls.com/en/soft-mobility/our-guarantee-of-quality.html>, consulted on the 05.12.2014.

these networks are probably worth studying, but, unfortunately, any given project only has a few investigators to analyse them. This basic, practical consideration leads to the obvious conclusion that a selection must be made when deciding which networks will be thoroughly investigated.

At the beginning, the decision was made to analyse two of them: Alliance in the Alps, a network of Alpine municipalities, and Alparc, the network of Alpine protected areas. These two are among the oldest ones to have been created in the Alps, Alparc in 1995 and Alliance in the Alps in 1997. Moreover, they are currently the most active networks and have not experienced periods of decreasing activity, as have for example the Alpine Towns of the Year or NENA networks.

Additionally, Alpine Pearls was included in the research, when, in 2009, a student, Nicolas De Groote, joined the research team and decided to write his master's degree dissertation (De Groote, 2009), under the direction of Prof. Debarbieux, on the Alpine Pearls network. This was a great opportunity to add material to the research. Alpine Pearls, contrary to the other two networks, is devoted to a single particular theme (sustainable tourism) and is working independently from CIPRA. Moreover, the network has a clear marketing goal: selling the image of "sustainable tourism resorts" to potential customers. This was seen as a good opportunity to analyse variations of discourses in a network that has a different purpose but chooses a similar way of acting: via the constitution of a pan-Alpine network of local political actors.

In that sense, a theoretical, and not a statistical, sampling strategy was chosen concerning the selection of the units of analysis. If a theoretical sampling is applied, units of analysis are chosen for theoretical, not statistical, reasons (Glaser and Strauss, in: Eisenhardt, 1989: 537). Hence, units of analysis are chosen so that they are likely to replicate or extend the emergent theory (Eisenhardt, 1989: 537). In the same way, for every unit of analysis, a non-fixed sample was established to conduct interviews.

The number of interviews for each unit of analysis and the kind of persons interviewed were decided according to the need for information and the ability of those persons to shine a light on the mechanisms of the process under investigation. As an example, after four months of research and 28 interviews, it became clear that the research could not be finished without interviewing Antonio Zambon, at that time Mayor of Budoia (Italy) and Vice President of Alliance in the Alps. Every person I interviewed and who knew him personally told me: "You should go to Budoia and meet Antonio." When on the 6th of July 2007 I finally met him, I understood why this suggestion was so valuable: Zambon embodies the soul of the network and could delight me with anecdotes and stories, which enriched the collection of information.

3.2 Choosing an interpretive method

In 2005 I attended a public conference held by the Italian philosopher Remo Bodei and organized by the Ticino Cantonal Library, in Bellinzona (Switzerland). Bodei was invited to speak about the traumatic experiences of war, in the context of an exhibition called *L'Histoire c'est moi*, a result of the most important research on the oral history of the Second World War in Switzerland. At this event, Bodei made the following memorable statement: "Scientific objectivity does not exist, but honesty does." This is certainly true for the sensitive theme of his speech:

> Trauma as wars separate what happens 'before' and what happens 'after' and induce people to reformulate, delete or reinvent memories. Memory and oblivion entwine and everyone selects and organizes, from his or her point of view, a personal heritage of experiences.[32]

This should be true for every research, as well.

Every researcher is called upon to select and organize, from his or her point of view, the material that he or she collects during the fieldwork. Certainly "scientific honesty" is what I endeavoured to follow

32 Presentation of the conference by Remo Bodei, "Traumi della memoria", Biblioteca Cantonale di Bellinzona, 27.10.2005. Conference cycle "Ricordare/Dimenticare" in the framework of the exhibition "L'Histoire c'est moi".

during the entire path of my research. I not only want to be "honest" towards the reader, but first and foremost I want to be honest towards all those who during this long process decided to dedicate some of their time and energy to me. Honesty is particularly important in qualitative research, which too often is accused of not being objective and reliable. The constructivist approach adopted here challenges the idea that research, qualitative or quantitative, is conducted by "impartial, detached, value-neutral subjects, who seek to uncover clearly discernible objects or phenomena" (Boyd in Mir, Watson, 2000:941).

The choice of a mainly qualitative approach is based simply on the assumption that the method chosen "depends on what is to be studied" (Mintzberg, 1979: 584), as Mintzberg already defended in 1979 in his *plaidoyer* for qualitative research. In other words, "choice of method should clearly depend on the problem under study and its circumstances" (Flyvbjerg, 2001: 75). Qualitative data analysis is, in the eyes of Eisenhardt, the best methodology to provide a good understanding of "why" something is happening and "what" that something is. Since it is the appropriate method to uncover the dynamics underlying relationships (Eisenhardt, 1989: 542), it is suitable for the purposes of my research.

Although qualitative research in general, and case study strategy in particular, is often seen as "unscientific" or "subjective", and qualitative researchers are time and again defined as "journalists" or "soft scientists" (Kohlbacher, 2006), in the past few decades, as Kohlbacher underscores in an interesting synthesis article, "the development of qualitative methods showed impressive advances and results" (Kohlbacher, 2006). It was undoubtedly the most suitable approach to answer the questions raised by the present research.

I agree with Reichenbach, who, already in 1938, stated that qualitative analysis is more a field of discovery, in opposition to the quantitative analysis that covers the field of justification (Reichenbach, 1939). Reichenbach defines the "context of discovery" as being an interpretative and exploratory research that aims to identify variables and the relationships among them. Interpretive research, as opposed to the positivist one, assumes that "people create and associate their own subjective and intersubjective meanings as they interact with the world around them" (Orlikowski & Baroudi, 1991: 5). In that sense, and as Orlikowski and Baroudi point out, interpretive researchers study phenomena by understanding the meanings that subjects (e.g. interviewees) assign to them. In contrast to descriptive studies, interpretive studies reject the prospect of an "objective" account of events and situations, seeking instead a relativistic, albeit shared, understanding of phenomena" (Orlikowski & Baroudi, 1991: 5).

Since the aim here is to uncover social facts in the Alpine region, in other words to answer the questions of "how" and "why" the Alpine regionalization is taking place in the contemporary European context, case study research using qualitative methods is the most suitable. The case study research, defined by Eisenhardt as a "strategy which focuses on understanding the dynamics present within a single settings" (Eisenhardt, 1989: 534), fulfils the aim of this research. Accordingly, to "choose case study analysis with an interpretive method it is just a matter of choosing the right tool for the right purposes".[33]

3.3 Data collection strategy: questionnaires, interviews and observation

The fieldwork started at an early stage of the research and continued throughout the process (see interview table). I interviewed and observed persons whom I call "local political actors". This term must be understood as comprising all the persons involved in the Alpine networks and Alpine institutions promoting a new way of thinking as well as a new way of governing what they define as the Alpine region. Their local action is supposed to have an impact at a regional and supra-regional level and therefore can be considered political.

Being in continual contact with the actors involved in the Alpine regionalization process has allowed me to continuously compare theory and data, which is as per Eisenhardt the central idea of

33 Personal notes of the course held by Michael Gibbert, *Case studies: design, methods and reporting*, Swiss Summer School 2009, Lugano.

a case study strategy (Eisenhardt, 1989: 541). In fact, the key feature of theory-building case research is rightly the "freedom to make adjustments during the data collection process" (Eisenhardt, 1989: 539). These allow researchers to explore emergent themes or take advantage of unique opportunities that present themselves in a given situation (Eisenhardt, 1989: 539).

Alliance in the Alps always opened its doors wide to me so that I could participate in all the events and meetings they organized. I quickly realized this was an opportunity that I should not miss; it was a chance given to me to observe the network in depth. As a result, I devoted more time to it than to other networks, something that was not calculated at the onset of the research. As a consequence of this choice, the reader will probably have the (correct) impression that the analysis of one network, Alliance in the Alps, is more in-depth than that of the others. I do not consider it a weakness, but rather a chance to better understand the functioning and challenges of one particular key actor of the regionalization process. To use Eisenhardt's words: "Flexibility is controlled opportunism in which researchers take advantage of the uniqueness of a specific case and the emergence of new themes to improve resultant theory" (Eisenhardt, 1989: 539).

Although the main methodology employed can be considered qualitative (interviews, first-hand observation and participation), the research started with a survey (questionnaire), distributed to the members of the three networks under investigation: Alliance in the Alps, Alparc and Alpine Pearls. The survey is to be considered a first exploratory research phase, while semi-structured interviews and observations compose the main core of the research (see summarizing Table 1).

Type of collected data	*Number of collected data*
Interviews	45
Participation at events	6
Questionnaires (answers received)	88 for Alliance in the Alps 68 for Alparc 9 for Alpine Pearls 77 for the Danilo Re event

Table 1: A summary of the collected data.

The multiple methods of data collection made it possible to triangulate data; in other words, it permitted the "adoption of different research angles from which to look at the same phenomenon, by using different data collection strategies and different data sources" (Gibbert *et al.*, 2008: 1468).

During the research process, I regularly took field notes that helped to "overlap data analysis with data collection" (Eisenhardt, 1989: 538). The journal mainly consists of notes that I wrote after the interviews, ideas about possible relations between interviewees' statements and links between those statements and the theory, in order to construct theoretical hunches. My notes cannot be regarded as what anthropologists consider a true field notebook. In my case, field notes primarily played the role of fuelling the theoretical reflection process.

Whenever possible, I also made note of anecdotal situations that I observed. The importance of recording anecdotes is elucidated by Mintzberg, who defends the idea that

> [...] for while systematic data create the foundation for our theories, it is the anecdotal data that enable us to do the building. Theory building seems to require rich description, the richness that comes from anecdote. We uncover all kinds of relationships in our 'hard' data, but it is only through the use of this 'soft' data that we are able to 'explain' them, and explanation is, of course, the purpose of research (Mintzberg, 1979: 587).

This can be compared to a young Sardinian author who sketches a delightful portrait of his island in his splendid novel, "for who, like me, is allergic to abstractions, does not believe in horoscopes and in ascendants and needs to stick to events, to words, anecdotes may be the only way we have to describe a person" (Soriga, 2008: 149). The story of the people and institutions building the Alpine region would not be as interesting and "graspable" if anecdotes were not recounted. That is why some of them will be related in this document.

3.3.1 Questionnaires for Alliance in the Alps, Alparc and Alpine Pearls leading to the choice of in-depth interviews

The survey was conducted in two stages, on the basis of the opportunities offered by three projects funded by the Swiss National Science Foundation (SNSF) and by the Boninchi Foundation (Geneva, Switzerland). They were all directed by Prof. Debarbieux at the Department of Geography (University of Geneva):

- "*Populations de montagne en réseau: Genèse, finalité et fonctionnement de réseaux de populations de montagne dans le monde depuis 1990*", 2006–2007, funded by the Boninchi Foundation;
- "Pan-Alpine Networks: *Analyse des réseaux d'acteurs et d'institutions à l'échelle des Alpes en rapport avec la Convention Alpine*", 2006–2008, funded by the SNSF;
- "*Réseaux pan-alpins: L'émergence des Alpes comme cadre de l'action collective*", 2007–2008, funded by the SNSF.

Marie-Anne Guérin started working as a researcher for the first SNSF project in 2006; she compiled the questionnaires for the members of two networks, Alliance in the Alps and Alparc, as planned in the two older projects. In February 2007 Petra Arnuš and I joined the research team "Environments and mountains: knowledge and policies" at the Department of Geography and Environment at the University of Geneva.[34] The three projects dealt with networks and the building of institutions at the pan-Alpine level and as such covered the main research questions of this thesis. The material collected during these projects was used in various publications by the members of the project and in the present book.

During the first stage of the research (January 2007 – June 2007), we sent questionnaires to the members of Alliance in the Alps and Alparc. The two questionnaires were elaborated using the same structure of questions, with some variations given the uniqueness of each network.

For Alliance in the Alps, 258 questionnaires written in four languages (German, Italian, French and Slovenian) were distributed,

34 Petra Arnuš left the project at the end of 2007.

yielding 88 usable responses, representing a 35 % rate of response. Because of the structure of the network, the survey was sent to the administration of the groups of municipalities (11 replies were received) and to the single municipalities belonging to such groups, that will later be called "regional members" (52 replies). For example, the questionnaire was not only sent to the administrator of the Pfynwald Region (Switzerland), but also to the three municipalities that are part of that region, i.e. Salgesch, Varen and Leuk. So, in the case of Pfynwald, I expected a total of four replies. It was also sent to municipalities that joined Alliance in the Alps directly (25 replies), that is municipalities independent of any such groups/regions.

For Alparc, a managers' survey was conducted. A questionnaire written in the four Alpine languages had been emailed in February 2007 to 170 protected areas managers. By means of additional phone calls, 68 answers, mainly written by directors and deputy directors of national, regional and nature parks, were collected. Altogether, they cover 83 protected areas of different categories, since some managers are in charge of several protected areas. A second questionnaire was filled in by 77 participants of the 12th Danilo Re Trophy (Val d'Allos, France, March 2007). This sporting event brings together teams of rangers from various protected areas throughout the Alps to compete against each other. The reason for participating in this event was to get another point of view, different from that of the park managers.

In 2008, when Nicolas De Groote joined the research team, he sent 20 questionnaires to the 20 members who, at that time, belonged to Alpine Pearls. He received nine answers.

The collected material was analysed and presented in a research seminar held in Geneva on the 28th of November 2007. The intermediate results of the research were presented at that time. The seminar was divided into two parts: The first was explicitly dedicated to the pan-Alpine networks, and the second to identities and collective action referring to mountains in general. A short document (four pages) summarizing the main results of the quantitative analysis was distributed to the participants (Del Biaggio, 2007; Arnuš, 2007). The seminar represented an opportunity for a constructive debate amongst selected academic and institutional partners. The representatives of Alparc, Alliance in the Alps and CIPRA were present, so that the first findings were discussed. That gave rise to clarifications and brought up new issues for further investigation. The analysis of the questionnaires, in addition to providing material for the first discussions between the research team and the partners, was central for two further reasons: First, it helped put forward themes to be discussed during in-depth interviews; and second, it facilitated the identification of locations (fields) where those interviews should be conducted.

For Alparc, for example, the differences of answers between managers and rangers, not only in the questionnaires but also in informal talks during the "Danilo Re" event, made it clear that some themes (like the introduction of a unique Alpine uniform) should have been discussed during the interviews with the managers of the parks. The answers given in the questionnaires made it clear, as well, that the decision by Alliance in the Alps to allow groups of municipalities, instead of only single municipalities, to become members of the network was not unproblematic, especially in terms of participation of the regional members. The latter, in fact, seemed to be less active than single members. Nevertheless, this did not appear to be the case for every region. For this reason, two locations were chosen: Ticino (Switzerland) and the *Grosses Walsertal* – Great Walsertal (Austria). For Ticino, I conducted interviews with the regional secretaries and some of the members of two groups of municipalities, Malcantone and Valli di Lugano.[35] For the Great Walsertal, I interviewed the regional manager and the mayors of four of the six municipalities composing the region. In fact, the single municipalities composing the different regions of these two areas already showed in the questionnaires a clearly different degree of identification with the network. The reasons for this divergence were worth investigating.

In the case of Alpine Pearls, Nicolas De Groote and I decided to conduct interviews with at least one member of every country belonging to the network (Austria, France, Germany, Italy and Switzer-

35 In 2011, these two regions decided to leave the network. Representatives of the network explained to me that this was due to territorial readjustments that took place in the canton following the introduction of the New Regional Politics. In this document they will be treated as members of the network.

land). We decided not to go to Slovenia because of time and financial limitations. Nicolas De Groote interviewed the members of the network, while I interviewed the head of the network, the President and the manager.

3.3.2 Interviews

The pertinence of conducting interviews as a way to obtain more in-depth and plausible information than is the case with questionnaires can be illustrated by one of the most symbolic examples with which I was confronted during my research: Before choosing which mayors I would interview during my fieldtrip to the Great Walsertal, I decided to first have a look at the answers that the different mayors of the valley had sent back to me. My curiosity was notably piqued by the answers given in one particular questionnaire. Part of the questionnaire, especially for questions concerning directly the network, was left blank. To all other questions a clear inclination for "no" and "never" answers could be detected. "No", the municipality does not have any project of sustainable development. "No", the municipality did not apply for Dynalp I or II projects. "No answer" to the question about whether the municipality takes part at events organized by the network. "No", the municipality did not try to convince other municipalities to join the network. "No", the employees of the municipality "never" consulted the Internet page of the network.

Before interviewing the mayors, I spent some time with Sabine (animator), who asked me whom I would interview. When I told her that I would meet the mayor of the "no municipality", she made a quite astonishing remark: "You have to know, Cristina, that this is the only municipality which does not have sunshine during the winter." This explanation brought back old memories of my first year as a geography student, when professors explained environmental determinism to us. In my eyes, at least, topography and exposition to the sun could not justify the answers I had read. While travelling to the small municipality by bus, I tried to imagine this undoubtedly gruff and cold mayor, who did not propose anything progressive for his municipality. Imagine how great my surprise was when I discovered the most enthusiastic and open mayor of the valley. I smile now as I recall this anecdote, but it led me to reflect a lot on the two different ways of collecting information: questionnaires and interviews. I used both methods, taking advantage of both types of information collected, but this anecdote shows why, in this specific inquiry, the data collected from interviews are typically more significant and, therefore, constitute the main research material upon which the results will be presented.

Table 2 shows the details of the 46 semi-structured interviews that my colleagues and I conducted during three and a half years of projects (the interviews are presented in chronological order). The interview guides covered a range of topics that were discussed during the interviews in a non-chronological order. It cannot be considered a standardized guide, since it was adjusted depending on the type of institution and the status of the interviewees, and of course, on the interviewing process.

The majority of the interviews were recorded and transcribed, in order to allow the coding process with qualitative data analysis software. Tommaso did not want to be recorded, because of the tense political situation linked to the introduction of the New Regional Policy in Switzerland. The interviews with Rainer, Gianluca and the second interview with Fabio were not recorded either, because of the inappropriate surroundings in which they were conducted. In those cases, notes were written down and entered in the fieldwork journal. The interview with Mario was conducted only per telephone, since the information he gave me during the call was sufficient for my purposes.

Name of the interviewee	*Institution*	*Function of the interviewee*	*Date of the interview*	*Interviewer*	*Language*
Robert	Alliance in the Alps	Animator	24.10.2006	Marie-Anne	French
Olivier	Alliance in the Alps	Animator	25.10.2006	Marie-Anne	French
Peter	Alparc	Ranger	17.03.2007	Petra	German
Laura	Alparc	Ranger	16.03.2007	Cristina and Petra	Italian

Andrej and his colleagues	Alparc	Rangers	17.03.2007	Cristina and Petra	Slovenian
Antonio and his colleagues	Alparc	Rangers	17.03.2007	Cristina	Slovenian
Giovanni	Alparc	Ranger	17.03.2007	Cristina	Italian
Dario	Alparc	Ranger	17.03.2007	Cristina	Italian
Ivica	Permanent Secretariat of the Alpine Convention	Employee	30.03.2007	Petra	Slovenian
Mirko	Alliance in the Alps	Ex-mayor of a single member of Alliance in the Alps	30.03.2007	Petra	Slovenian
Fabio	Alliance in the Alps	Animator and manager of a group of municipalities	02.04.2007 and 23.09.2009	Cristina	Italian
Branko	Alparc	Ex-director of a protected area and former member of the Alparc board	12.04.2007	Petra	Slovenian
Dragan	CIPRA	Member of the board of CIPRA	12.04.2007	Petra	Slovenian
Goran	Alparc	Director of a protected area	12.04.2007	Petra	Slovenian
Jean	Alparc	Director of a protected area and member of the board of Alparc	10.05.2007	Petra and Bernard	French
Julien	Alparc	Director of a protected area and member of the board of Alparc	31.05.2007	Petra and Bernard Debarbieux	French
Claudio	Alliance in the Alps	Manager of a group of municipalities	06.06.2007	Cristina	Italian
Rainer Siegele	Alliance in the Alps	President of Alliance in the Alps and mayor	11.06.2007	Cristina and Bernard Debarbieux	English
Sabine	Alliance in the Alps	Animator	11.06.2007	Cristina	German

Maximilian	Alliance in the Alps	Mayor of a municipality belonging to Alliance of the Alps via a group of municipalities	12.06.2007	Cristina	German
Ulrike	Alliance in the Alps and Alparc	Manager of a group of municipalities that also belong to a protected area	12.06.2007	Cristina	German
Henrik	Alliance in the Alps	Mayor of a municipality belonging to Alliance of the Alps via a group of municipalities	13.06.2007	Cristina	German
Jakob	Alliance in the Alps	Mayor of a municipality belonging to Alliance of the Alps via a group of municipalities	14.06.2007	Cristina	German
Niklas	Alliance in the Alps	Mayor of a municipality belonging to Alliance of the Alps via a group of municipalities	14.06.2007	Cristina	German
Leonardo	Alliance in the Alps	Former employee of a group of municipalities, in charge of finding new members for Alliance in the Alps	25.06.2007	Cristina	Italian
Nicolò	WWF	European Alpine Programme of the WWF	27.06.2007	Cristina	Italian
Manuele	Alliance in the Alps	Employee of a municipality belonging to Alliance of the Alps via a group of municipalities	28.06.2007	Cristina	Italian

Tommaso	Alliance in the Alps	Employee of the Section of the economic promotion for the Canton of Ticino (Switzerland). Member of RegioSuisse, network of regional development.	29.06.2010	Cristina	Italian
Antonio Zambon	Alliance in the Alps	Mayor of a single member of Alliance in the Alps and member of the board of the network	06.06.2007	Cristina	Italian
Christian and Giorgio	Centre of Alpine Ecology (Trieste)	Director and ex-director of the centre	24.10.2007	Cristina	Italian
Giulia	WWF	Ex-employee for the WWF Alpine programme	26.10.2007	Cristina	Italian
Gianluca	Alliance in the Alps	Mayor of a potential member of Alliance in the Alps	12.06.2008 (and email exchanges, August 2010)	Cristina	Italian
Erik	Alparc	Employee	09.10.2008	Cristina	French
Paolo	Alpine Convention	Member of the board	10.10.2008	Cristina	Italian
Peter Brandauer and Karmen Mentil	Alpine Pearls	President and manager	29 and 30.01.2009	Cristina	German
Mathieu and Philippe	Alpine Pearls	Employees of two municipalities	02.04.2009	Nicolas	French
Hans	Alpine Pearls	Mayor of a municipality and member of the board	06.04.2009	Nicolas	German
Antje	Alpine Pearls	Employee of the tourist office of a municipality, member of Alpine Pearls	07.04.2009	Nicolas	German

Normann	Alpine Pearls	Employee of the tourist office of a municipality, member of Alpine Pearls	08.04.2009	Nicolas	German
Viktoria	Alpine Pearls	Employee of the tourist office of a municipality, member of Alpine Pearls	28.04.2009	Nicolas	German
Edy and Patrizia	Alliance in the Alps	Employees of a potential member of Alliance in the Alps	04.11.2009	Cristina	Italian
Mario	Alliance in the Alps	Manager of a group of munici-palities, potential member of Al-liance in the Alps	04.11.2009	Cristina	Italian
Giovanna and Martina	Alliance in the Alps	Managers of a group of munici-palities, potential member of Al-liance in the Alps	04.11.2009	Cristina	Italian
Enrico	Alliance in the Alps	Manager of a group of munici-palities, potential member of Al-liance in the Alps	06.11.2009	Cristina	Italian
Andreas Götz	CIPRA	Director of CIPRA International	21.06.2010	Cristina	German

Table 2: Summary of the interviews conducted.

As the table shows, interviews were conducted mainly with members or heads of the networks under consideration (Alliance in the Alps, Alparc and Alpine Pearls). Two interviews were carried out with employees of CIPRA, given the importance this organization had and still has in the creation and management of the networks. For this research, it was decided to study in depth the two oldest pan-Alpine networks (Alliance in the Alps and Alparc), as well as Alpine Pearls. The latter introduces a new point of view, since, unlike Alliance in the Alps and Alparc, it is clearly theme-orientated (tourism). We constantly kept an eye on each other, thus allowing us to enhance the analysis and to better generalize some facts.

Two additional interviews were done with members of other organizations (the WWF and the Centre of Alpine Ecology in Trento, Italy) that have a pan-Alpine strategy for their projects. The WWF's office in Bellinzona (Switzerland) runs the European Alpine Programme. In 1999, some national WWF's offices decided to adopt the ecoregional conservation approach in the Alps. It is now an official partner, together with Alparc, CIPRA and ISCAR (International Scientific Committee on Research in the Alps), of a consortium for the Ecological Continuum initiative[36]. An interview was conducted with two employees of the CEALP (Centre of Alpine Ecology in Trento, Italy), which, already in 1993, set a goal to create a network of protected areas in the Alps. Though the project did not really take off, it can still be considered the first attempt to link all the Alpine protected areas in a network. These additional interviews were undertaken in order to provide me with a broader picture of the major institutions working at the Alpine range level. Time and financial means were insufficient to make in-depth studies of these institutions as well. As in every research undertaking, material circumstances condition the concrete research possibilities.

During the project, it was decided also to interview some representatives of Alpine municipalities that are not members of Alliance in the Alps. The decision was taken to interview members in Ticino (Switzerland), knowing that they were in contact with some representatives of Alliance in the Alps and already knew of the existence of the network. An additional interview was held with Gianluca, mayor of a municipality that is not a member of any network, but whom I met during an event where all the Alpine networks were present (Forum Alpinum 2008 in L'Argentière-la-Bessée, France). Contrary to the intent with the other interviews, here the idea was not to find out why it is important for a municipality to join a pan-Alpine network, but to know why a municipality decides not to take advantage of membership to a network. I mainly wanted to know whether this was due to non-identification with the Alpine region.

---

36 For more information: <www.alpine-ecological-network.org>, consulted on the 10.12.2014.

3.3.3 First-hand observation and participation to events

The main *corpus* of the research consists, as the length of Table 2 demonstrates, of interviews. Nonetheless, the opportunity to participate in events organized mainly by Alliance in the Alps constituted for me yet another possibility to collect first-hand observations on the workings of the networks. For example, participation at the General Assembly of the Swiss section of Alliance in the Alps was for me particularly enlightening, since it showed a discrepancy of the political vision inside the association. Some of these events will therefore be broadly discussed in this book.

Attending those events was also essential to build trust between myself and the persons gravitating around the networks. My regular presence, certainly at first considered "exotic", with time came to be seen as "normal". Thus, friendliness and mutual knowledge could be cultivated through acquaintance, which gave rise to informal and sometimes confidential discussions that would not have occurred without regular presence in the field.

Date	*Place*	*Event(s)*
15.06.2007	Flühli (CH)	General Assembly of Alliance in the Alps
30.03.2007 and 06.–08.08.2007	Bovec (SL) and Budoia (I)	Events on the occasion of the 10th anniversary of Alliance in the Alps
06.12.2007	Schaan (LI)	Meeting of the "Reflexion team Dynalp2"
31.01.2008	Schaan (LI)	Meeting of the animators of Alliance in the Alps
01.02.2008	Schaan (LI)	Meeting of the steering committee of Alliance in the Alps

Table 3: My participation at events.

3.3.4 What has changed since 2007

Networks are dynamic institutions, and so many things have evolved since 2007 when I started my research. Some changes are important to mention, since they involve persons and institutions that are central to my argument.

The first changes concern the number of members of the networks. When I started researching Alliance in the Alps, the network comprised around 200 members. It now totals around 300. Alpine Pearls also saw an increase in membership. Some municipalities, attracted by what the networks have to offer them, decided to join. This aspect did not influence the results.

Four other changes more directly affected the results I present in this document. They concern the following persons I met and the following municipalities on which I worked:

- Firstly, Andreas Götz, Director of CIPRA International since 1996, decided to leave his post in 2012. Claire Simon, at the time the research assistant director of CIPRA International and coordinator for Alliance in the Alps, became Director of CIPRA International;
- Secondly, the presidency and vice presidency of Alliance in the Alps changed. After 13 years, Rainer Siegele, quit his function, and Marc Nitschke, mayor of Übersee am Chiemsee (Germany) and since 2009 Vice President, was chosen to replace him as President. Antonio Zambon, vice president of the network at the time of the research, is no longer the mayor of Budoia (Italy), since the regional law in Friuli does not allow a mayor to sit for more than three terms. He occupied this position from 1995 to 2009. Roberto De Marchi took over the post of Mayor of Budoia after Antonio Zambon. Consequently, Zambon changed position within the network as well: He left the position of vice president and became animator for Italy;
- Thirdly, Dominik Siegrist, after having been the President of CIPRA International for 10 years, made way for his successor, Katharina Conradin;
- Finally, the two regions in Ticino (Switzerland) that were the members of Alliance in the Alps that I had chosen for my fieldwork decided to leave the network in 2011. They based this decision on the territorial reforms taking place in the canton and affecting the way municipalities are grouped into regions.

In this book, the persons I interviewed and the institutions I analysed are presented with the status they had at the moment of the interviews.

The aim of this research is to show how networks work and how people are involved in them. In that respect, the fact that some persons changed does not affect the analysis on the functioning of networks.

3.4 Discourse analysis: from coding to analysis

Rudaz (Rudaz, 2005: 33) states in his PhD dissertation: "Discourse contributes to the very process of objectivation." The same could be said for another process: identification. The discourse produced by the pan-Alpine actors is analysed in order to uncover these two processes: the objectivation of and the identification with the Alps (chapter 5). It will be analysed as well to portray the complex Alpine articulation of institutions (chapters 6 and 7).

As we have seen, the socio-constructivist approach defended in this research suggests that mountains are not merely a way to designate and comprehend a reality (Gerbaux, 2001: 21). This means that it is crucial to find strategies to grasp these ways of designation and comprehension. As Petite says, "discourse is fundamentally a producer of social reality" (Petite, 2009: 108). In order to understand these ways of designation and comprehension, researchers have to listen to those discourses and analyse them.

The aim of any research is the understanding of "the ways in which discourses ensure that certain phenomena are created, reified, and taken for granted and come to constitute 'reality'" (Phillips & Hardy, 2002: 21). It is thanks to discourse that we can have access to the territorial reality that individuals and social groups are building (Petite, 2009: 108). And, as Petite continues, discourse allows us to establish facts, retrace a process and grasp representations; it translates the articulation between individual and collective identity (Petite, 2009: 10). What I classify under the category of discourse is in reality a variety of material. Mainly interviews, but also Internet sites, discussions, maps, observations, images, opinions, etc.

Once the research material is collected, the "real" analysis and interpretation can start. This is constituted by the coding process, which Holsti defines as a "process whereby raw data are systematically transformed and aggregated into units which permit precise description of relevant content characteristics" (Holsti, 1969: 94). As Bergmann told us during a methodology seminar,

> [...] coding is an analysis, because the selection of codes is already an analytic process. In all the following stages, analysis is made: definition of the codes, text coding, analysis of the coded text, interpretation of the sense of the analysis.

Just as Rudaz did in his dissertation, I privilege the most restrictive approach of content analysis, which focalizes on the manifest sense of a communication, a discourse (Rudaz, 2005: 33). The qualitative content analysis opted for here can be defined as

> [...] an approach to documents that emphasizes the role of the investigator in the construction of the meaning of and in texts. There is an emphasis on allowing categories to emerge out of data and recognizing the significance for understanding the meaning of the context in which an item being analysed (and the categories derived from it) appeared (Bryman, 2004: 542).

The active role of the researcher as a "competent knower" (Campbell, 1975: 191) is therefore needed if this approach is favoured. Content analysis thus consists in identifying different types of information, extract its sense, classify and interpret its content (Gumuchian, 2000). In my case, a thematic coding was adopted. This permits the classification of texts according to themes that the researcher identifies as providing answers to the research questions. Coding text is a process consisting in two non-chronological steps: creating codes and assigning them to parts of the text.

3.4.1 Use of qualitative data analysis software

Atlas.ti, a frequently used software programme for qualitative content analysis in social sciences, helps the researcher ensure a systematic coding process. What, some decades ago, was done with scissors, glue and a lot of paper on a table, can nowadays be done sitting in front of a computer. Although this method can be seen as less romantic, it is undeniably more efficient.

The software provides many more features than the strict coding process, especially if used as the adherents of the grounded theory do. I, however, used it almost exclusively for the purpose of coding. I exploited two main characteristics of the software: the attribution of codes to parts of the text and the production of *memos*, notes that can considered as first analyses directly generated during coding.

After coding all the interviews, quotations were collected in themes that were already partly identified in the theory and partly emerged from the codification process itself. Once the important themes had been identified, I could start reconstructing the discourse of the interviews, according to those themes. The result of this reconstruction is included in the following chapters.

3.5 How to cope with five languages and one dialect?

A crucial decision had to be made about the choice of the language for conducting interviews. Since I speak three of the four Alpine languages well enough, I let the interviewees speak in their own mother tongue. This choice was mainly dictated by three reasons:

- Firstly, being myself part of a linguistic minority (Italian speaker in Switzerland), I consider it a sign of respect to let people speak in their own language;
- Secondly, the institutions with which I worked attach great importance to the use of the four Alpine languages and to everyone having the possibility during pan-Alpine events to communicate in their own language. Ethically, it would have been misplaced to use English;
- Thirdly, the possibility given to interviewees to express themselves in their own language allowed them to speak more easily and more comprehensively.

In one setting, namely the interviews in the Ticino Canton (Switzerland), I let interviewees speak in the local dialect, if they so wished. Because I myself know and use the dialect, I am aware of the potentiality of this spoken language, capable of adding a colourful and friendly tone to the conversation. I am also quite conscious of the sensitive feelings some people harbour on the issue of the dialect; they were always positively surprised that a person working for a "university" could discuss with them in the dialect. Even this was a challenge during the transcript process, since dialect is, by definition, a spoken and not a written language, but it had allowed me to interview people in a much more relaxed and pleasant atmosphere.

Petra Arnuš followed the same rules for Slovenian. Unfortunately, the consequence was that I could not read the original transcripts and had to rely on Petra's English translations. In this book, Slovenian extracts of interviews are presented only in the English translation prepared by Petra.

Quotations reported in this book are presented in a double column (except for Slovenian): On the left side, the original transcript, and on the right side, my own English translation. This sort of *bricolage* partly reflects the "coloured", spontaneous and improvised mixed conversations that I witnessed during the Alpine events.

As regards the extracts taken from Internet sites, I have, obviously, chosen to quote the English version whenever it was available. In those cases where an English version did not exist, I have chosen to take French as a reference and translated it into English myself.

3.6 Some reflections on my positionality

Research on social relations is also constituted by social relations. This means that the researcher, his or her personal experiences, his or her beliefs, and his or her academic choices influence how data are collected, analysed and presented. In my case as well, the "me researcher" is partly the result of the experiences of and beliefs I have formed during my life. As regards this research, those experiences and beliefs influenced my way of imagining what identities could be, and thus also the way of collecting data and analysing them. In the following pages, I tell two personal stories that can help the reader understand this point: The first shows how I have been involved in a project on mountain identities; the second how I became aware that identities are constructed and not given.

3.6.1 How I became a "mountain specialist"

I grew up surrounded by mountains, but, since I was a small child, I always dreamt about Africa and Latin America and never paid much attention to the reality around me. Rarely, and with great reluctance, I went walking in the mountain, and then only when my parents forced me to go "up there". The sheer "pleasure of the effort" never attracted me. Mountains always represented closure for the small girl I was, always in search of exotic and distant travels, which would let me discover new people, new food, new sounds, new colours, new smells. Of course, I supposed that those distant cultural features were undoubtedly more interesting than the ones close by. Too close, too high, I had the impression that mountains obstructed my view; they prevented my eyes from seeing the horizon and my soul from flying beyond them.

The pride of being a mountaineer only appeared after my teenage years had passed and on just one occasion: after the (rare) victories of "my" ice hockey team, the Hockey Club Ambrì-Piotta. Ambrì and Piotta are two small villages at the foot of the Gotthard Pass. They both belong to the municipality of Quinto. The latter counts slightly more than 1000 inhabitants, who are really proud that their hockey team, the poorest but also the most popular in Switzerland, has since 1985 played in the main national league without interruption, even though, in the past few years, it has been a struggle for Ambrì-Piotta to continue to play at that level.

Every sport event has its rituals, and the matches in the Valascia, the stadium that can accommodate 7000 supporters, are no exception. When the local team wins, its supporters, filled with enthusiasm, sing *La montanara*, a song dedicated to the mountain. It had been the only moment in which I felt emotionally attached to the northern part of Ticino (also called Sopraceneri), whose inhabitants are taunted for being mountaineers. The rivalry with the other big team from Ticino, the Lugano Hockey Club, known for its urban and bourgeois supporters, whipped up with more vigour, if possible, this mountaineer identity during the crowded derbies. As Perruchoud reports in an article published in *Le Temps*, *La montanara* is a teasing refrain when the Lugano supporters sing it, but it becomes the expression of irrefutable pride, the pride of people that one cannot humiliate, when it is sung by the Ambrì supporters (Perruchoud, 2010).

Years later, when I joined the Department of Geography and Environment at the University of Geneva, Bernard Debarbieux asked me if I wanted to conduct research in the context of a project funded by the Swiss National Science Foundation called "Pan-Alpine networks: the rise of the Alps as a frame for collective action". After an initial moment of true perplexity, wondering whether I was really ready to work long years in a region that I considered uninteresting, I decided to answer positively to Debarbieux's proposition. And, actually, thanks to this project and thanks to my fieldwork I have been able to discover a world much more open and innovative than the one I could ever have imagined. The research allowed me to reconcile myself with the Alps and with the mountain dwellers.

This book therefore is not only the history of this discovery, made of projects, hope and people who open up to their neighbours to make the Alps a pilot-region within Europe, but it also represents my personal story of renewed attachment to my region, Ticino, and to my mountaineer roots. I am sincerely thankful to Prof. Debarbieux and to all the people I met during these five years; it is them who have permitted this to happen. Thanks to all these people, I am no longer ashamed to say that I "come from the mountains", because I know, now, that behind this word, there is a fascinating world.

Thanks to this research, I experienced how identities are, indeed, constructed. But, still and ironically, even now that I am claiming that identities are socially constructed as much as mountains, once inside the *Valascia* I forget to be a geographer and simply become a supporter, singing *La montanara* and believing that the mountains are really there, in the heart of every *Biancoblù*. I do this even as I am aware that claiming to be a mountaineer during a sport event is just another way to dig a trench between us, the "Ambrì supporters", and them, the "Lugano supporters". The mountain is just a pretext to reinforce one's identity, an identity built upon constructed stereotypes.

3.6.2 My double identity: becoming an Italian citizen in Switzerland

I was born in Bellinzona, Ticino, a Swiss Italian-speaking canton situated in the southern part of the Alps. I was proud of my red-with-white-cross passport and had always hidden the fact that my maternal grandparents were from Southern Italy. They immigrated to Switzerland in the 1960s in order to find work, as so many did at that time. In the 1980s, when I was a child, Italians were still stigmatized, as Seiler's documentary film *Siamo italiani – Die Italiener* (1964) dramatically reminds us. In the 2000s, when the political destiny of Switzerland and that of the European Union took different paths, a decision by the Italian government made it possible for me, the daughter of an Italian woman, though born outside of Italy, to claim Italian citizenship.

Simply and practically, I, a non-EU citizen with only my Swiss citizenship, could become, thanks to my grandparents, part of the European Union and benefit from the advantages given me by my new nationality. In 2002, a couple of documents sent to the Italian consulate in Lugano and 45 euros were sufficient to obtain Italian citizenship. This does not even come close to what immigrants to Switzerland have to endure in order to get Swiss citizenship, as Lyssy's 1978 film *Die Schweizermacher (The Swissmakers)* satirically points out. This procedure of naturalization taught me that one can construct one's identity and become part of another country than the one of birth during one's life. My mother, through marriage, did the opposite: She became a Swiss citizen by receiving her second nationality through my father. Thus, already by observing what happened in my family, I could appreciate that one can be a citizen of a country without the feeling of belonging and without identifying with it. By contrast, one can strongly feel part of a territorial entity without any "written proof", as I learned while interviewing the Alpine political actors for my thesis.

4. The actors of the Alpine region

In this chapter, I will present the institutions that helped me understand the pan-Alpine regionalization process. The main objective here is to show how these institutions were established, their history, and how they function, i.e. their structure. As I go through the institutions, some personalities will emerge: the thinkers and builders of the pan-Alpine region, whom I qualify as the leaders.

The chapter starts with a reflection on the ways the Alpine Space and the Alpine Convention envision the Alpine region. The discussion will make it clear that these two institutions do not share the same designation of the Alpine perimeter.

4.1 (At least) two ways of drawing the Alpine territorial unity: the Alpine Space and the Alpine Convention

The recognition of the Alps as a single unique region is part of a long history in which mountains and mountaineers played different roles. In the chapter called "L'Union Européenne: la montagne introuvable?" ("The European Union: the mountain that cannot be found?"), Debarbieux and Rudaz trace the history of how the European Union has regarded the mountains (Debarbieux, Rudaz, 2010). They point to the 2000s as the decisive period during which the mountains in general, and the Alps in particular, finally acquired increased significance in public action in Europe (Debarbieux & Rudaz, 2010: 289).

The European Economic and Social Committee, the European Parliament, the Committee of the Regions and the European Council, as well as the Council of Europe Conference of Ministers Responsible for Spatial/Regional Planning (CEMAT), agreed on the necessity of a policy in favour of mountains (Debarbieux & Rudaz,

2010: 289–293). The real impact of this recognition at the European level is nevertheless mitigated, according to the two geographers, by the impression that the definition of the mountain in Europe is far too heterogeneous to impose a unique vision and a common model for political strategy (Debarbieux & Rudaz, 2010: 299). Although a common model for the European mountain has not yet been found, the Alps certainly play a pioneering role for other mountain ranges to claim mountain-specific strategies.

4.1.1 The Alpine Space

At present, different actors are supporting a pan-Alpine strategy and politics. Debarbieux identifies the following: the Alpine national states, which are asked to coordinate the initiatives undertaken within the framework of the Alpine Convention; the sub-national entities (regions, cantons, *Länder*, etc.), which are active in Alpine politics; the municipalities, which are building a wide range of coalitions; the associations and clubs, which are defending their idea of the Alps (Debarbieux, 2008a: 40). All those actors demonstrate the existence of a veritable "pan-Alpine activism" (Debarbieux, 2008a: 40).

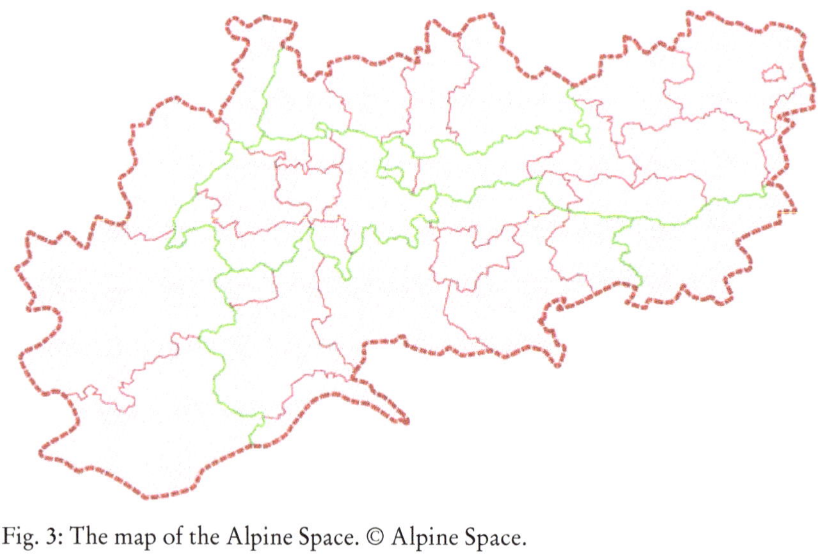

Fig. 3: The map of the Alpine Space. © Alpine Space.

Additionally, the role of the European Union must be highlighted. In fact, the constitution of the Alpine Space, promoted by the European Union, produced a map of the Alps giving them territorial unity (see fig. 3). Thanks to this map, not only do the Alps exist but simultaneously acquire a legal status, since legal and administrative procedures are fixed inside of these borders.

The area of the Alpine Space covers a population of some 70 million inhabitants, a surface of 390 000 km² and some of the biggest European metropolitan areas, such as Munich and Milan. It was established in the third generation of the Interreg programmes created to stimulate cooperation between member states of the European Union. The Alpine Space was integrated into the Interreg IIIB programme. The "B" programmes involve national, regional and local authorities striving to promote better integration within the European Union through the combination of large groups of European regions.[37] The first phase of the Alpine Space (2000–2006), defined in the *Executive summary of the Alpine Space prospective study* as "the mountainous area of the Alpine core space, the mountain cities and the peri-Alpine area with the metropolitan cities" (Bausch *et al.*, 2005: 2), aimed to increase the attractiveness of the area, promote sustainable development, improve accessibility to the Alps and protect its cultural and natural heritage. The second phase (2007–2013) is pursuing the objectives of the first by focusing on growth, employment and sustainable development in accordance with the EU's strategic goals set up in the Lisbon and Gothenburg agendas. The group of experts that wrote the prospective study identified four priorities for the Alpine Space Programme (Bausch *et al.*, 2005: 7):

- Promoting the Alpine Space as an innovative, competitive and attractive living and economic space in the scope of a polycentric spatial development;
- Improving accessibility of the Alpine Space;

37 See: <http://europa.eu/legislation_summaries/regional_policy/provisions_and_instruments/g24204_fr.htm>, consulted on the 10.12.2014.

- Supporting wise management of nature, landscapes and cultural heritage, promotion of the environment and prevention of natural disasters;
- Encouraging Alpine innovation and ensuring an equitable repartition of factors of competitiveness.

Thus, as a result of this designation, EU funds could reach the Alpine territory. These funds were mainly used to finance transnational projects inside its core area, which the Alpine Convention defined.

4.1.2 The Alpine Convention

The Alpine region, as defined in the Alpine Convention, is home to some 13 million people. The 190 912 km² contained within the Alpine arc comprise the territory of eight countries, 83 NUTS 3 regions and about 6 200 municipalities (see fig. 4).[38] It is hence clearly smaller than the Alpine Space region. Above all, the Alpine Convention's territory does not include the big metropolitan areas surrounding the Alpine arc. The perimeter of the Alpine Convention can be considered the "core area" of the Alps, where the partly EU-funded projects are implemented. The Alpine Convention does not give out funding for large-scale projects, as Church pointed out, and in fact the Alpine Space Programme is the major funder (Church, 2010: 48).

The Alpine Convention is undoubtedly the result of, on the one hand, lobbying work done by activists and NGOs and, on the other hand, the desire of national states to create new partnerships between them. Nevertheless, its implementation could not have been undertaken to such a degree without the financial intervention of the European Union, which can be considered a facilitator in the European regionalization processes.

38 <www.cipra.org/fr/politique-alpine/les-alpes,> consulted on the 10.12.2014.

Fig. 4: Perimeter of the Alpine Convention. © Alpine Convention.

The history of the Alpine Convention starts in the 1950s with a founding document drafted by the *Commission Internationale pour la Protection des Alpes* (CIPRA), a non-governmental organization that will be presented in the next section. The document stipulates that CIPRA's most important task is the elaboration of an international treaty for the protection of the Alps.[39] Indeed, CIPRA is the first institution to "call for the creation of a convention to protect the Alpine environments and corresponding natural resources" (Fall, 2007: 109). Despite the considerable efforts deployed by CIPRA's first members, it took 40 years to finally have, on the 7th of November 1991, the Alpine Convention signed by the representatives of the respective Ministries of the Environment of Germany, Austria, France, Italy, Liechtenstein, Switzerland and the European Union.[40] With this agreement, the signatory states commit themselves to orientating

39 <www.cipra.org/fr/nouveautes/4286>, consulted on the 10.12.2014.

40 For an historical overview, refer as well to Price, 2000.

the development of their territories in the direction of the convention (Rudaz, 2005: 213–214). The convention is aimed at promoting sustainable development in the Alpine area and at protecting the interests of its inhabitants. It embraces the environmental, social, economic and cultural dimensions.[41]

The present structure of the Alpine Convention can be summarized as follows: The primary decision-making body is called the Conference of the Contracting Parties and meets every two years. The chairmanship and location of the meetings change after each ordinary meeting. In addition to the contracting parties, the United Nations and its specialized agencies, the Council of Europe, all the European countries, cross-border associations of Alpine territorial authorities and specified NGOs can attend the meetings as observers.[42]

In 2002 the contracting parties decided to set up a Permanent Secretariat that would become a junction allowing to effectively coordinate and provide support to the bodies established by the Alpine Convention. They resolved that the main office would be located in Innsbruck (Austria). The main tasks of the Permanent Secretariat are to provide administrative and technical support to the bodies of the Alpine Convention; public relations; the coordination of the Alpine research projects; the facilitation of the implementation of the convention and the compliance with its protocols; the setting up and management of an observation and information system for the Alps (SOIA).[43] In addition to a general secretary and a vice-general secretary, eight other persons work for the Permanent Secretariat.

One of the most important successes of the Alpine Convention has been to institute, for the first time in history, the Alpine states as a single actor espousing a common Alpine policy, which until then had been applied separately by every national state (Götz & Rohrer, 2011: 5). Thus, with the signature of the Alpine Convention, the Alpine arc was framed within the bounds of a common political and administra-

41 <www.alpconv.org/en/convention>, consulted on the 10.12.2014.

42 For more information about the functioning of this body: <www.alpconv.org/en/organization/conference>, consulted on the 10.12.2014.

43 <www.alpconv.org/en/organization/permanentSecretariat>, consulted on the 10.12.2014.

tive structure (Norer, 2002:8) with the "recognition that the Alps are a region with distinct environmental and cultural characteristics, of considerable significance at the European scale" (Price, 1999: 88).

The Alpine Convention can thus be considered the first international treaty seeking to coordinate a variety of public policies at the mountain massif scale; it is even cited as a model inside and outside Europe (Debarbieux, Rudaz, 2010: 316; Siegrist: 2010: 3). The particularity of this convention is the fact that it does not focus on narrow sectors, as most other multilateral environmental agreements do. It consists instead in a framework treaty aimed at promoting sustainable development, i.e. cross-sectorial policies (Balsiger, 2008: 15).

Notwithstanding these successes, the title chosen for issue number 95 of CIPRA's official journal *Alpenscène* does not conceal the difficulties encountered by the treaty. The authors ask "who will wake the sleeping beauty",[44] i.e. the Alpine Convention. In this 17-page document, different authors, mainly employees of CIPRA, assess the successes and failures of the international agreement. In his editorial, Siegrist, the then-President of CIPRA International, does not deny the problems experienced by the Alpine Convention. He regrets, for example, that the Alpine states do not promote innovative initiatives, for example on the topic of climate change (Siegrist, 2011: 3).

The Alpine Convention was conceived as a large instrument for the promotion and implementation of sustainable development. As Götz and Rohrer remark, considering that the Alps are not only a natural area, but also an economic and social space, the Alpine Convention could not be limited to only one theme. Ecological balance, economic security and social justice, the pillars of sustainable development following Götz and Rohrer, were included in the text (Götz & Rohrer, 2011: 5).

Despite this, local authorities and the broader population still view the convention with a critical eye, as an instrument for the protection of nature (Messerli, 2010: 121; Simon, 2011: 18). An indication of that point of view is the fact that the convention is signed by the Ministries of Environment and not by the heads of states, as

44 The journal can be downloaded online in the four Alpine languages: <www.cipra.org/fr/publications/4586>, consulted on the 10.12.2014.

Debarbieux and Rudaz intimate (Debarbieux & Rudaz, 2010: 314). That point, together with the official title, *Convention on the Protection of the Alps*, clearly emphasizing protection, raises criticism, as the convention allegedly puts the Alps "under a bell jar" (Götz & Rohrer, 2011: 5–6). It is also one of the reasons mentioned by Price for Switzerland's reluctance to ratify the convention, since it was perceived by mountain cantons as promoting environmental protection rather than economic development (Price, 2000: 193). Switzerland and Italy eventually ratified the convention seven years after the other countries did.

Three other main criticisms are levelled against the Alpine Convention:

- Firstly, the most important means to reach the objectives of the Alpine Convention are constituted by its protocols. Eight were elaborated in the 1990s; others, planned at the beginning, were never written (Götz & Rohrer, 2011: 6). Götz and Rohrer regret that the wording of the protocols is far too evasive, as it complicates their legal implementation (see as well Bätzing, 1994: 200; Gerbaux & Zuanon, 1995: 108–109). The protocols are the result of too many compromises, which, in the end, diminish the impact of the final text by avoiding controversial aspects (Gerbaux & Zuanon, 1995: 109). Furthermore, there are no sanctions planned for states acting against the principles of the Alpine Convention (Norer, 2002: 24; see as well: Götz & Rohrer, 2011: 6). This serves as a pretext to the majority of the states for not adapting their legislation in accordance with the protocols. In general, the Alpine Convention is criticized for being a "piece of paper" with no real implementation (Simonis 1997: 13; Balsiger, 2008: 5). Not only are the protocols considered too vague, but, as Price underlines, no measures are in place to resolve potential conflicts or disputes (Price, 2000: 192). Gerbaux and Zuanon sum up these thoughts well: The model is appealing on paper, satisfying at the intellectual level, but difficult to enforce in practice (Gerbaux, Zuanon, 1995: 110);
- Secondly, two Alpine states, Italy and Switzerland, which possess 40 % of the Alpine territory (Wülser, 2011: 9), did not ratify the protocols of the Alpine Convention. This can be called a failure of the process, says Hasslacher, President of CIPRA Austria,

in an interview with Wülser (Wülser, 2011: 9), and a weakening of the convention itself (Simon, 2011: 17);

- Additionally, some scholars have noted, not without some criticism, the "colonial manner" (Bätzing, 1994: 186) in which the convention and the different protocols were established. Scholars point out that the convention was formulated in Rome, Bonn and Paris, but not in the Alps (Bätzing, 1994: 186; Norer, 2002: 11). An activity report from the *Groupement de la population de montagne du Valais romand* (Switzerland), quoted by Rudaz in a doctoral thesis on mountain lobby groups in Switzerland, expresses the frustration of some local political actors in this regard. The *Groupement* writes that the philosophy behind the Alpine Convention and its protocols strongly resembles a new form of colonialism. The city dwellers try to legislate the way of life in the Alps, while the mountain inhabitants do not have any rights regarding the urban world, Rudaz says (Rudaz, 2005: 216). Consequently, some associations defending mountain people distance themselves from the project of the Alpine Convention (Rudaz, 2005). Fourny observed the same attitude concerning the network *Communauté de travail des villes des Alpes*. She notes that the community expresses clear opposition to the Alpine Convention, not because of its content but because of the modalities of its institution. For the community, it represents an interference in local decisions (Fourny, 1999: 174).

Beyond those justified criticisms, however, the Alpine Convention constitutes the milestone of a new Alpine policy (Götz & Rohrer, 2011: 5). Until then, as Gerbaux and Zuanon underscore, each Alpine state had elaborated separate specific policies in response to the problems that the Alps were encountering; even if some initiatives presented a number of common features, they reflected sectorial and national visions and political philosophies (Gerbaux & Zuanon, 1995: 102–102). Therefore, the Alpine Convention can be seen as the most concrete attempt so far at implementing the concept of sustainable development in a broad and multi-lingual region in Europe. This is true even though, as Price concludes in his article, "few of the expected impacts on environmental policy-making and implementation have been realized" (Price, 2000: 194).

4.2 CIPRA and its pivotal role in the creation of a transnational Alpine entity

The two maps drawing the borders of the Alpine region, the one of the Alpine Space (see fig. 3) and the one of the Alpine Convention (see fig. 4), and the parallel idea of considering the Alps as a united entity, are the result of a long process involving different actors. One actor stands apart: the International Commission for the Protection of the Alps (CIPRA). In fact, CIPRA is the first organization concerned with treating the Alps in their entirety (Debarbieux & Rudaz, 2010: 313), and this already from the 1950s. CIPRA International has its seat in Schaan (Liechtenstein) and is financed by the fees paid by its members. Its structure is composed of three bodies, namely an assembly of delegates, a presiding committee and an executive board:

- The assembly of delegates is the supreme body; it is made up of the delegates of the ordinary and extraordinary members. The ordinary members are CIPRA representatives of the national sections from Austria, France, Germany, Italy, Liechtenstein, Slovenia and Switzerland. Extraordinary members are defined by the CIPRA's statutes as follows: "Any international governmental and non-governmental organisation or national representative of states other than those listed as ordinary members which support the objectives and tasks of CIPRA International; regional (as for example CIPRA South Tyrol) and/or supra-regional representatives may also be admitted as extraordinary members by agreement with the relevant national representatives."[45] The assembly meets at least once a year;
- The presiding committee consists of the executive board (which comprises the president, the first and second vice presidents, and the treasurer) and two representatives of each of the national delegations, as well as one representative of each extraordinary member. It may discuss and vote on all matters to be adopted by

45 The statutes can be found here: <www.cipra.org/en/cipra/about/mission/2013_Statutes_e.pdf>, consulted on the 10.12.2014.

the assembly of delegates. The presiding committee determines the content, date, time and place of performance of important events;
- The executive board, elected every three years, features a president, two or more deputy presidents and a treasurer. It assumes decision-making powers in urgent cases when such powers are reserved to other bodies. Additionally, it oversees the management and supervises the administration of finances.

CIPRA was founded in 1952 in Rottach/Egern (Germany) under the initiative of the International Union for Conservation of Nature (IUCN).[46] The Director of the Gran Paradiso National Park, Renzo Videsott, reported to the general assembly of the IUCN in 1950 various problems concerning the protection of nature in the park and its transboundary neighbouring region. Notably, Videsott was concerned by the construction of hydroelectric power plants. At the general assembly of 1951, Videsott proposed a transboundary collaboration with his French neighbours for the protection of endangered species. That year, it was acknowledged that, in times of great economic growth, all the Alpine states were facing similar problems.

Since no quick solution to those problems could be found within the IUCN, two German organizations, the *Bund Naturschutz* in Bayern and the *Schutzgemeinschaft Deutsches Wild*, invited the concerned countries to a preparatory meeting for the foundation of an international commission for the Alps, on the 5th of May 1952 in Rottach/Egern (Germany). CIPRA's activities in the 1950s and 1960s concentrated on the organization of yearly assemblies. The members of the association were, at that time, almost exclusively natural scientists. The focus on "naturalist" themes, which characterized the first decades, was subsequently expanded to more socio-political topics by embracing the concept of sustainable development.

This tendency is clearly recognizable in the last *Report on the state of the Alps* published in 2007 by CIPRA and called *We Alps! Women and men shaping the future* (Michel, 2007). The main charac-

---

46 The history of CIPRA has been taken from this Internet page: <www.cipra.org/fr/cipra/nous-connaitre/histoire>, consulted on the 10.12.2014.

ters portrayed in the report are women and men "committed to being master builders of the future". Nature is no longer a central theme, as it was in the first two reports (1998 and 2001). At the end of the 1970s, CIPRA opened its doors to a more heterogeneous group of actors, as well as to private associations or representatives of public administrations. Today, CIPRA counts more than 200 members, including single members, organizations and federal states.[47]

Götz, the former and "historical" director of CIPRA International, whom I interviewed in June 2010, identifies two turning points in this process of increased importance to social issues: the beginning of the awareness of sustainable development, and collaboration with municipalities. Indeed, it would have been impossible to work with local authorities in terms of "pure" protection of nature. Time has proved Götz right and, aside from some internal criticisms for this new direction, the path to a more open association was traced and became even more apparent when CIPRA decided to found a network of small and medium-size businesses, NENA.

At the same time, CIPRA took on the role of an information platform, notably via the publication of *CIPRAInfo*. CIPRA, as is stated on its Internet site, "communicates in a multitude of languages, providing information and public relations for experts, politicians, the media and anyone with an interest in the Alps".[48] CIPRA's Internet site contains a large collection of information and documents; it also regularly publishes reports, journals and different articles concerning the Alps. As Church and Ramcilovic highlight, CIPRA's publications are among the most accurate and abundant sources of information dealing with the Alps (Church, Ramcilovic, 2010: 14).

On a political level, from the end of the 1980s, CIPRA established contacts in order to realize its main project, i.e. the establishment of an international treaty for the Alps, the Alpine Convention. This became reality at the beginning of the 1990s. The convention would not exist without CIPRA's preliminary work and lobbying. In that sense, CIPRA has been, and still remains, the decisive driv-

---

47 The list of the members can be found here: <www.cipra.org/fr/cipra/nous-connaitre/membres>, consulted on the 10.12.2014.

48 <www.cipra.org/en/cipra/about/mission>, consulted on the 10.12.2014.

ing force of the Alpine Convention (interview with Töpfer, in: Fitzthum, 2011: 15). Gerbaux and Zuanon declare without hesitation that CIPRA's current major concern is to find a solution for the implementation of the Alpine Convention (Gerbaux, Zuanon, 1995: 111). Concurrently, CIPRA plays a central role in the promotion of citizen participation in the pan-Alpine activities (interview with Töpfer, in: Fitzthum, 2011: 15).

Additionally, CIPRA launched and supported most of the pan-Alpine networks, which makes it somewhat of a "network of networks" and the pivot of the "new collaborative regional approaches and initiatives" (Price, 1999: 88) taking place in the Alps. I will present three of those networks, the ones that I have chosen as units of analysis, in the next section.

4.3 The pan-Alpine networks: Alliance in the Alps, Alparc and Alpine Pearls

As mentioned previously, some of the funds provided by the European Union via its Interreg programme were used by Alpine actors to support transnational cooperation. The networks used the opportunity offered them by the European Union funds to carry out projects aimed at the implementation of the Alpine Convention's principles. Therefore, as Götz says, they became the "ambassadors of the Alpine Convention" (Götz, 2011: 10). They promote sustainable development and a horizontal *modus operandi*. Together, they make the Alps exist as promoters of a collective action supporting the pan-Alpine project. Those networks are one of the *tesserae* of the Alpine mosaic, made up of an increasing number of actors and levels of action that are building a complex system of governance (Debarbieux, 2008: 40). This is quite a unique phenomenon, the geographer says, since only a few transnational regions know such public activity.

Thus, the Alpine networks contribute to the creation of new forms of expression for democratic legitimacy and accountability along ecoregional boundaries (Balsiger, 2007: 4–5). Local political actors, such as managers of protected areas or mayors of Alpine municipalities, were, and still are, key persons for the smooth operation of these networks. The most important networks active in the Alps at present are Alliance in the Alps, a network of municipalities, the Alpine Network of Protected Areas, the Alpine Town of the Year and the network of businesses, NENA.

All of them have been created to find innovative ways to anchor the principles established in the Alpine Convention at the local level. As discussed above, one of the major criticisms directed at the Alpine Convention is its weak power of implementation. The pan-Alpine networks were created precisely to overcome this problem and to put into practice what the framework of the Alpine Convention promised: sustainable development in the Alpine region. In the next sections, I present the three networks that I analysed for this research: Alparc, Alliance in the Alps and Alpine Pearls.

4.3.1 Alparc: connecting Alpine protected areas

The Alpine network of Protected Areas[49] was created in 1995 by the French government. Philippe Maigne, Director of the Ecrins National Park, convinced of the importance of networking in order to infuse life to the Alpine Convention, set up a partnership between three national parks in three Alpine countries: Ecrins (France), Hohe Tauern (Austria) and Triglav (Slovenia). He persuaded the French Ministry of the Environment to create the Alpine network of Protected Areas (Lalande, 2004: 127–128). The concretization of this network thus represents one of the rare initiatives taken by one of the contracting parties to implement the Alpine Convention (Götz, 2011: 11). As written on Alparc's Internet site: "The main goal [of the network] is the concrete application of the protocol of the Alpine Convention 'Nature conservation and landscape management'."[50]

---

49 Most of the information presented in this section is taken from the Internet site of Alparc: <www.alparc.org>.

50 <www.alparc.org/the-alparc-network>, consulted on the 10.12.2014.

Alparc is composed of three major bodies[51]:

- The General Assembly, which brings together the ALPARC members. All key decisions and the choice of subjects for cooperation are submitted to the Assembly every year. The General Assembly also elects the ALPARC Board, which is the representative body of the protected areas within ALPARC and is made up of managers of protected areas;
- The missions of ALPARC are defined by the Council, which is composed of managers of the Alpine protected areas, representing the diversity and number of protected areas by Alpine country. It establishes the main directions that define the international work according to the priorities and requests of the protected areas. The Permanent Committee of the Alpine Convention is informed about the ALPARC working programme;
- The Board is one of the ALPARC executive bodies. It is composed of five members: the President, two Vice Presidents, the Treasurer and the General Secretary. Board members are elected every four years by the ALPARC Council. The board is responsible for the administration and for the finance management in coordination with the Alparc coordination unit.

It is from the coordinating unit, located in Chambéry (France), that the network is managed. The team was first attached to Les Ecrins National Park, then in 2006 to the Permanent Secretariat of the Alpine Convention with the name of "Task Force Protected Areas". Since 2014 the Coordination unit is employed by the ALPARC Association.

The network formally brings together all the Alpine protected areas (more than 900 areas corresponding to more than 100 hectares of land). Altogether, these protected areas cover some 25 % of the territory defined by the Alpine Convention. Alparc's main purpose is to "to promote the exchange of expertise, techniques and methods among the managers of Alpine protected areas".[52] The promotion of interchange is thus one of the essential elements of its mandate. Co-

51 <www.alparc.org/the-alparc-network/association-alparc>, consulted on the 10.12.2014.

52 <www.alparc.org/the-alparc-network/alparc-s-objectives>, consulted on the 10.12.2014.

operation with Alparc is entirely voluntary and free of charge, so that all Alpine protected areas are automatically members of the network. This constitutes a major difference with Alliance in the Alps and Alpine Pearls, as we will see later.

The most important project carried out by Alparc is the Alpencom project (2005–2008). The "ALP ENvironmental COMmunication's" objective is to establish an information and communication exchange system for the sustainable management of natural resources and public relations for protected areas.[53]

4.3.2 Alliance in the Alps: connecting local authorities

Alliance in the Alps[54] was set up in Bovec (Slovenia) in 1997, under CIPRA's impetus. It was created by the Alpine Convention as an instrument for implementing the principles of the convention at the local level. The idea of creating a network of local authorities goes back to the beginning of the 1990s, when Dominik Siegrist, the former President of CIPRA International and at the time an active member of the association, reflected, along with others, on how to implement the Alpine Convention, which was experiencing difficulties of local acceptance (Petite, 2009: 312). Siegrist came up with the idea to cross the Alps on foot with some friends in order to discuss the convention with locals. Thus, the TrasALPedes project was born.

In his book *Sauver la montagne*, François Lalande recounts this walk (Lalande, 2004). From June to October 1992, the delegation walked from Vienna to Nice. With this walk they wanted to publicize that the Alpine massif's nature was under threat. In the project's presentation they condemn the "crazy mobility that wreaks havoc in all of Europe" (in Lalande, 2004: 123). That is why the group decided to walk. Walking permits one to recapture the "rhythm of things" and to find time to speak with people (in Lalande, 2004: 123). Four main

53 <http://www.alpencom.org/pages/index.php>, consulted on the 10.12.2014.

54 Most of the information presented in this section is taken from the Internet site of Alliance in the Alps: <http://alpenallianz.org/en/about-us> and from the interviews I conducted.

themes were discussed with local authorities and inhabitants: transport, energy, tourism and mountain farming (Lalande, 2004: 124). In fact, this was the second objective of the walk: taking the opportunity to speak with local authorities, who could thus be motivated to form a network (Petite, 2009: 312). The group hoped to "contribute to the creation of a Europe built from the base" (in Lalande, 2004: 123). The third objective was to overcome national barriers and to show that the Alps were a single entity (Lalande, 2004: 123).

As Lalande reports, in the end, the walk was considered a positive experience. The small delegation walked almost 2000 km and engaged in more than 100 meetings with local authorities. Indeed, contacts were established and experiences exchanged (Lalande, 2004: 127). The notes taken by the delegation (and reported in part in Lalande's book) show their collective impressions about the problems that worried the local authorities and population. This walk, thus, reminds us of old times, when messengers crossing the Alps brought news from one corner of the Alpine arc to the other. By reporting each municipality's concerns to the other, they contributed to the building of a common consciousness. After the delegation's visit, local authorities were doubtlessly more aware that other municipalities along the Alpine arc shared the same problems. In the same spirit of sharing knowledge, Siegrist and CIPRA officially created the Alliance in the Alps network in 1997 (Lalande, 2004: 127).

Siegrist reflects on the beginnings of Alliance in the Alps by underlining the fact that everything started with a European project.[55] He found six Swiss potential members who were prepared to form a network of municipalities. Other people related to CIPRA were in charge of finding more interested municipalities in the other Alpine countries. Contacts were established via personal conversations or per telephone, since Internet did not yet exist. Siegrist remembers that he had to travel several times to the Malcantone Region (until 2012, a Swiss member of Alliance in the Alps) to convince them to participate in the project. Apart from the personal links, an intern

55 Notes taken during a discussion between Bernard Debarbieux, Dominik Siegrist and me in the restaurant coach during a train trip between Schaan and Zürich, coming back from a meeting organized by Alliance in the Alps (06.12.2007).

was hired to find all the addresses of every Alpine local authority. A letter was then sent to each of them. Around 30 municipalities expressed an interest. CIPRA subsequently started a pilot project with a group of 27 municipalities. Each had to prepare and carry on a project of sustainable development. In the interview I conducted with him, Antonio Zambon recalls receiving that letter. He speaks of a happy coincidence: At the exact moment he was thinking about an innovative project for his municipality, he received CIPRA's letter.

From these first contacts and projects, the decision was made to establish a network of municipalities. Currently this network includes over 300 local authorities from seven of the eight Alpine countries.[56] The Alpine Convention provides a basis for the network's activities and a guiding line towards sustainable development in the Alps. The implementation of a sustainable development policy constitutes one of the network's goals. Petite describes the main objectives of the alliance in his dissertation. The geographer highlights that, firstly, the network fulfils the role of a political lobby, notably in pressing the national states to promote the Alpine Convention. Secondly, the network is conceived as a platform, where members can exchange experiences, know-how and good practices. Thirdly, the network is deemed to be a material support for the implementation of the Alpine Convention, notably though the realization of concrete projects (Petite, 2009: 77).

The current structure of the network is constituted by an executive board composed of seven local representatives of the Alpine states. Additionally, a secretariat is based in Übersee am Chiemsee (Germany). The international association is funded by membership fees, while projects are mainly funded by EU programmes or private foundations.

In 1998, Switzerland decided to found a national section of the network. This allowed the Swiss members to get additional funding thanks to extra national subscriptions from the members and above all from the Federal Office for Spatial Development (ARE). In 2002, Austria followed the Swiss path and decided to implement an Austrian section of the network. As in the Swiss case, this allowed Austria to get more money from subscriptions and apply for extra federal funding.

56 Monaco does not have a member.

To become members of the network, municipalities have to submit a project in the field of sustainable development and must pay an annual fee.[57] This constitutes a major difference with Alparc. In fact, membership to Alliance in the Alps is neither automatic nor free. The amount of the yearly fee is proportional to the number of inhabitants.[58] Moreover, membership must be motivated and justified by concrete projects, as potential members are reminded on the network's Internet site: "You can become a member of the network if you approve the principles of the Alpine Convention and if you show that you want to implement environmental-friendly objectives in your municipality."[59] It is important to underline, since it will be discussed later in this book, that there are two ways of becoming a member of the network, as a single municipality or as a group of municipalities, which I will later refer to in this document as "regional members". While at the beginning of its history the network accepted almost exclusively single memberships, in more recent years the number of groups of municipalities increased. Later, I highlight the advantages and the problems of this evolution.

Alliance in the Alps included in its structure a group of people, which I will call, as per Alliance in the Alps' official translation of the French *animateur*, animator. At the moment 12 animators work for the network; they can be considered the intermediaries between the executive board and the network's members. Their role is to pass information from the executive board to the members and to convey to the board propositions emanating from the municipalities. The rule is that there is to be one animator per country, more in multilingual settings (Switzerland had three when members from all linguistic areas were present) or territorially large countries (as Italy). All animators have different backgrounds (some are mayors, but others are technical and scientific consultants) and are attached to dif-

57 The amount of the fee is proportional to the number of inhabitants of the municipality and can be found here: <www.alpenallianz.org/fr/sur-alliance-dans-les-alpes/devenir-membre>, consulted on the 13.07.2015.

58 Amount of fees can be found here: <www.alpenallianz.org/fr/sur-alliance-dans-les-alpes/devenir-membre>, consulted the 13.07.2015.

59 <www.alpenallianz.org/fr/sur-alliance-dans-les-alpes/devenir-membre>, consulted on the 13.07.2015.

ferent kinds of institutions (municipalities, private institutes, NGOs, etc.). Their task is to advise members, to give support to local authorities in the establishment of projects and to facilitate links between the different municipalities all around the Alpine territory.

The role of the animators is particularly central in the managing of projects, especially for the most important in terms of the amount of money allocated, as has been the case for Dynalp. The Dynalp Interreg IIIB project brought together 52 towns and villages with the aim of initiating and exchanging information on local projects in the areas of tourism, conservation of nature and the countryside, mountain agriculture and regional planning. Dynalp2 (2006–2009), which followed the first Dynalp, had a budget of 1.8 million euros, almost entirely covered by the MAVA foundation, a foundation established by Luc Hoffmann under Swiss law in 1994. Most of the budget was allocated to support projects of sustainable development undertaken by Alpine municipalities. The municipalities received between 10000 and 30000 euros. The most illustrative projects, 12 accounts in all, were collected on DVD. The latest project initiated by the network is called DynAlp-climate. As its name suggests, it focuses on the theme of climate change. The DynAlp-climate programme, started in 2009, aims to help municipalities deal with a sustainable approach of climate change. It is also funded by the MAVA foundation, with a budget of around 800000 euros in support of 20 projects.

4.3.3 Alpine Pearls: associating touristic towns for sustainable mobility

The network Alpine Pearls was born out of the Interreg IIC programme. It is a "pilot project for environmentally sound travel logistics linked with electronic booking and information systems in Alpine tourist regions" (abbreviated as "Alps Mobility") (1998–2001).[60] Its aim was to "create practicable, environmentally compatible transport strategies for travel to the Alps and mobility at the visitors'

60 More information can be found in the brochure "Alps Mobility" (downloadable online: <www.alpsmobility.net>, consulted on the 12.12.2014).

holiday destinations".[61] In the pilot phase, partners from only three Alpine countries (Germany, Italy and Austria) joined the project. A subsequent three-year project called "Alps Mobility II" began in 2003 and had two additional partners from Switzerland and France.[62] This second project took place within the framework of the Interreg IIIB programme and focused on the creation of a package of eco-tourism offers called "Alpine Pearls".

As the President of the network, Peter Brandauer, suggests, the objective of the Alps Mobility project was to establish an association built upon the existing structures through the pilot project. That is why, in 2001, the Alpine Pearls association was founded.[63] The network was created in Vienna on the 30th of January 2006, while Alps Mobility II was still ongoing, under the motto "holidays with soft mobility in the most beautiful areas of the Alps". It is designed "for all guests who combine leisure with a desire to preserve nature", so the brochure presenting the network declares.[64] In January 2011, four new members joined the pearl necklace that now comprises 27 municipalities[65] from six countries.

This network's specificity is that it concentrates on a single particular theme: environmental friendly transport and tourism solutions. This is what distinguishes it from the other two networks presented earlier, which are multi-thematic and not oriented specifically towards tourism. As suggested in the presentation of the project, in order to get the "pearl" designation the partners have to fulfil "certain mobility and tourism standards in the sense of sustainability according to a fixed criteria catalogue".[66] The president and the manager of

61 <www.alpsmobility.net>, consulted on the 12.12.2014.

62 More information can be found in the brochure "Alps Mobility II" (downloadable online): <http://alpsmobility.net/downloads/am_II/downloads/amII_folder-0106-kl.pdf>, consulted on the 12.12.2014.

63 Interview conducted by Cristina Del Biaggio with Karmen Mentil and Peter Brandauer (Werfeneweg, 29.01.2009).

64 Brochure downloadable online: <http://alpsmobility.net/downloads/am_II/downloads/ap_folder_en.pdf>, consulted on the 12.12.2014.

65 At the time of the interview they were 20.

66 See p. 2 of the brochure "Alps Mobility II": <http://alpsmobility.net/downloads/am_II/downloads/amII_folder-0106-kl.pdf>, consulted on the 12.12.2014.

the network personally travelled to a potential member's location in order to check that the motivation and criteria were satisfactory for it to become member.[67] Criteria are detailed in an exhaustive 19-page catalogue, and they concentrate on different types of mobility: mobility upon arrival, regional mobility, mobility on-site and so on.

Alpine Pearls is constituted by a General Assembly, which meets every year, and in which every member has one vote. Additionally, there is a Board of Directors, composed of one president and six directors, one from each country plus one director from South Tyrol. An external controlling body, comprising two persons, verifies that the criteria of soft mobility are fulfilled in every pearl. The association is overseen by a manager, who appoints the General Assembly. This structure is financed by the annual fee of 12000 euros paid by each member.[68]

To conclude this chapter, one has to acknowledge that the Alpine Convention agreement made it possible to draw a map of the region. Thus, local political actors received a means to recognize the Alps as a region with distinct environmental and cultural characteristics (Price, 1999: 88). This map is one among several that could have been drawn. However, this particular map presents local political actors with a perimeter within which to implement their projects. Thanks to the convention, the Alpine region is being institutionalized and represents a case, in Balsiger's words, of "ecoregional institutionalization" (Balsiger, 2007: 4). It is precisely inside the area sketched by the Alpine Convention that a new space of action has been created. So, as Rudaz points out, with the signature of the Alpine Convention, a new scale of action is emerging, and new insider actors came to light (Rudaz, 2005: 214). Three networks of local political actors, Alliance in the Alps, Alparc and Alpine Pearls have been portrayed, as well as the non-governmental organization CIPRA. It is exactly the role and significance of those networks as promoters of a "new space of action" that will be the focus of the following chapters.

---

67 Interview conducted by me with Karmen Mentil and Peter Brandauer (Werfeneweg, 29.01.2009).

68 Members with fewer than 100000 overnight stays per year pay a reduced fee, and the admission charge for new members is 6000 euros.

5. Regional objectivation and identification

In this chapter, I start presenting evidence from the fieldwork to support the theoretical framework fleshed out in chapter 2. The structure of this chapter will have two parts that follow my conception of identity as a double feature comprising processes of objectivation and identification.

In the first part, I show how local political actors objectivize the Alpine region. The aim is to detect and analyse the different kinds of images that local political actors use to frame the Alpine region. Firstly, I analyse how local political actors think of the Alps as a unique mountain range. Then, I underline the intrinsic qualities that the local political actors attribute to this continuous space: its aesthetic, nostalgic, economic and ecologic values. In the second part, I frame the objectivation process presented in the first part with the ways in which local political actors identify with this objectivized Alpine space.

5.1 Regional objectivation

In this research, the map of the Alpine Convention sets the territorial framework, as the projects of the pan-Alpine networks are taking place within its borders. I will therefore show how local political actors assume the identity of this Alpine region's framework, namely the features of nature, culture and inhabitants that distinguish one region from others (Paasi, 2009: 469). In other words, I illustrate which social image of the Alps (Avanza and Laferté, 2005: 143) the promoters of the pan-Alpine initiatives espouse and convey to the broader population. How mountains in general, and the Alps in particular, have been objectivized is the result of a long historical process that, although captivating, does not have a place in this research, which concentrates on contemporary events.[69] The local political actors under investigation here refer, directly or indirectly, to this objectivation process that took place in the *longue durée*.

Scholars identify the United Nations Conference on Environment and Development, also known as the Earth Summit, held in 1992 in Rio de Janeiro, Brazil, as a turning point at which the international community started to recognize mountains as a common good (on this point, see the critical analysis by Crivelli & Raffestin, 1992; Debarbieux & Price, 2008). It was the first time, Debarbieux and Rudaz assert, that mountains became part of the global political sphere (Debarbieux & Rudaz, 2010: 219). With that year's Earth Summit, mountains were recognized for playing a major role in issues of sustainable development (Debarbieux & Price, 2008: 153). This idea of mountains as guarantor of sustainable development has been one of the most important images internalized by the local political actors I interviewed.

It is in this context that the idea emerged of considering the Alps as one continuous region that should be preserved. The operational programme of the European Commission that was approved in 2007 states that the Alps constitute a common space, regardless of national borders and administrative barriers (European Commission 2007: 10). This programme assumes that the national borders inside the Alpine region should be more permeable and that the Alps should be considered a unique entity. In the European Commission's view, the fact of considering the Alps as a unique entity is a condition for developing shared strategies for sustainable development.

The establishment of the Alpine Convention is, thus, the best example to demonstrate these socio-political evolutions. As Debarbieux points out, the convention contributes to a social process – one that started with the Enlightenment – of a conversion from a negative to a positive image of the Alps and their inhabitants. Additionally, it is the outcome and the origin of a political process that takes shape during the creation of new institutions (Debarbieux, 2008b: 101), as will be shown later in this book.

69 See Bozonnet, 2002 or Debarbieux & Rudaz, 2010 for a historical analysis of the image of the Alpine region.

As some scholars have underscored, in the first phase, which corresponds to the establishment of the Alpine Convention, predominantly scientists and politicians (Bätzing, 1994; Debarbieux 2008b: 43) undertook the objectivation process of the Alpine region. In this phase, the Alpine non-governmental organization CIPRA, and the International Scientific Committee on Research in the Alps (ISCAR) played a key role in defining what the Alpine region actually is, where it starts, and where it ends.

Later, during the implementation phase, when the pan-Alpine networks were created, other actors participated in the conception of an image (or many images) of the Alps. It is this second set of images – of the Alps as conveyed by the networks and its members – that will be considered here. In this section, the objectivation process will be investigated in order to find an answer to the following questions, with a focus on the three networks that I analyse: How do local political actors and their institutions qualify this "thing" they call "the Alps"? What territorial shape is the Alpine region taking for local political actors? What is it behind the word "Alps" that these actors use as a common point of reference to implement their projects? And, last but not least, how are local political actors integrating (or not integrating) the way of objectivizing the Alps that scientists and politicians have proposed?

5.1.1 Recurrent images of the Alps that local political actors use

The present section is structured according to the categorization established by Petite in his research about the territorial and material dimensions of collective identities in the Alps (Petite, 2009). Petite, while analysing the role of representations in collective projects dealing with material objects, suggests that there are five recurrent images that local political actors use to refer to the mountain in general, and to the Alps in particular (Petite, 2009: 37–39):

- The aesthetic image: mountains thought of as "naturally beautiful";
- The patriotic image of mountain landscape: the use of the mountain to increase national feelings;

- The touristic image: the image of the mountain as a suitable environment for sports and leisure activities;
- The ecological image, in which the mountain environment is seen as a natural resource;
- The nostalgic image, which tends to idealize the link between mountain societies and the environment in which they live.

I use the same categorization to understand how the pan-Alpine networks and their members think and objectivize the Alps. Of course, for the purposes of my research, I adapted the grid proposed by Petite to my case. For example, I will not use the second image, since there is no evidence, based on the interviews I conducted and in the presentation of the networks' Internet sites, that the mountain is used as a "patriotic means". It is difficult to explain why this is the case. This is possibly because the interviews were not conducted with an emphasis on the meaning of the mountains as a symbol of nationalism. In fact, the emphasis was placed almost exclusively on trans-nationalism, so that the interviewees were not guided to stress this point. Instead, and this is probably also a result of the questions that were asked, many persons put forward the idea of a "unique Alpine region", and it will, therefore, be analysed here first.

5.1.2 The definition of the Alpine arc

Not just in their actions, i.e. the establishment of pan-Alpine networks linking different points inside the Alpine region, but also through their words, local political actors stress the idea that the Alps constitute a unique structure. This unity is underlined in different ways, notably through the geomorphologic structure of the Alps, and also through some homogeneous socio-cultural characteristics. The two elements that interviewees emphasize to signify the unique Alpine structure will be presented in the following paragraphs. I will conclude this part by underlining what could at first be seen as a paradox: the constant reference to Alpine diversity in order to stress Alpine unity.

A unique natural structure

For some, the particular topographical and geographical Alpine structure is the key element to define the region, which sets it apart from others. This image ties up with what Sacareau identifies as the common-sense definition of the mountain: an evident visual reality, corresponding to a dramatic landform (Sacareau, 2003: 5). This attitude corresponds well to Alparc's definition of the Alps, which can be read on its Internet site. Alparc lists some quantitative features, linked to topography (the surface and the altitude of the highest point), in order to describe the biogeographical Alpine region:

> The Alps are a biogeographical region extending over approximately 190 000 km² in the centre of Europe. Mont Blanc is the highest point, at 4 810 m above sea level (Alparc webpage).

All these characteristics can be summarized in one German word: the Alps are a *Gebirgszug*, a mountain range, as the president of Alpine Pearls underlines. He adds that the Alps are simply known by the average Joe as a mountain range, with all its geomorphologic qualities. Even more laconically, to my question "What, for you, are the Alps?" Paolo, an employee of the Centre of Alpine Ecology, Trento, Italy, answered: "They are an elevation of the tectonic plates." This answer underlines the same geological origin, and with that the similarity characterizing the mountain range.

This particular topographic characteristic of mountains and valleys eroded by rivers is seen as something unique in the European context:

In den Bergen und Flüssen [herrschen] einfach ganz spezielle strukturelle Bedingungen [in den Alpen]. Was für mich in Europa sicher ein Alleinstellungsmerkmal ist. (Sabine, Betreuerin)	Through mountains and rivers, the Alps have special structural conditions. It is, from my point of view, a distinctive feature in Europe. (Sabine, animator)

Although the Alps seem to be easily recognizable, as the German word *Merkmal* (which comes from the verb *merken*, i.e. to perceive, to notice) suggests, questions about the limits of the Alpine region are not uncontested, as has always been the case in the history of natural science. Consequently, the debate on the definition of the mountain's boundaries is still a reality in contemporary discussions. When Karmen, manager of Alpine Pearls, says that for her the Alps go all the way to Vienna, her view indicates that there is not a consensus on where the Alps begin and end, which leaves room for personal and flexible definitions. However, as Karmen first declared, she does share with others the idea that the Alps can be distinguished by their topography. So, in a certain way, they should be easily identifiable:

Für mich ist das auf Grund der Topographie ein vorgegebener Lebensraum. Zusammenhängend vom Mittelmeer, den Bogen über Frankreich, Schweiz, Österreich, Deutschland. Bis Wien, für mich [*Lächeln*]. (Karmen, Alpine Pearls Manager)

The predetermined and resulting living space of the Alps is caused by its topography. A contiguous arc from the Mediterranean Sea, to France, Austria and Germany. All the way to Vienna, for me [*smiling*]. (Karmen, manager of Alpine Pearls)

This small comment about the Alps reaching Vienna, probably added as a provocation, as evidenced by the smiling, seems to take a more concrete contested dimension in the interviews conducted with two other Alpine protagonists. The Alps, which seem to be a reality that everybody can easily define, is in fact a matter of choice. I will prove this with one example. Sabine, who lives in the northern part of the Austrian Alps, summarizes the idea in a single sentence: "the boundary between the foothills of the Alps and the Alps is fuzzy *(verschwimmend)*".

The definition of the Slovene Alps is an eloquent (and contested) example to illustrate this issue. Markus, an employee of the Berchtesgaden National Park, while remembering the beginning of Alparc, exposes the case of the Škocjan Caves Park, situated in the south-eastern part of Slovenia. Markus speaks about the difficulties, at the beginning, to get the big parks involved in creating a network of protected areas. He stresses that the project was able to become reality only when the most important parks decided to join. What is interesting here is the justification he gives for one small regional park to belong to the Alps:

Und erst als die großen und bekannten Parks dann gesagt haben "Ja, wir arbeiten zusammen", war das eigentlich der Durchbruch, weil man ab dann auch die Nationalparks hatte... und dann hat man auch die kleinen Regionalparks und Naturparks langsam dazu genommen – aber es war eigentlich eine schwierige Geburt... Man hat gerade dann an Škocjan denken müssen. Škocjan liegt so an der äussersten Ecke der Alpen, also eigentlich liegt es nicht ganz in den Alpen, aber es gibt sehr viele Gemeinsamkeiten... Der ganze Karst und so, und deshalb hat man gesagt, "Ja, Škocjan eigentlich gehört dazu!". (Markus, Mitarbeiter eines Schutzgebietes)	The breakthrough happened only when the biggest and best-known parks said, "Yes, we want to work together", because then the national parks decided to join. And then even the small regional and natural parks slowly got around to joining. But it was a difficult birth... Then we had to think about Škocjan. Škocjan lies at the outermost corner of the Alps, so, it does not lie entirely in the Alps, although there are a number of similarities: the whole karsts and so. That is why we said, "Yes, Škocjan actually is part of the Alps". (Markus, employee of a protected area)

The decision to include the Škocjan Caves Parks in the Alpine perimeter was therefore dictated by geomorphologic reasons. Owing to the presence of the karst rock formations, the Škocjan Caves Park has been classified as belonging to the Alpine geomorphologic structure and therefore to the Alps *tout court*.

Dragan, a CIPRA employee, in discussing the designation of the Slovene Alps, also uses the karsts to validate the argument that a larger area of Slovenia than the one currently defined should be considered as Alpine and therefore be part of the Alpine Convention:

> We try to extend the definition of the mountain in Slovenia. In our country, when people speak of mountains, they think of Triglav and of "things" higher than 1 500 m. This is a problem [...]. According to some international definitions, more than 70 % of Slovenia has to be considered mountain area, and if you look at the profiles of Slovene rivers [...]. We don't have lowland rivers, I mean, we have some, but only high in the mountains, where some plains lie. The Sava, Drava, Mura and Soča rivers all have a mountainous profile in the geographical sense. That is why the definition of the territory as per the Alpine Convention is stupid in Slovenia, because it is somehow restricted to a certain geographical middle area of the Alps. The definition leaves out large areas and practically most of Slovenia. In Italy, the whole karst is inside the Alpine Convention, up to Trieste, but in Slovenia the karst is not inside. (Dragan, employee of CIPRA)

Dragan also cites the authority of vague "international definitions" to give weight to his argument. This, added to what he calls the "geographical sense", or the "scientific definition of mountains", is for him strong enough elements to bring into question the Alpine Convention's official definition of the Alps.

All these remarks suggest that what people think the Alps are, how they objectivize them, is still anchored in the naturalist tradition that started in the 18th century. Natural elements (here mostly geological), thought to be objective, are used to signify the pertinence of the Alpine perimeter. The dispute, as it relates to Slovenia for example, is also based on naturalist considerations, i.e. the karst as synonymous with "alpinity".

A unique natural structure as support of unique socio-cultural features

The natural (geological) structure of the Alpine mountain is one way – albeit one that, as the examples before have shown, can be contested – to represent Alpine unity. Many more interviewees see an Alpine hyphen in socio-cultural elements. With a suggestive image, Paolo remembers his passion for taxonomy when he was younger:

Io da piccolo facevo il tassonomo, se dovessi catalogare un montanaro, un trentino, un austriaco o uno svizzero, farei la stessa specie, con qualche sottospecie, [...] però, ecco, c'è sicuramente qualcosa che unisce. (Paolo, impiegato al Centro di Ecologi Alpina)	When I was a child, I was a taxonomist. If I had to place a mountaineer in a category, a person from Trentino, an Austrian or a Swiss, I would put him or her in the same species, probably within some sub-species, but, well, there is obviously something in common. (Paolo, employee of the Centre of Alpine Ecology)

Paolo recognizes a similar Alpine culture. As Sandro said, this shared cultural background builds the space called "Alps", and in this space there is a shared culture:

Una cultura straordinariamente simile. Una cultura immateriale che è fondamentalmente simile. Tu hai gli stessi segni culturali in Slovenia, in Val d'Aosta, nel Sud della Francia, perché c'è una "cultura alpina". Le Alpi sono lo spazio di queste persone. (Sandro, impiegato della Convenzione delle Alpi)	An extraordinarily similar culture. An immaterial culture that is essentially similar. One sees the same cultural signs in Slovenia, in the Aosta Valley, in southern France, because an "Alpine culture" exists. The Alps are the living space of these persons. (Sandro, employee of the Alpine Convention)

This "same living space", according to Hans, gives rise to a similar "mentality", built up by a way of interacting with the Alpine environment that is shared by all Alp dwellers:

> Gerade die Verwurzelung, auch durch das Leben mit der Natur, ist sehr ähnlich. Die Arbeit auf den Bauernhöfen zum Beispiel, die ganz ähnliche Probleme, ganz ähnliche Herausforderungen haben, ob das jetzt in Südtirol ist, in der Schweiz oder bei uns. Die Mentalität, die sich herausbildet, die ist ähnlich, egal wo ich hinschaue. Und das ist spannend. Und ich glaube, das lässt sich nicht so sehr an nationalen Grenzen festmachen. (Hans, Bürgermeister)

> Precisely the roots, stemming from living with a similar landscape, make us analogous. Farming, for example, which knows similar problems, similar challenges, is the same, whether you are in South Tyrol or Switzerland, or here. Everywhere I look, a similar mentality develops. And this is fascinating. And I believe that this cannot be fixed along national borders. (Hans, mayor)

The way Hans thinks the mountains and the influence on the people living there is reminiscent of the way natural scientists in the 18th century tried to characterize the "mountaineer". As did the naturalists, Hans recognizes a peculiarity of the mountain inhabitants, because of the geographical characteristics of their environment (the "landscape", as he says). What is different from the 18th century is that people whom scientists at the time described as "mountaineers" did not think of themselves as such (Debarbieux, 2010). Yet, in this case, it seems that local political actors (Paolo, Sandro and Hans, but also others) do not hesitate to describe themselves as "mountaineers with a similar mentality". For most of them, this "mentality" has a positive connotation today, so that it is easier to identify with it.

In a contemporary world where it is common to affirm that national borders are losing significance, it is not surprising that Alpine political actors emphasize a "common culture" overlapping the national ones. Charlotte Penel, animator of Alliance in the Alps for France, speaks openly of an existing Alpine culture[70] that, in her eyes, forms another layer that must to be added to the national cultures. The Alps, as she says, "are a big mountain chain, with gorgeous landscapes, a cultural

70 The interview can be found on YouTube: <www.youtube.com/watch?v=L13Vzsm7FCc&feature=related>, consulted on the 04.03.2015.

richness and a heritage that should be promoted. But above all, what I think is outstanding, is the fact that the Alps are beyond boundaries. It is a whole that crosses borders going beyond language and national culture. The Alpine culture is added on top of these cultures".

Paolo, while answering my question about the significance of the national Alpine borders in comparison with the Alpine boundary, emphasizes not only the political issues (the introduction of the Schengen Area) to signify the diminished importance of the national boundaries, but also the cultural similarities that characterize the Alps. In Paolo's words, there is an allusion to a collective awareness process that took place in the contemporary Alps, in which people started thinking about the Alps as a hyphen, and no longer as a barrier. Interestingly, Paolo opens his reflection by referring to cultural characteristics that identify not only the Alpine populations specifically but also mountaineers in general:

> E' indubbio che le Alpi venivano viste come un ostacolo in tutti i modi, e visto che non ci sono più questi ostacoli [con l'introduzione di Schengen], psicologicamente ha favorito [la visione di una regione alpina unita]. Assieme a cose ben più importanti che sono di tipo evidentemente culturale, ideologico, cioè le Alpi come fonte di unione anziché come divisione. Io ricordo che avevamo organizzato, questo sì, insieme agli scozzesi, una conferenza internazionale sulle montagne d'Europa. Una sessione in Scozia, ad Edimburgo, e una sessione al Centro di Ecologia Alpina. [...] Effettivamente c'è una similarità nelle persone [di montagna], che poi ho ritrovato anche con i montanari scozzesi, quella volta. (Paolo, impiegato del Centro di Ecologia Alpina)

> Obviously, the Alps were seen as a barrier. The removal of these barriers with the introduction of the Schengen Agreement encouraged [the vision of a united Alpine region]. This, in conjunction with more important cultural and ideological phenomena, helped see the Alps as a source of union instead of division. I remember that we organized, together with the Scots, an international conference on European mountains. One session in Scotland, in Edinburgh, and one in Trento. In fact, there is a similarity between [mountaineers], which I discovered with the Scottish mountaineers, on that occasion. (Paolo, employee of the Centre of Alpine Ecology)

One of the elements that mountain inhabitants have in common is that they share the same problems, distinct from the ones that people living in the lowland have to face. Henrik, a mayor, highlights that those problems are "geographic", as he says, i.e. they are caused by the geomorphologic structure of the mountain. Structurally, he continues, it is impossible for lowland municipalities to suffer from avalanches.

The interviews show, as will be discussed later in this chapter, that this consciousness of being part of the same "Alpine structure", and therefore sharing the same problems, is one of the major sources of identification with the Alpine region.

Diversity as foundation of the Alpine unity

The Alps are a particularly complex socio-cultural area due to the coexistence of four commonly spoken languages (Italian, German, Slovene and French) with a number of derived dialects, and of manifestly differing national political attitudes. Paradoxically, this diversity is one of the main elements cited by the pan-Alpine actors to underline the Alpine uniqueness and unity. As Crivelli and Raffestin underscore, CIPRA does not underestimate the natural, cultural and economic diversity of the Alps; nonetheless, the global vision of problems and threats leads to a homogenizing representation of the mountain range (Crivelli & Raffestin, 1992). Interestingly, the same arguments used to characterize the unity of the Alps are also used to point out the differences within the same perimeter: The natural and socio-cultural elements that characterize the Alps are exactly the same ones upon which the Alpine unity is built.

This point is already underscored in the preamble to the Alpine Convention, where mainly socio-political aspects are used to indicate Alpine diversity:

> The seven Alpine states and the European Union,[71] aware of the substantial differences existing between national legal systems, natural conditions, population distribution, agriculture and forestry, the state and development of the economy, the volume of traffic and the nature and intensity of tourism, have agreed as follows [etc.] (Preamble to the Alpine Convention)

71 Monaco, the eighth state to sign the protocol, became a party on the basis of a separate additional protocol.

Alparc's Internet site dedicates more space and details to underscore the local differences identified in the Alps than do the sites of the other networks. Alparc gives importance to the natural (geomorphologic as well as biological) differences, but also to the socio-cultural facets of the Alps, producing an almost exhaustive portrait of all the elements that constitute the Alpine mosaic. By doing so, Alparc gives the Alps a distinctive character within the European context:

> The Alps are extremely diverse in terms of biology, landscapes, languages and cultures. [...] The mountains of the Alps have many different faces: the limestone plateaus of the pre-Alps, mounds of schist, granite needles, moraine and scree, farmed terraces, marshes and wetlands, dark pine forests and vibrant larch forests, flower-strewn meadows and grassy pastures, lofty cliffs, breathtaking caves and valleys, tumbling waterfalls, roiling mountain streams and gentle rivers, etc. The variety of scenery is the product of geology, morphology and climate, but is also the result of over one thousand years of human habitation. That scenery has also produced a peerless variety of flora and fauna.
>
> The Alps are home to a wide range of languages and cultures which in turn have produced many tales and legends. Some legends can be found in different guises throughout the Alps. In addition to the four main Alpine languages used in Alparc, namely French, German, Italian and Slovenian, the Alps are home to a very diverse linguistic heritage. Languages range from Slavonic to Germanic, with Alemannic German, Bavarian and the Walser dialect still being spoken in some regions, to the Rhaeto-Romance languages (Romansch, Ladin, Friulian) and Gallo-Romance and Italo-Western dialects. Many of these languages and dialects are now only spoken in remote areas and are gradually dying out.
>
> The architectural heritage also bears witness to the diversity found in the Alps. It is fascinating to examine both the differences and also the similarities in the solutions used to cope with the steep gradients and harsh climate. The same is true of the culinary heritage, costumes and traditions, not to mention the skills of local craftsmen and artists, etc. (Alparc internet site)

Alliance in the Alps, by contrast, limits itself to a couple of short and dry, albeit essential, sentences on this topic: "The Alps represent a region of great diversity in terms of landscape and the flora and fauna. The Alps are a region of cultural and linguistic diversity."[72]

72 Text published on the page "The Alps and the Alpine convention" of the old website of Alliance in the Alps but no longer accessible. Consulted the last time on the 04.03.2015.

The attitude of Alpine Pearls towards the Alpine diversity is encapsulated by the inclusion of local (and therefore diverse) "treasures", as they call them. These are advanced almost exclusively for the purpose of marketing to tourists, as witnessed in the tone of the following quotations:

> The Alpine Pearl communities pledge in the following ten areas that they will offer their guests an environmentally sustainable, culturally diverse, qualitatively high and adventure-packed holiday experience.
>
> Diversity is our goal, and that does not stop at the local culture and cuisine. In fact, we support Alpine culture, traditions and cooking.
>
> The Pearls conserve the esthetic character and local beauty of our Alpine villages. We want you to discover our local treasures; that is why we celebrate local culture and cuisine, foster traditions and support the environment. Each one of our "Pearls" enriches your experience in a unique and special way. (Alpine Pearls Internet site)

The interviews with two of the Alliance in the Alps animators, Francesco Pastorelli for Italy and Andreas Güthler for Germany, which are available on YouTube, also accentuate cultural diversity as one of the major characteristics of the Alps:[73]

Le Alpi per me sono un territorio affascinante per la diversità culturale e per la varietà di paesaggi naturali e culturali che rappresenta (Francesco Pastorelli, animatore)	For me, the Alps are a fascinating territory because of their cultural diversity and because of what they represent in terms of the variety of natural and cultural landscapes. (Francesco Pastorelli, animator)
Die Alpen sind für mich ein ganz faszinierender Raum, in dem ich sehr gerne lebe, und wo einfach auf relativ kleinem Raum die totale Vielfalt, sowohl der Natur als auch der Kultur der Leute die dort leben, existiert. (Andreas Güthler, Betreuer)	The Alps are for me a truly fascinating space, where I am happy to live, and where there is, in a relatively small space, an incredible variety of nature and cultures of people living in them. (Andreas Güthler, animator)

73 The interview with Francesco Pastorelli can be found here: <www.youtube.com/watch?v=aTOSlc7Wmy8&feature=related>; the interview with Andreas Güthler here: <www.youtube.com/watch?v=TGisL61tdGE&feature=related>, both Internet sites were consulted on the 04.03.2015.

These quotes highlight the importance accorded to the socio-cultural diversity of the Alps. However, Ivica's words probably best reflect the idea that I want to express in this section, namely the construction of an Alpine unity based on diversity:

> In this diversity I see the specificity of the Alpine space, and this diversity has to be cherished, has to be reinforced, and this is exactly what forms the basis for a common identity: We are all aware of how different we are in many respects, but we have common values, like to live in a beautiful environment, [be conscious] of all this, what the Alps represent. Unique nature, multiculturalism ... the Alps represent all of this. (Ivica, Alpine Convention)

Not only does Ivica observe local specificities inside the Alpine perimeter, but he deems it important to, as he says, "cherish" them. This view is shared by the networks that lend importance to the promotion of projects that preserve local knowledge and heritage.

This recognition of Alpine unity through diversity can be surprising and raise the following questions: How can these people and institutions stand for diversity if their aim, in the end, is to promote *one* Alpine region? How can their call for respecting diversity be credible in the eyes of those listening and reading their arguments? How is it that the claim for diversity does not destroy the design of a united Alpine region they want to build? In his article about Finnishness, Häkli borrows an interesting concept from the philosopher Perelman: the root metaphor.

The root metaphor is defined as "an interpretative framework that is automatically applied without ever recognizing that it is only a metaphor". In other words, Häkli continues, "when something is described through a 'root metaphor' it is seen as the reality – the only way of looking in things" (Häkli, 2005: 18). The Alps, and by the Alps I mean the "stuff" constituting them, probably have to be considered as a root metaphor. They are built on such solid rock (rock is understood here in a metaphorical sense but simultaneously as the mineralogical thing), that they "transcend difference and unite what is seen as different" (Häkli, 2005: 18). They cannot crumble if somebody puts forward their differences, instead of their unifying elements, because they are "obviously there".

A comparison can be made with the opposite position (other examples, taken from other parties calling for regional-ethnic independence in Europe, could have been used as well), namely that of the constitution of the Padania Region, the Po Valley, for which the Lega Nord party claimed independence from Italy, especially in the 1990s. Without entering into the details of this political movement, it is worth recalling that, as specialists on the question have stated, "Padania" is a pure invention, a region that was unknown before the 1994 elections and has no past political unity, no religious specificity, no common language and no feeling of common belonging (Avanza, 2007).

In order to create this inexistent union, the thinkers of Padania took the Po River as a basis for unity. In one of the first editions of the *Quaderni Padani*, the intellectual voice of the party published by the *Libera Compagnia Padana*, an association that promotes the culture, language, traditions and autonomist aspirations of Padania,[74] the unifying element is well identified. Indeed, Oneto, one of the most important intellectual exponents of the party, recognizes the Po River as this element: It is the unifying sign of this land; it unifies two seas and the people who quench their thirst thanks to it; for everybody, the Po River is the symbol of unity and freedom (Oneto, 1996: 3).

The river, as is the mountain in the case of the Alps, seems to be used because it is "incontestable": It is "matter" that is there and that everybody (supporters and opponents) can see and touch. Therefore, it cannot be denied. It shows in both cases, although probably in a more spectacular way for the Padania than for the Alps, the political construction of natural objects, the mountain and the river. The "water ritual", which Bossi, the president of the Lega Nord party, invented in 1996, shows this attempt to give, on the one hand, materiality, and, on the other, political weight, to this idea. The yearly "water ritual" consists of a long march of party supporters, which starts on the Monviso Mountain, in Piedmont, and ends in Venice. At the top of the mountain, a representative of the party collects some water from the source of the Po and puts it in a cruet that he pours, some hours later, into the Mediterranean Sea: a highly symbolic gesture that signifies the unity between the eastern and western parts of northern Italy.

74 <www.laliberacompagnia.org/statuto.php>, consulted on the 04.03.2015.

This ritual, and the intellectual production on the Po River as a unifier of the region, is not necessary with regard to the Alps, because they are, even in popular beliefs, a real "thing" that exists and is observable. People both inside and outside the Alps have an image of what the Alps look like, something they would never question. Because this belief is so strong and anchored, the supporters of a regionalization process do not have to worry if they call for diversity. Calling for diversity in Padania could result in the collapse of the entire project; that is why justifications, in this case the river, have to be used and defended.

5.1.3 Landscape features and an aesthetic image of the Alps

In the collective imagination, the Alps are associated with spectacular natural and cultural landscapes: high peaks, sloping mountainsides, water-falls and rocky plateaus that were sources of fear during the Middle Ages, inspired naturalists, writers and painters from the Enlightenment onwards. This fascination for the aesthetic value of the Alpine landscape can be found in the words and images produced by the networks investigated here. The image of the postcard "Together for the Alps" published by Alparc (see fig. 5) probably best summarizes this idea.

Fig. 5: Postcard produced by Alparc, "Together for the Alps". © Alparc.

These sorts of pictures, showing the "gorgeous Alpine region", as defined by Alparc, can be found on the websites of the three networks. Of course, the networks highlight the aesthetic beauty of the Alpine landscapes to varying degrees. Alliance in the Alps, whose aim is to connect Alpine municipalities on the field of sustainable development, is the network with the less idyllic images. Alparc, in its photo gallery of the protected areas,[75] exposes almost exclusively natural "untamed landscapes and legendary peaks",[76] without any human beings or traces of their activities. A different outlook is displayed by Alpine Pearls, which shows the presence, albeit a "sustainable presence", of human beings in the gorgeous, natural, mountain landscapes. The main objective of the network (the promotion of sustainable and slow tourism) is to combine these two factors: the possibility of tourist activities in a wonderful landscape.

In interviews, some also refer to the beauty of the mountain scenery as one of the Alps' characteristics. This was in answer to the question: "What are the Alps for you?" Short keywords, not structured into elaborate sentences, can be found in the answers: "gorgeous mountains" (Antje, employee of a tourist office), "a nice living space" (Henrik, mayor), "a unique nature and a beautiful environment" (Ivica, Alpine Convention).

Andreas provided a reason why local political actors do not underline this aspect of the Alps:

Für mich sind die Alpen wirklich diese Netzwerke. [...] Objektiv sind die Alpen ein Gebirgszug und ich könnte auch [sagen, dass] für mich die Alpen die Kraft und die Natur ist, die ich spüre. Das ist auch so, aber vor allem sind die Alpen für mich diese Netzwerke und diese Leute, die etwas tun und die sich besuchen, und daran glauben, dass man etwas verändern kann. Das ist eigentlich für mich das Wichtigste in den Alpen.

For me, the Alps are really these networks. [...] Objectively, the Alps are a mountain chain, and I could also say that, for me, the Alps are the power and the nature that I can feel. But above all, the Alps are these networks and these people that do something, who meet each other and believe in change. I consider this the most important thing going on in the Alps. The Alpine Convention is an important foundation, but

75 <www.alparc.org/the-protected-areas/the-apa-in-images>, consulted on the 04.03.2015.

76 <www.alparc.org/the-alps/legendary-scenery>, consulted on the 04.03.2015.

Eben wie gesagt, diese Alpenkonvention ist ein ganz wichtiges Fundament, aber wenn die einmal nicht mehr existieren würden, diese Netzwerke, die sind so gut, dass sie weiter funktionieren dürfen. Das ist für mich das Allerwichtigste. (Andreas, Direktor CIPRA International)

should it not work out anymore, these networks are so good that they will continue working. This is for me the most important thing. (Andreas, director of CIPRA International)

The sentiment behind Andreas' words is probably shared among local political actors. Without anyone denying the beauty of the Alpine landscape, they are aware of their role inside the Alpine community, which is to act to promote their projects. Through their efforts to implement sustainable development, they "see beyond" the beautiful features that the Alps represent in the collective imagination and concentrate their efforts on preserving this beauty instead of only talking about it.

5.1.4 Nostalgic image of the Alps

Probably exactly because the networks are almost by definition turned more towards the future than towards the past, action rather than admiration, there are only a few allusions to the nostalgic image of the Alps. The cover of the third report on the state of the Alps published by CIPRA and named, emblematically, *We, the Alps! Women and men shaping the future* (Michel, 2007), is probably the best illustration of the way in which CIPRA and the networks conceive the Alps. Although the image shows (see fig. 6), in the background, a "classical" Alpine landscape with mountains and a church, the foreground clearly gives another perspective, notably the presence of a laptop and a young boy in the very centre of the picture.

Fig. 6: Cover of the third report on the state of the Alps.
© CIPRA International.

Moreover, even in the evidence collected during fieldwork, there is almost no presence of a "real" nostalgic image of the Alpine region. Only a couple of examples of nostalgic images can be found on the networks' websites and in interviews. And even in those cases, one can debate whether they really qualify as nostalgic examples. They refer to a way of coping with the mountain environment, judged as being "sustainable". In some ways, it recalls activities of the past, but these activities are thought of more as an investment for the future than as a nostalgic remembering of "what it was but is no longer". Let us take a look at a few examples in the category "nostalgic image of the Alps".

Alparc, for example, refers to traditional agriculture as one important factor for the preservation of Alpine biodiversity:

> Traditional agricultural and pastoral activities have shaped the farmed landscapes in large sections of the Alps for over a thousand years and are a significant factor in maintaining biodiversity. (Alparc website)

Alliance in the Alps remembers the capacity of adapted Alpine economic activities to preserve the environment. Mountaineers, often seen in history as the cause of overexploitation of resources (Debarbieux & Rudaz, 2010: 109), are now considered to be the most skilled persons to implement sustainable development by adapting traditional farming to the new context:

> The Alps can also play a pioneer[ing] role in the creation of a convincing design for a sustainable economy. Experience with economic activity adapted to the needs of the natural environment has been preserved longer in the Alpine space, so that the measures required for sustainable development can often be introduced more easily. (Alliance in the Alps Internet site)

Although it is possible to find slight references to a "nostalgic image" of the Alps on the networks' websites, I could not detect any in the interviews I conducted. This is probably a sign that local political actors are even more forward-looking than their umbrella associations in thinking about the Alps. As Luc admits, there is no need to look backwards; quite the contrary is true, since it is in the Alps that one can find the "ideals of life":

On retrouve dans la montagne et dans les Alpes aussi, un petit peu un idéal de vie. (Luc, employé d'une municipalité)	In the mountains, and in the Alps, you can find a bit of the ideals of life. (Luc, municipal employee)

5.1.5 The Alps as an economic and tourism capital

Alpine Pearls is, among the three networks, the one that exploits in a more direct way the image of the Alps as a location suitable for leisure and tourism activities. This position is obvious in the presentation of the network on its website:

> Holidays are your precious time off-work. You look forward to relaxation, fun-filled activities and cultural diversions. Yet you still desire to do your bit to sustain nature and to protect the environment and climate. With the Alpine Pearls, you'll have both – gorgeous Alpine regions and a soft environmental footprint. (Alpine Pearls website)

Antje, the representative of a member of the Alpine Pearls network, clearly manifests this idea by stating that the Alps are "their capital":

[Die Alpen sind] unsere Basis für alles. Wenn wir die nicht hätten, dann könnten wir unser Gebiet nicht in dieser Art bewerben, vermarkten. Das ist für mich eine Voraussetzung. Eine wunderbare, wunderschöne natürliche Voraussetzung. (Antje, Mitarbeiterin eines Fremdenverkehrsbüros)

[The Alps are] our basis for everything. If we did not have them, then we could not market our region in the same way that we are doing now. For me, they are a condition. A delightful, gorgeous natural condition. (Antje, employee of a tourist office)

Claudio, whose municipality is a member of Alliance in the Alps, also speaks of the tourist capital that the Alpine territory represents. He starts off by saying that "they" (referring to the people of the Ticino Canton in Switzerland) are not aware of the resources that the Alpine territory represents. He regrets this, especially because it seems that this potential is well mobilized by the other members of the Alpine Pearls. Alpine Pearls is indeed basing its entire project for promoting tourism on the awareness of the capital that the Alpine environment represents. However, the people working in collaboration with Claudio do not seem to be sensitive to this point, as he affirms:

Non ci rendiamo conto del capitale che è lo spazio alpino, o pre-alpino, o lo spazio rurale, nel senso che non viene monetizzato e invece sono convinto che, d'altra parte lo conforta anche le ricerche che ha fatto il Politecnico, il territorio è fondamentale per il turismo, loro calcolano che l'indotto del turismo, su 15 miliardi di franchi, un miliardo e mezzo è generato dall'attrattività del paesaggio gestito, mantenuto, ben conservato. (Claudio, segretario regionale)

We are not aware of the capital that the Alpine, or pre-Alpine, or the rural space, represents. Therefore, it is not monetized, but I am convinced that (and this is supported by the research undertaken by the Swiss Federal Institute of Technology Zurich) the territory is essential to tourism. The technical university calculates as follows the economic outcome of the tourist industry: Out of 15 billion Swiss francs, 1 billion is generated by the attractiveness of a well-managed and preserved landscape. (Claudio, regional secretary)

The accuracy of the numbers and the sources of the study are not important here. What must be noted is the awareness of the capital that the Alpine territory represents for this individual. Claudio was the only member of Alliance in the Alps whom I interviewed that openly emphasized the network's potential to actively promote tourism. Tourist marketing, notably via the creation of a common website where potential tourists could see the offers on rural tourism, is one of the most important challenges that Claudio sees for the network. It is, in his eyes, a way to economically exploit the capital represented by the Alpine territory; an approach that was identified and put into practice by the Alpine Pearls.

5.1.6 Ecological image of the Alps: the shared concept of sustainable development

One of the shared concepts among local political actors is that of "sustainable development". This principle is enshrined in the Alpine Convention, as described by CIPRA: "The Alpine Convention is an agreement between various countries for the protection and sustainable development of the Alpine region."[77] Sustainable development, thus, is constantly used as reference by the local networks, whose members, "together with their citizens, strive to develop their Alpine living environment in a sustainable way".[78] Thereby, the Alpine Convention seeks to generalize sustainable development at the entire mountain range level, via the networks that are expected to implement sustainable development principles.

The mountain environment is thus seen as one of the three pillars (the others being economy and society) of sustainable development. The reference to sustainable development, the wish to implement projects capable of finding a balance between these three pillars, is clearly present in the discussions I have undertaken with local political actors, as well as on the networks' websites.

Nevertheless, the decision to collaborate with municipalities and integrate themes other than protection issues has not gone uncontested. Götz remembers the intervention of one of CIPRA's members, who declared: "You now work together with the municipality of Saalbach-Hinterglem, in the Salzburg region, a big tourism resort.

77 <www.cipra.org/en/alpine-politics/alpine-convention?set_language=en>, consulted on the 04.03.2015.

78 <http://alpenallianz.org/en/about-us>, consulted on the 13.07.2015.

They do these foolish things, we appeal against them, and you support their point. This is crazy." However, now that the concept of sustainable development has increased dramatically in popularity and the mountains have become the model of sustainable life, collaboration with municipalities is no longer in dispute. Güthler says the following in a video published on YouTube,[79] in answer to the question "What are the Alps for you?":

Die Alpen sind für mich ein wunderschöner Lebensraum, in dem leider auch nicht alles perfekt ist, in dem aber auch trotzdem viele Beispiele von Kreativität, von guten Ideen und ein Austausch von Kultur- und Sprachgruppen zu finden ist. Und deswegen kann ich gut sagen, dass die Alpen eine Modellregion für andere Gebiete auf dieser Welt sind. (Andreas Güthler, Betreuer)	The Alps for me are a wonderful living space in which everything is unfortunately not perfect but many examples of creativity, good ideas and exchange among cultural and linguistic groups can be found. Therefore I can say that the Alps are a model region for other places in this world. (Andreas Güthler, animator)

Sustainable development, thus, acts as a "regional conceptual shape" (Paasi, 1986: 125) among local political actors, and it is seen as, on the one hand, the aim of the actions carried out by the networks, but also, on the other hand, as the reason for undertaking these actions. Local political actors use the concept of sustainable development as "taken for granted" without clearly explaining (neither on their websites nor in the interviews I have conducted) what this concept consists of and how exactly it is implemented.

Despite full adherence to sustainable development, as I previously discussed, when objectivizing the Alps, the "natural" character of the mountain environment seen by the networks is closer to a definition of nature conservation than one of sustainable development. Alparc's description of the Alpine flora and fauna and its celebration of the number of species living in the Alps do not merely seek to show the beautiful species of animals and plants in the Alps. It also implies a duty to protect them and safeguard them from extinction:

79 <www.youtube.com/watch?v=FhNynnOUnGw&feature=related>, consulted on the 04.03.2015.

> The edelweiss, alpenrose and gentian are undoubtedly among the best-known Alpine flowers, but are just a drop in the ocean of the 4 500 vascular plant species found in the Alps – that's 39 % of all European plant species. Of these 4 500 plant species, 350 are classified as native to the Alps or certain Alpine regions and are often found at high altitude. These native plants include the Zois' bellflower (*Campanula zoysii*), round-leaved St. John's wort and the carniolan lily to name but a few. (Alparc website)

This quotation has closer links to protection than to sustainable development. The idea of nature being intact and in need of preservation is also mentioned in the preamble of the Alpine Convention:

> [We, the contracting parties of the Alpine Convention], aware that the Alps are one of the largest continuous unspoilt natural areas in Europe, [are] convinced of the need for economic interests to be reconciled with ecological requirements. (Alpine Convention website)

Of all three networks, Alliance in the Alps seems to be the one that uses the concept of sustainable development in a more univocal way. Nevertheless, on its website, it emphasizes its focus is on both sustainable development and protection:

> The future of the Alpine space is being seen more and more in terms of sustainable development, with equal attention afforded to the twin aspects of protection and use. (Alliance in the Alps website[80])

Networks, even those that more openly favour protection, constantly refer to sustainable development. Pointing to the Alpine Convention, which "contains general measures for the sustainable development in the Alpine region",[81] social and political Alpine actors believe that the implementation of its principles will preserve the nature of the Alpine territory, while simultaneously allowing economic activities to take place.

---

80 Text published on the old website of Alliance in the Alps but no longer accessible. Consulted the last time on the 04.03.2015.

81 <www.alpconv.org/en/convention/framework>, consulted on the 04.03.2015.

Through this conception of nature, the Alpine region emerges. Debarbieux describes the process of internalization of the representation of the Alpine nature in an article published in the *Journal of Alpine Research*:

> For a long time, the Alps only represented an entity relevant in the eyes of the scientific world, and secondarily a geopolitical object identified and worked by the states concerned. Today the introspection that is found at the heart of some of the territorial initiatives identified here – associations of Alpine communes, the Alpine Convention, networks of managers, ski resorts or companies, etc. – is also based on just as many representations of the Alps, representations that orient the action, and on the action itself. The Alps as such emerge under the double effect of shared representations and competing or concerted actions that refer to these representations (Debarbieux, 2008: 51).

Thus, there are indeed some "shared representations" that circulate among Alpine social and political actors and define their projects.

Sustainable development is one of the shared concepts that shape action. These actions, which take the form of different projects,[82] are mainly funded by the Interreg IIIB programme; others are financed by private initiatives, such as the dynAlp-Climate programme started in 2009 and funded by the MAVA foundation.[83] The concrete expression of these actions, which are conditioned by a certain representation of the Alpine territory, have an impact on the Alpine materiality by modifying the modes of management and the ecological character of the Alps, as Debarbieux highlights (Debarbieux, 2008: 102).

To conclude this section and to respond to the questions asked in the introduction of the chapter, some general remarks can be made. First of all, it seems that there is a consensus about the material evidence structuring the Alps. The Alps are characterized primarily by their geomorphology. Secondly, it is from this solid foundation that sociological and natural diversity can develop. What binds the different socio-cultural conditions is the Alpine inhabitants having to

82 See, for example, the Dynalp and Dynalp-climate projects, the Ecological Continuum Initiative (<http://www.alparc.org/it/our-actions/biodiversity-and-ecological-connectivity/ecological-continuum-initiative>) and the Alpencom programme (<www.alpencom.org/pages/index.php>), consulted on the 04.03.2015.

83 <http://en.mava-foundation.org/>, consulted on the 13.07.2015.

face the same problems, the same challenges. Finally, the solution that the networks propose, namely sustainable development, is seen by local political actors as one of the most efficient ways to face these challenges in unison. In that sense, local political actors make the definition of the Alps proposed by scientists and politicians their own – notably by accepting the Alpine Convention's map as the perimeter for their actions, but also by incorporating the concept of sustainable development – in order to conceive their projects.

5.2 Regional identification

Having analysed, by using the four images proposed by Petite, how local political actors are objectivizing this space called the Alps, I will emphasize in the following sections the role that this object, the "Alps", is playing in the construction of a common regional identification. Before entering the debate on identification, I want to make two preliminary remarks on the link between objectivation and identification. Firstly, these two processes need not be considered as chronological and causal, one being the consequence of the other. Secondly, we must draw attention to the fact that it is possible to create an image of the Alps and implement projects at the pan-Alpine scale without identifying with it, as will be demonstrated in the following paragraphs.

This section will be structured as follows: Firstly, I analyse how local political actors internalize the idea shaped by others that they belong to the particular social category of "mountaineers". Secondly, I take the example of two of the most active persons in local networks to show their different paths to identify with the Alps. Their personal stories of involvement in the pan-Alpine project show that it is impossible to generalize on the basis of the involvement and the identification process. However, these two persons and the majority of the other local political actors share a common reason for working at the pan-Alpine level: the awareness of being part of one "community of problems", as I will highlight in the third section. Thirdly, I introduce two types of actors who use different modes of identification with the Alps, although working in the same regional context: rangers and managers of protected areas. Finally, I suggest that the Alpine region is only one among many other scales of identification of the Alpine local political actors.

5.2.1 Proud to be a mountain inhabitant

The objectivation process of the Alps – the identification of the inhabitants with their region, both from outside and from inside – has a long history. The figure of the "mountaineer" takes shape in the 18th century. Its positive and idealized figure starts with Rousseau's writings. The philosopher of the Enlightenment glorifies the mountaineer as the incarnation of the *bon sauvage* (Bozonnet, 2002: 341–342). As Bozonnet underlines, the ideal of Rousseau's mountaineer constitutes a social model, capable of rehabilitating the contemporary urban societies (Bozonnet, 2002: 341–342). But while thinkers such as Rousseau were rediscovering in a new light people living in the mountains, the latter still did not consider themselves as mountaineers (Debarbieux & Rudaz, 2010: 34–35; Poche & Zuanon, in Rudaz, 2005: 27).

The identification, intended as an endogenous process of "collective construction of recognition" (Poche, 1996: 229), of Alpine inhabitants as mountaineers is recent. It is parallel to the promotion of the mountain as a "local development laboratory" by national public policies in the 1970s (Debarbieux & Rudaz, 2010: 150). At that time, some groups were formed that defended the interests of people living in the mountains and declaring themselves to be "mountaineers". Mountain inhabitants, organized in lobbies, took charge of their own destiny and even became the example to follow in terms of sustainable development (Debarbieux & Rudaz, 2010: 155–156). These developments can be seen as an endogenous appropriation of external images and symbols, which designate mountaineers as the guardians of a renewed link between society and nature, under the aegis of sustainable development.

One anecdote perfectly encapsulates this development of a renewed identification with the mountain. It was related by Antonio, mayor of Budoia,[84] during the interview I conducted with him in July 2007.[85] The story dates to 1999, when Antonio had newly been elected mayor and was searching for new ideas and projects to implement in his municipality. He was trying to find opportunities to link Budoia to the Alpine region. At the same time, Antonio discovered that Budoia and three other municipalities in the region (Polcenigo, Caneva and Aviano) had been excluded from the European funds "Objective 5b Programme".[86] As Antonio notes, "even some peripheral urban areas could benefit from the funds". In that sense, he could not understand the decision taken by the then *Comunità Pedemontana del Livenza* concerning Budoia.[87] In the same time period, the Friuli–Venezia Giulia Region was joining the "Objective 2 Programme" (Redevelopment of regions severely affected by industrial decline) promoted by the European Union. The region decided to apply to this programme with the same territory of the 5b objectives, thus excluding Budoia from the funds once more. With a colourful expression, Antonio writes that, when he found out about this decision, "a protest was triggered in me".[88]

Judging from the pictures (fig. 7), August 8th 1999 was a quite cold and windy day in Budoia. Antonio and some other people decided to hold a symbolic protest that day to show that Budoia was indeed a rural and mountain area and therefore had a right to the funds.

---

84 Antonio was mayor of Budoia from 1995 until 2009, thus, also at the time of the interview. He is no longer mayor, because the regional law in Friuli does not allow mayors to stand for re-election after three terms. Since he was mayor during my research period, I have retained his title in the current document.

85 Antonio hinted at this event during the interview I conducted on the 06.07.2007 in Budoia, and he gave further details of the event in an e-mail exchange between us in November 2010.

86 These funds are supported by the European Regional Development Fund (ERDF). Objective 5b corresponded to the objective of "development of rural regions".

87 The *Comunità Pedemontana del Livenza* no longer exists, because it has since been integrated into the *Comunità Montana del Friuli Occidentale*: <www.cmfriulioccidentale.it>, consulted on the 09.03.2015.

88 E-mail from the 23.11.2010.

Antonio organized a walk from the lowest point of the municipality of Budoia (90 m) to the highest, at Monte Forcella (1 900 m). Antonio recalls the event in his e-mail: "During five hours of walking, we passed villages, strode through hamlets, then through high meadows, until we reached the mountain top."

The small group of people emblematically carried two symbols during their walk: the flags of the municipality of Budoia and the flag of the European Union. The choice of these two flags suggests the following message: "Look, Budoia is indeed a mountain municipality, and therefore it should benefit from the European funds allocated to mountain regions."

Fig. 7: Images of the walk. Monte Forcella (1900 m), Budoia (Italy), 08.08.1999. © Franco Del Maschio.

In his account, Antonio says that despite this emblematic action, the area of the "Objective 2 Programme" was not modified: Budoia and the other three communities remained outside the perimeter. Nevertheless, representatives of the Friuli–Venezia-Giulia Region, surprised by the actions taken by the group, visited the municipality of Budoia. Thanks to this visit, the regional representatives discovered Budoia's commitment to the mountain cause and its projects to implement sustainable development at the local level. Consequently, the region decided to allocate special funds to the municipality of Budoia in order to sustain its projects. This anecdote shows three aspects linked to the identification process.

Firstly, it illustrates how individual identity, i.e. the recognition by one person (Antonio) that Budoia could build stronger links to the adjacent mountain, could be used to create a collective identity.

This provides content to the theory of Di Méo, who writes that identity shows the consequences, experienced by individuals, of common problems. It expresses shared issues and objectives of a collectivity (Di Méo, 2004: 345). When Antonio realized that Budoia was no longer considered a mountain municipality, he brought together a group of committed persons to demonstrate the opposite.

His walk had been transformed into a collective walk, not only because he was not alone in climbing the Monte Forcella, but also because the flags were used as a metonymy for the entire Budoia community. The capacity of Antonio to unite the community around a concept, a narrative, i.e. the promotion of sustainable development in a mountain area, corresponds to the power that Martin ascribes to identity narratives. The latter has the task to define the group and transform it from a latent status to one of a "community" in which the members are convinced they have common interests that are worth defending collectively (Martin, 1994: 23).

Secondly, it demonstrates the shift from an exogenous to an endogenous identification with the figure of the mountaineer. With his symbolic walk, Antonio clearly chose to be part of the mountaineering community. In fact, every individual, as stated by Gervais-Lambony, belongs to different identity communities, but he or she affirms one or the other, depending on the moment (Gervais-Lambony, 2004: 470). Undoubtedly conscious of the positive image that this identification with the Alps implies, Antonio chose to identify himself and his municipality with the mountain. He knows that the mountain represents, as Debarbieux and Rudaz declare, an exemplary value for the promotion of sustainable development (Debarbieux & Rudaz, 2010: 150).

Thirdly, it proves that the adherence to a community is often motivated by some interests, as Gervais-Lambony suggests (Gervais-Lambony, 2004: 470). In that sense, it shows that individuals and groups mobilize identity as a category of thought in order to achieve their aims and to initiate new dynamics (Petite, 2009: 20). The first motivation for Antonio's reaction was indeed dictated by the concern that Budoia could no longer have access to funds. The walk was considered a way to demonstrate that Budoia had the right to apply for those funds. The decision to belong to the Alpine community is, ultimately, a more or less conscious choice, based on the interest that this affiliation signifies.

This last aspect will be more deeply developed in the following examples. I will first take the example of two municipalities: Budoia, as well as Mäder, which is situated in the eastern part of Austria, in the *Land* of Vorarlberg. These examples will bring the reader closer to the core of the research and highlight more specifically the role of the mountain as justification for the common work at the pan-Alpine scale. It will answer the question regarding the interest of adhering to the pan-Alpine project. Budoia and Mäder are taken as examples because they are emblematic in two aspects: Firstly, although they are both pre-Alpine municipalities, their identification with the Alpine region is based on different rationalizations; secondly, these two municipalities are among the most active in the Alliance in the Alps network.

5.2.2 *To choose to be part of a mountain project*

The mayors of Mäder and Budoia municipalities (see map, fig. 8) Rainer and Antonio, respectively, were pioneers in the creation and development of Alliance in the Alps.

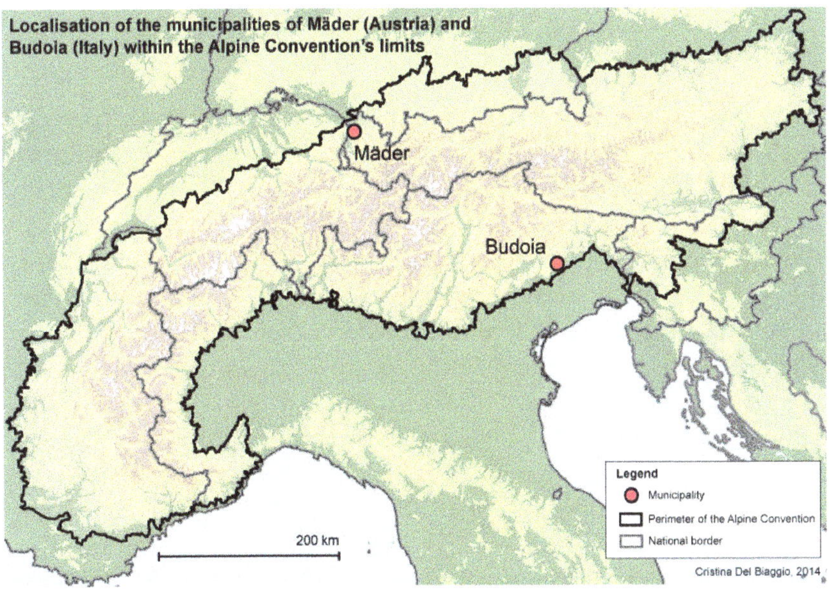

Fig. 8: The municipalities of Mäder (Austria) and Budoia (Italy). © Del Biaggio 2014.

At the time I interviewed them, Rainer was the president of the network, and Antonio was its vice president. Mäder and Budoia are among the most active municipalities of the Alliance in the Alps network, and their mayors are amongst its most charismatic personalities. Their role in the network will be discussed later, but one interesting point is worth mentioning here: Besides holding similar positions within the network, both Mäder and Budoia share a similar geographical situation in relation to the Alps. A glimpse at a topographic map (see fig. 8) makes it clear that these two municipalities do not have much in common with the typical, idyllic Alpine villages we are accustomed to seeing on postcards, calendars or tourism journals: They have no wildflower meadows, no wooden houses, no slopes and no mountain farmers looking after the cattle. I was probably not the only one surprised (and, I have to admit, a bit disappointed) by the landscape these municipalities offer, as their commitment to the Alps is well known. While walking around in either of the two municipalities, the following question spontaneously comes to mind: "Why and how do these municipalities feel they are part of the Alps?" The interviews conducted with the mayors pointed to an explanation.

Mäder: suffering the consequences of the Alps

Rainer decided to govern his municipality according to the precepts of sustainable development. In January 2007, he organized a guided tour of the municipality for

Giraud, the mayor of L'Argentière-la-Bessée (France) and freshly elected vice president of Alliance in the Alps. Since I was already in the region for another event planned by Alliance in the Alps, the network asked me whether I would interpret from German to French. The tour lasted an entire morning, quite a feat, considering that the municipality has just over 3000 inhabitants living on only 3.4 square kilometres and does not even have a historical centre where stories of a glorious past could be recounted. Yet, in every corner of the town, in every public building, Rainer had a story to tell, a contemporary story of big and tiny ecological and/or social projects for implementing sustainable development. With this visit, Rainer's commitment to sustainable development was obvious to see. Still, the reasons for the municipality's identification with the Alpine region were not evident in the discussions between Rainer and his French colleague.

The significance of the Alps for Mäder became clear during the interview that Debarbieux and I conducted with Rainer and Elke in June 2007, in their office. Rainer shared with us that his engagement with sustainable development was above all dictated by the municipality's struggles with the consequences of being situated at the foot of the mountain: "What happens up there, in the mountains up there, has important implications here", says Rainer.[89] Flooding and trans-Alpine traffic were two examples he mentioned of the concrete impact that the mountain has on Mäder, which is situated in the plain. Thus, for the mayor, a logical reaction was to partake in the decision-making process concerning the Alpine perimeter and to actively promote sustainable development as a possible solution to these problems. Indeed, according to Rainer, the implementation could help prevent problems such as trans-Alpine traffic. Joining an Alpine network, whose aim is precisely to promote sustainable development at the Alpine scale, was seen as a significant opportunity.

The interview with Sabine, who lives in Mäder, was in harmony with the mayor's answers:

Für mich liegt Mäder sehr wohl im alpinen Bereich, weil wir in diesem klimatischen Raum sind, in diesem strukturellen Raum "Alpen". Für uns sind z.B. diese starken klimatischen Veränderungen im Alpenraum sehr wohl gravierend, wenn ich da nur an den Rhein denke. Dafür müssen wir keine Berghänge haben, wir haben wahrscheinlich nicht die Muren, die Rutsche, aber die klimatischen Auswirkungen haben wir auch. (Sabine, Betreuerin)	For me, Mäder belongs without any doubt to the Alps, because we are part of this climatic area, this structural space called the "Alps". For us, these dramatic climate changes, for example, have important consequences, if I think about the Rhine River. Even if we do not have mountain slopes or avalanches or landslides here, we still suffer from the consequences. (Sabine, animator)

The reasons that Rainer and Sabine raised reinforce the slogan of the International Year of the Mountains (2002), "We are all mountaineers!", as interpreted by Debarbieux and Rudaz. Indeed, it suggests

89 Mäder, 11.06.2007.

that all persons in the region, including people living in the plain, are influenced by how mountains are managed (Debarbieux & Rudaz, 2010: 230). This is undoubtedly one of the main reasons for Rainer to take part in activities and projects concerning the Alps, and it is the reason he feels part of the mountain's community. Yet, in order to pursue his primary scope, i.e. the implementation of sustainable development, the Alps are for Rainer one amongst many other scales. Sustainable development is the "engine" that leads him to take action at different scales of involvement.

The map of the scales of projects and Rainer's involvement (see fig. 9) shows, on the one hand, the variety of the projects and, on the other, the diversity of the scales and links that the small municipality of Mäder has with other municipalities and regions around the world. Mäder is involved in the following projects:[90]

- The *Am Kumma* region, a group of four municipalities situated around the Kummen Mountain (667 m) that carries out common projects in the economic sector. The most emblematic project has been the creation of a local currency;
- The *Abwasserverband Region Hohenems*, which includes the municipalities of Hehenems, Altach, Götzis, Mäder and Koblach. Together, they manage drainage systems and sewage disposal;
- The *Umweltgemeindeverband Vorarlberg*, an association of municipalities in Vorarlberg concerned with environmental issues. They first created the association with the scope of regionally organizing the management of waste, like sorting out and recycling rubbish, and then integrated other aspects. For example, it reflected on the possibility of changing the way of organizing and carrying out public festivities. This consideration led to the concept of "*Ghörig feschta*",[91] which allows municipalities to organize parties in a sustainable way: organization of public trans-

90 Information received on the 01.11.2010 via e-mail by Elke Klien, former secretary of the Mäder municipality.

91 See the explanation of the concept by Rainer Siegele online: <http://video.vol.at/vorarlberg/ghorig-feschta-mehr-spab-beim-feiern?autoPlay=true>, consulted on the 09.03.2015. *Ghörig* is a local expression that means "correctly" or "harmoniously".

port, sharing of material between the municipalities, use of local products, management of waste, promotion of prevention campaigns, etc;

- The creation of the *Landbus Unterland*, a network of some municipalities in the Rhine Valley that promotes and manages the public transport offer in the region;
- The *Gemeindeverband Vorarlberg*, an association that assembles all the municipalities in the *Land* of Vorarlberg. It represents the interests of the municipalities towards the *Land* and the Austrian federal government;
- The *Wasserverband Rheintal*, an association that links municipalities along the Rhine River and organizes common water supply;
- The *Vision Rheintal*, a project based on the identification of the potentialities of the Rhine Valley region and on the subsequent development of a concept for the future of the region;
- The *Klimabündnis Vorarlberg*, which is member of the European network *Klimabündnis*, the largest network of municipalities dealing with climate change in Europe. Inside this network, Mäder carried out a development project with the municipality of Chocó, in Colombia;
- The e5 municipalities network, which assists municipalities in developing renewable energy policies.

The "Map of the scales of projects and involvement of Rainer Siegele" (see fig. 9) shows that the link, which some decades ago was considered evident and univocal, between one single person and one single place/scale, does not have any basis in the reality of social interactions in the contemporary world. In order to achieve a goal, i.e. the implementation of sustainable development in his municipality, Rainer plays with different scales. The adherence to networks is seen as a possible innovative medium to promote his ideas and is an important way to "seek strategic advantage" (Häkli 2008: 475). The result is a "mosaic of unevenly superimposed and densely interlayered scalar geometries" (Brenner, 2001: 606).

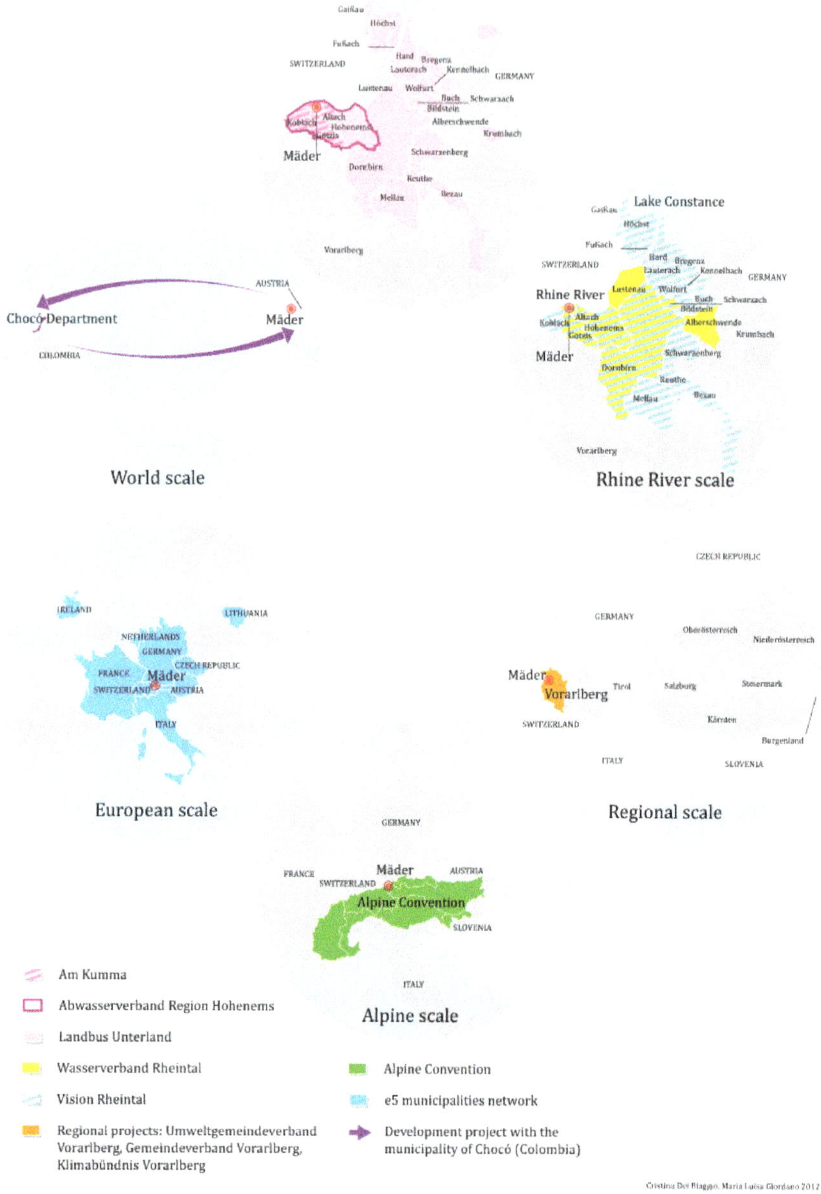

Fig. 9: Map of the scales of projects and involvement of Rainer Siegele, mayor of Mäder. © Del Biaggio & Giordano, 2012.

Budoia: opening the municipality to the mountain

Budoia's membership in Alliance in the Alps is the result of the will of one person, Antonio Zambon, who was mayor at the time. He later became vice president of the network and was elected to the steering committee from 1999 to 2009. As I mentioned before, I met Antonio in June 2007, and on that occasion he explained the motivations that led him to join the network. The account was completed by an exchange of e-mails between us in November 2010.

In the e-mail,[92] Antonio mentions two main factors that motivated him to join the network. Firstly, since Budoia was excluded from the European funds, Alliance of the Alps represented an alternative to get funds to implement projects of sustainable development in his municipality. Secondly, a strong link exists between Antonio and the Alps. His forefathers immigrated to Switzerland, to a town on Lake Maggiore, in search of work, as others from the same region had done in large numbers from the end of the 19th century onwards, and especially in the 1920s. When, later in the century, the Po Valley became an important industrial centre, municipalities such as Budoia, situated in the Italian Eastern pre-Alps, experienced a significant demographic decline. Antonio realized that the municipality of Budoia was becoming a commuter town without much local life. Therefore, for Antonio, the Alps represented something close to the history of the municipality, something "natural and aggregating", as he says in his e-mail, that links the people to their territory, since until recently the lives of Budoia's residents always took place in the mountain:

Nella Convenzione delle Alpi si potevano riconoscere filosofia ed obiettivi utili a far rivivere la storia del paese.	In the Alpine Convention, one could recognize a philosophy and objectives that are useful for making people relive the history of the town.
Anch'io figlio di emigranti in Svizzera, sentivo (per cultura appresa sul Lago maggiore) l'attrazione verso la montagna e quando ne ho parlato alla gente, mi sono sentito ascoltato. (Antonio, sindaco)	I am also the son of migrants in Switzerland. I feel (because of a culture learned on the Lake Maggiore) the attraction towards the mountain. When I spoke to the people about that, I sensed that they understood. (Antonio, mayor)

92 Information received via e-mail from Antonio on the 23.11.2010.

Antonio's reflection shows that he wanted to intensify the inhabitants' connection to the mountain, which was once strong but had weakened in the past few decades, and which he considers important for the inhabitants of the municipality. As much as his words, his strong commitment to the Alps made it possible for the rest of the municipality to share his personal story. In so doing, a personal identity was transformed into a collective one. This argument motivated the search for and creation of stronger links with the mountain range at the collective level. Thanks to his encounter with CIPRA, Antonio's project of promoting and waking up the mountaineer feeling in the population became real:[93]

CIPRA e Alleanza nelle Alpi, mi hanno dato la possibilità di promuovere e di risvegliare sentimenti che comunque la gente covava dentro di se. (Antonio, sindaco)	CIPRA and Alliance in the Alps gave me the possibility to promote and wake up feelings that the people were harbouring in their hearts. (Antonio, mayor)

During the interview, Antonio recalls the fruitful encounter with CIPRA as a good opportunity for his municipality to join a network interested in mountain issues. It all started in 1995, with the election of a new municipal administration in Budoia. Emulating Antonio, the elected persons wanted to forge a new path for the municipality based on its history. They felt that Budoia was no longer the "open town" it had once been, a town with strong links to other countries, built thanks to many of its immigrant residents.

As Antonio told me,

> [when I became mayor, my colleagues and I] thought that the town's identity, at that moment, was not that of an open municipality; this was a closed municipality. We were wondering what to do in order to place Budoia in a larger context, to somehow open the mentality of the inhabitants.

At the precise moment that the deputies were pondering how to execute their ideas, CIPRA sent them (as to all other municipalities of the Alpine arc) a letter asking if any municipalities were interested in participating in a network. Antonio's positive response led Moroder,

93 E-mail of the 23.11.2010.

the then-president of CIPRA Italy, to visit Budoia. On that occasion, Antonio explained the motivations and objectives of Budoia's wish to belong to the network and decided to accept CIPRA's proposition.

With that "happy coincidence" (the reflection on new projects in the municipality and the arrival of the letter), Antonio realized that Alliance in the Alps corresponded to a concrete expression of the vision that he was trying to establish in his municipality. The following part of the interview shows how this happened. It is in answer to my question whether Alliance in the Alps affected the creation, in Budoia, of an Alpine identity:

Questa è una domanda difficile! Io dico che forse, innanzitutto, c'è stato un riconoscimento di qualcosa che si era perso, cioè, la montagna non era più una cosa del paese, perché finché era abitata erano storie vecchie, che era bello sentire da qualche anziano, mentre aderendo all'associazione si è di nuovo riscoperto che questa è un'area pre-alpina, quindi dentro nelle Alpi. La montagna è cominciata a diventare, per lo meno emotivamente, è stata riconosciuta come una cosa appartenente al paese, non una cosa che non c'è più. Quindi si è guardato un po' verso la montagna piuttosto che verso la pianura. Quindi questa reazione c'è stata!

[...]

La montagna è tornata ad essere un elemento di discussione, e sono sorti i progetti europei, che abbiamo fatto. Noi siamo nel progetto *River Basin Agenda*, che è stato un progetto molto piccolo per noi, però abbiamo portato i giovani a seguire un percorso dell'acqua, abbiamo scelto una montagna, che è molto bella, il Col Cornier, che si trova a 1700–1800 metri, e abbiamo seguito il percorso dell'acqua, cioè la goccia che cade in cima al Col Cornier passa attraverso le nostre zone carsiche, negli alpeggi, dove l'acqua viene recuperata, poi arriva giù, abbiamo un

This is a difficult question! First of all, I can say, that there was the recognition of something lost. The mountain was no longer part of the town. The mountain, as long as people were not living there anymore, was the subject of old tales told by some elderly persons; by joining the association, we rediscovered that this is a pre-Alpine area and belongs to the Alps. Mountains became, at least emotionally, something that belonged to the town, not something that does not exist anymore. Consequently, we looked a bit more at the mountain instead of at the plain. This reaction really happened.

[...]

Once again, the mountain became a subject of discussion, and some European projects were carried out. We participate in the River Basin Agenda project, which is a really small project for us, but we brought young people to follow the course of the water... We chose a really nice mountain, the Col Cornier, which is situated at 1700–1800 m, and we followed its course of water, i.e. from the drop that falls on top of the mountain that goes through our karsts areas, where water is recovered, then runs into the Artugna Torrent, [...] then into the Liv-

torrente, si chiama Artugna, [...] e poi va nella Livenza, che qui poi va fino al mare. Quindi con i ragazzi abbiamo fatto una "settimana di avventura", conoscendo il territorio dalla cima della montagna fino ad arrivare in canoa fino al mare. [...] Dentro questa cosa gli anziani hanno raccontato le storie di quando vivevano nelle Alpi, nella montagna. (Antonio, sindaco)

enza River, and so to the sea. With the young people, we organized an "adventure week", so that they could know the territory from the top of the mountain and reach the sea by canoe. [...] And old people told them stories of the time when they lived in the Alps, in the mountain. (Antonio, mayor)

So, as Antonio explains, the consequences of this choice, on the one hand, helped to create a stronger bond between the inhabitants and the mountain. On the other hand, it brought European funds to the municipalities. Budoia was able to benefit, for example, from a Dynalp fund for the project "Development of a sustainable model for rural tourism" (January 2005 – June 2006). These projects, in turn, helped awaken the relationships with the mountain that had disappeared over time. The new way of conceiving the town contributed not only to initiating new identity processes but also to developing a new socio-economic dynamic in the town:

Oggi nelle scuole elementari i bimbi sono sopra i 20 ogni anno, è stata ampliata la scuola materna oggi ha tre sezioni e circa 80 bambini, anche alla scuola elementare si sono dovute costruire nuove aule. Oggi i residenti a Budoia sono 2500. Molte famiglie hanno scelto di venire qui perché è migliorata la vivibilità. (Antonio, sindaco)

Today, in the primary schools, we have more than 20 children every year. We had to expand the kindergarten to three sections, and we had to build new classrooms for the primary school. Today, Budoia has some 2 500 residents. Some families decided to live here, because the liveability has improved. (Antonio, mayor)

This example shows the strong link that exists between identification and action, where neither is the consequence of the other, but both grow side by side. Antonio's story, contrary to Rainer's, proves a stronger attachment to the mountain in general and to the Alps in particular. The networks are there to support and develop this attachment. However, both are strategic identities in the sense that both mayors put in place a strategy to achieve a precise aim and ideal: the creation of a sustainable Alpine space as defined by the Alpine Convention. The two accounts, however, illustrate that there are different strategies to reach this aim. I would qualify Antonio's strategy as more affective, while Rainer's is more functional and pragmatic.

5.2.3 To choose to identify with the Alps

As Mäder's and Budoia's commitments within the Alpine perimeter attest, identification with the Alps is a personal choice that the two mayors proposed to the residents of their towns and that later became a collective strategy. Both municipalities are situated in a pre-Alpine area and, through their involvement, they chose to be part of the Alps, to "look up", instead of "looking down at the plain", and to contribute to the pan-Alpine project.

The town of Mendrisio (Switzerland) is an example of another point of view. In order to demonstrate my argument, I will first show three photographs of the three municipalities: Mäder, Budoia and Mendrisio (see fig. 10, fig. 11, fig. 12).

Fig. 10: Municipality of Mäder (Austria) © Gemeinde Mäder.

Fig. 11: Municipality of Budoia (Italy) © Angelo Balla.

Fig. 12: Municipality of Mendrisio (Switzerland) © Alberto Campi.

These pictures illustrate the three very similar geographical situations of the three municipalities. They all lie at the foot of a mountain in a pre-Alpine setting, as can be seen on the map (see fig. 13).

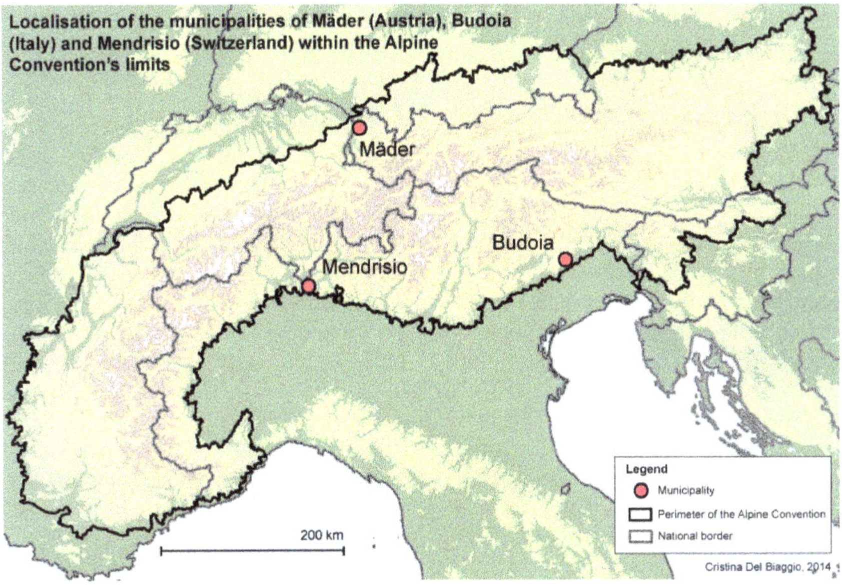

Fig. 13: Map of the municipalities of Mäder (Austria), Budoia (Italy) and Mendrisio (Switzerland).

Despite the similar geographical situation with respect to the Alps (all three are pre-Alpine towns and inside the Alpine Convention's perimeter), the municipality of Mendrisio did not join Alliance in the Alps, even though employees of the municipality have regular contacts with Fabio, who is trying to promote Alliance in the Alps in the Ticino Canton. I interviewed Edy (an employee of the municipality) and Silvia (employee of the *Ufficio aggregazioni urbane*, the office in charge of the municipal incorporations)[94] on November 4, 2009. They both knew about the existence of the Alliance in the Alps network. Mendrisio has a 10-year-old history of commitment to sustainable development, and the projects it implemented at the local level to promote sustainable development could have been used to join the

94 In Ticino, the reform of the municipalities, which includes the incorporation of small entities into bigger municipalities, is one of the major institutional projects carried out by the canton.

network. In this excerpt, Edy explains the town's projects and the reasons why it did not join Alliance in the Alps:

Edy (segretario municipale): Mendrisio sul discorso dello sviluppo sostenibile è attivo da una decina di anni. Questa visione di cercare qualche cosa di diverso per quanto concerne la gestione del territorio è sicuramente partita con il progetto pilota VEL, andando a toccare con mano uno dei nostri problemi ancora oggi non risolto che è quello della mobilità, del traffico. [...] Man mano che il VEL andava in avanti, il comune di Mendrisio cercava sempre di approfondire questa sua visione dello sviluppo sostenibile. Il passo seguente è stato "Città dell'energia". E poi una serie di cittadini si sono attivati nel far proprio i movimenti di Agenda 21. E poi, un po' per caso, siamo venuti a conoscenza del discorso *Città slow*, questo negli anni 2006. Ripeto, veramente per caso. Per noi era importante di aderire a questa rete internazionale, anche con lo scopo di scambiare le varie esperienze nell'ambito della ricerca dello sviluppo sostenibile.	Edy (employee of the municipality): Mendrisio has been active for about 10 years in the area of sustainable development. This idea of searching for something different for the management of the territory started with the VEL project.[95] With that project, we gained experience first-hand with problems that have not yet been solved, such as mobility and traffic. [...] While VEL was going ahead, the municipality of Mendrisio tried to advance its sustainable development vision. The following step was to become a City of Energy.[96] Next, some citizens adopted the Agenda 21. Then, by chance, we learned about the *Città slow* network.[97] I repeat, really by chance. For us, it was important to join this international network, because it allowed us to exchange experiences in the area of sustainable development.
Cristina: Per capire il discorso alpino...	Cristina: In order to understand the Alpine discourse...
Edy: Forse sbagliando, sicuramente sbagliando. Molto probabilmente non abbiamo, pur consapevoli della forza dell'arco alpino, forse non ci siamo immedesimati in modo corretto in questo tipo di realtà, perché siamo forse più prealpini che alpini; e forse ci siamo detti "Non abbiamo forse le stesse cose in comune come potrebbero avere i comuni o	Edy: Maybe it was a mistake, certainly a mistake. Even though we are conscious of the strength of the Alpine arc, perhaps we did not correctly identify ourselves with this kind of reality, because maybe we are more pre-Alpine than Alpine; and maybe we thought, "We do not have the same things in common with other municipalities or other re-

95 For more information about the project: <www.infovel.ch/en/>, consulted on the 16.03.2015.

96 <www.citedelenergie.ch>, consulted on the 16.03.2015.

97 <www.cittaslow.net>, consulted on the 16.03.2015.

le regioni del Sopraceneri, che vivono in modo diverso la vicinanza e il territorio dell'arco alpino". Questo non vuol dire che non guardiamo questo tipo di discorsi, li abbiamo guardati, le mail ci arrivavano, le abbiamo girate comunque al Dicastero ambiente e territorio, il quale molto probabilmente non è stato colpito nella maniera corretta, altrimenti non riesco a trovare altro tipo di giustificazione del perché non abbiamo aderito a questa iniziativa.

gions of the *Sopraceneri*[98], because they experience the proximity with the Alpine territory in a different way." This does not mean that we do not contemplate this kind of discourse; we looked at them. When we received e-mails, we sent them to the Office of the Environment and Territory, but they were probably not properly understood; otherwise, I cannot find another justification for why we did not join this initiative.

This excerpt is interesting for four reasons. Firstly, there is a similarity between Edy's discourse and that of Rainer. Both, obviously, believe strongly in sustainable development. Yet Mendrisio found a way to implement sustainable development without a discourse linked to the mountain. Secondly, as for Budoia, there is some happenstance in the decision to join this or that network: "We learned about *Città Slow* by chance", says Edy. Antonio received an unexpected letter from CIPRA, while somebody, the son of a colleague of Edy, brought up the idea of *Città slow*. The two coincidences ended in different paths, with only one connected to the Alps. Thirdly, the Mendrisio municipality employees did not seem to identify strongly with the Alpine region with respect to the strategic projects proposed for the city. Edy clearly mentions that, for him, pre-Alps do not mean Alps, and that the problems in these two territories are different. Finally, Edy was unable to transform his personal identification with a collective movement into commitment for the Alps. These last two points can be identified in this excerpt as well:

Le Alpi sono sicuramente un patrimonio importante del nostro territorio, sia per quello che rappresenta ancora oggi dal punto di vista dell'opportunità di lavoro e di ricchezza dal punto di vista economico. Possono essere fonti di ricchezza

The Alps are certainly an important part of the heritage of our territory, both in what they represent in terms of work opportunities and wealth from the economic point of view. They can be a source of wealth concerning the energy

98 The *Sopraceneri* corresponds to the northern part of the Ticino Canton, the most mountaineer one. It is situated to the north of the Monte Ceneri Pass (554 m.).

per quanto concerne il discorso energetico, per il turismo, per lo sviluppo del loro patrimonio che hanno, non so per esempio per quanto concerne il discorso produzione del legno. Ma sono anche patrimonio per quel che concerne quelle aree che oggi sono delle piccole rarità naturalistiche della nostra Svizzera. Io adesso sto parlando un po' egoisticamente. Il fatto che partono dalla Francia e che arrivano fino in Slovenia, è vero sì, geograficamente... Quindi le vedo anche come patrimonio e ricchezza perché hanno ancora oggi delle particolarità forti che devono essere salvaguardate per garantire la diversità della nostra fauna, a garantire magari ai nostri figli di andare a vedere un giorno che le Alpi non sono solo sci lift. Lo dico perché forse vivo in una regione pre-alpina, ma di origine sono proprio un alpino al 100 % visto che la mia famiglia viene dalla Val Calanca e dalla Val Mesolcina. Quindi proprio attaccati sotto le montagne. (Edy, impiegato comunale)

question, for tourism, for the development of their heritage, for example the wood production industry. But they are also a heritage of what are today the small naturalistic rarities of Switzerland. I am now speaking a bit selfishly. The fact that they begin in France and end in Slovenia is true, yes, geographically... I see them as heritage and richness, because at present they still have some strong peculiarities that should be preserved, in order to guarantee the diversity of our fauna, to guarantee perhaps that our children, when in the future they look at the Alps, will not only see ski resorts. I say that, because I live in a pre-Alpine region, but I am originally 100 % Alpine, since my family comes from the Calanca Valley and from the Mesolcina Valley.[99] Therefore, from villages really close to the mountains. (Edy, employee of a municipality)

Three arguments can be highlighted from this quotation, to partly explain Mendrisio's non-adherence to the mountains projects.

Firstly, Edy begins speaking using the collective "we" when mentioning the Alps: "The Alps are an important heritage of *our* territory", he says. Yet, shortly afterwards, he says: "[The Alps] can be sources of wealth concerning the energy question, for tourism, for the development of *their* heritage". There is a transformation of the discourse from a "we" to a "they" form, which is probably a sign of an incomplete identification with the Alps.

The second point is the hesitation he displays while thinking about the Alpine arc. Edy affirms that "[The fact that the Alps] begin in France and end in Slovenia is true, yes, geographically..." These

99 Both valleys are situated in the Italian-speaking part of the Canton Grisons (Switzerland).

last two words ("yes, geographically...") suggest doubt, as if to say that geographically the Alpine arc is an entity, but that other elements do not link the different parts of the Alps. As such, the Alps for Edy probably do not exist the same as they do for Antonio and Rainer.

In the third and probably most important argument, Edy separates his private life, his mountaineer roots, from his collective mission in the municipality. Unlike Antonio, Edy did not transform his individual identity into a collective one capable of building a common project from his personal history. This shows that identities, both individual and collective, are built with the help of resources that are chosen freely (Petite, 2009: 28). It emphasizes that nowadays identities are built upon an adherence mode (Dubar, 2000: 5); they are not given in advance and depend on personal and collective choices.

5.2.4 Feeling part of a "community of problems"

The first interview for this research project was conducted on October 24, 2006 by Marie-Anne with Robert, an animator of Alliance of the Alps. This interview already provided indications as to why mayors decide to take part in the network: "[The mayors] are happy, among elected persons, to be able to share their worries" (Robert).

This capacity of the network to create a community of people sharing the same problems was detected in several other interviews that my colleagues and I conducted. In terms of numbers, members of Alpine Pearls and Alliance in the Alps mentioned this theme more often than did the members of Alparc. All seem to strongly identify with the Alps, because they all share the same mountain environment and the problems this entails for mountain communities. In fact, members of the pan-Alpine networks can be considered, as I like to define it, as a "community of problems". Joining a network, therefore, facilitates concrete solutions to common problems (mobility, climate change, etc.), an observation that has also been made by Price, who asserts that amongst people living in the Alpine arc it is clear that "many issues cannot be solved only through national legislation; coordinated regional approaches and initiatives are essential to solve common problems" (Price, 1999: 88).

The "sense of community" that I could feel by interviewing local political actors can be defined by using four qualities, as proposed by McMillian and Chavis (McMillian & Chavis, 1986: 9–11). For the scholars, the sense of community entails four elements that will be explained here with help of interviewees' responses.

Firstly, the membership, which is the feeling of belonging to a group, demonstrated by one having invested part of himself or herself to become a member of a community, affords the right to share a sense of personal relatedness. As explained in chapter 4, in order to become member of Alpine Pearls and of Alliance in the Alps, municipalities have to send a request and pay a fee, which attests to belonging to the network. One of the most concrete results of the membership is that municipalities are thus informed about the activities of the networks. Henrik explains that receiving journals from the network increased his feeling of being part of a community of problems:

Das interessiert mich schon, wenn Zeitschriften [von Alpenallianz] kommen, in denen man auch andere Länder sieht und dass diese zum Teil auch solche Probleme haben. Das ist gut für uns. (Henrik, Bürgermeister)	I am interested when I receive journals [from Alliance in the Alps]. Thanks to them, we can see that in other countries they also have some of the same problems. This is good for us to know. (Henrik, mayor)

And, in those documents that the networks or CIPRA sends out, this notion of "common problems" is constantly mentioned, as a recent publication from CIPRA demonstrates: "The problems are the same in Manosque, in France, or in Mürzzuschlag, in Austria" (Götz & Rohrer, 2011: 5). The history of CIPRA dates to the 1950s with the recognition that the Alps are one entity, because the people living in them face the same problems. This recognition was the basis for the establishment of CIPRA, as already mentioned in a previous chapter.

The second element that characterizes a sense of community, as underscored by McMillan and Chavis, is influence. This is a dual process whereby, firstly, a member of the group hopes that he or she can have influence over what the group does; secondly, cohesion inside the group is built on the group's ability to influence its members (McMillian & Chavis, 1986: 9). This is also, as already discussed in previous chapters, the aim of the networks, whose objective is the implementation of the principles of sustainable development at the local level. The networks, thus, influence the behaviour of the Alpine municipalities on their territory. This should be the objective of every municipality joining the network, as is written in the charter:

> We commit to implementing the founding principles of the Alpine Convention. [...] We ensure and stimulate participation. Citizens, associations and local institutions actively participate in the decision-making, production and implementation process.[100]

This document, signed by the members, acknowledges that a single member can influence the entire group and the entire territory.

The excerpt of an interview with a mayor member of Alliance in the Alps reported by Petite in his Ph.D. thesis supports this idea of a win-win situation (Petite, 2009: 313). The network acts as a discussion forum in which solutions can be found; at the same time, the municipalities that have decided to join the network consciously do something for the entire Alpine community: They work to impede the deterioration of the Alpine environment:

> L'idée de l'Alliance des Alpes était une très bonne idée. C'était un peu un forum. Et de prendre comme unité les Alpes, ça veut dire, ça invitait tous les gens à raisonner au-delà des frontières politiques, au-delà des mentalités. C'était plus une question de discuter avec la commune voisine. [On s'est ainsi rendu] compte que tout le monde avait les mêmes problèmes et que si à travers les Alpes tout le monde faisait la même chose, c'est-à-dire un développement intensif et ne se préoccuper pas d'équilibrer ça, tout le monde était perdant indirectement. (maire)

> The idea of Alliance of the Alps was really good. It was a bit of a forum. And taking the Alps as a unit means inviting people to think beyond political borders, beyond mentalities. It was more a question of being able to discuss with the neighbouring municipality. We realized that everybody had the same problems, and that if throughout the Alps everybody was equally promoting an intensive development without caring about balance, everybody was indirectly loosing. (mayor)

This sentiment introduces the idea that it is not only the fact of living in the Alps that makes municipalities of the pan-Alpine networks identify with a community, but the community is constituted only by those

100 <www.alpenallianz.org/fr/sur-alliance-dans-les-alpes/charte>, consulted on the 13.07.2015.

municipalities that do not "promote an intensive development" and instead support sustainable development. It is the identification with a territory, the Alps, but also with a concept, an ideal, that is important. The community exists only if these two requirements are fulfilled.

Thirdly, the integration and fulfilment of needs (also called reinforcement by the McMillian and Chavis) is the feeling that members' needs will be met by the resources received through their membership in the group. Henrik's final sentence, "this is good to know for us", probably best reflects this feeling. Later in the interview, he twice uses the expression, "it already brings a lot", in order to convey this feeling:

Ich bin erst vor zwei Jahren hier zum Bürgermeister gewählt worden. Ich muss sagen, für mich ist das schon eine große Hilfe, dass ich weiß, dass viele Gemeinden auch mit unseren Problemen zu kämpfen haben. Das bringt eigentlich schon viel. Nicht dass ich denke: "Ok, ich bin alleine, ich habe alleine mit dem Schutzwald da oben ein Problem". Und das bringt, das muss ich sagen, eigentlich viel. (Henrik, Bürgermeister)	I was elected mayor two years ago. I have to say, for me it is indeed a big help to know that many other municipalities have to deal with the same issues. It already brings a lot. Not that I think: "OK, I am alone; I am alone with the problem of protecting the forest up there." And, I have to admit, this actually brings a lot. (Henrik, mayor)

This integration can be also illustrated by the efforts on the part of the head of the networks to provide information and input to members. For example, through the establishment of a good practices database, the information service and the various seminars offered over the year and all over the Alps are organized by the networks with the aim of creating synergies between members. This idea is well expressed by Olivier:

Vous [en tant que maire] ne partez pas de zéro, vous n'allez pas réinventer la roue comme on dit. Et ça ira beaucoup plus vite, ça sera beaucoup plus facile, vous aurez à votre disposition des personnes ressources, vous pourrez appeler les maires. (Olivier, animateur)	You [as mayor] are not starting from zero; you do no re-invent the wheel, as we say. And everything will happen much faster, and it will be much easier because you have resource persons at your disposal, and you can call other mayors. (Olivier, animator)

Fourth among these elements is shared emotional connection, which is, for McMillian and Chavis, the commitment and belief that members have shared and will share common history, places, time together and similar experiences. I would add, in the case of the pan-Alpine networks, the strong sentiments resulting from the awareness of shared problems, which is the main point here and in the interview with Mayor Mirko:

> One of the interesting things [that we discovered we share] was traffic, especially the dormant traffic in towns. We all were simply finding out that we have difficulties, both in small and in large towns. (Mirko, mayor)

Even those mayors, such as Maximilian, who realize that the members of the networks are heterogeneous, find some similarities. For instance, one commonality is the capacity to deal with the marketing of local products or the preoccupation with the preservation of cultural characteristics. This last point is particularly important for Maximilian, who is also an active member of the *Vorarlberger Walservereinigung*, an association whose aim is, among other things, to support cultural activities linked to the preservation of the Walser heritage:[101]

Es ist schon ganz unterschiedlich. Jetzt sind meiner Meinung nach auch Gemeinden dabei, die größer sind, fast schon Städte. Die haben sicher nicht dieselben Probleme wie wir hier oder dieselben Vorgaben, die haben vielleicht andere Probleme. Aber irgendwie hat man schon das Gefühl, wenn man näher hinsieht, [dass man dieselben Probleme hat] mit den ganzen Sachen, da bemüht man sich auch mit der Kultur und der Tradition, die irgendwie aufrecht zu erhalten, und vor allem versucht man jetzt die Produkte, die man vor Ort hat, besser zu vermarkten und eine bessere Wertschöpfung zu erzielen. (Maximilian, Bürgermeister)

It varies a great deal. In my opinion, there are some municipalities inside Alliance in the Alps that are much bigger; they are almost cities. They certainly do not have the same problems we have here, or the same characteristics; they probably have other things. But, somehow, we have the feeling, if one looks deeper into it, [that we share the same problems] in various fields. We deal with culture and tradition, we try to preserve it, and, above all, nowadays, we try to sell local products better and to create added value with them. (Maximilian, mayor)

101 <www.vorarlberger-walservereinigung.at>, consulted on the 16.03.2015.

The words of Hans, of Alpine Pearls, reveal not only the recognition of the existence of similar problems in the Alpine arc, but they also bring out the strong bond that this awareness creates amongst people who are living with those problems on a daily basis. Using another example, Hans admits that he "feels close" to Interlaken (also a member of Alpine Pearls), while he experiences a feeling of distance when he thinks about Nice, which is just outside the perimeter of the Alpine Convention and, in his eyes, faces completely different problems:

Ich glaube, dass sich die Lebensweise und die Art der touristischen Ausrichtung sehr gut auf den Alpenbogen, auf die gesamten Alpen ausdehnen lässt. Wenn ich in Val d'Isère bin, ist der Unterschied nicht so groß, was die touristische Ausrichtung angeht. Die Probleme, die es mit den Straßen, dem Abwasser, mit der Wasserversorgung gibt, die sind sehr, sehr ähnlich, bzgl. der saisonalen Nutzungen. [...] Die Problemstellungen, die Aufgabenstellungen sind sehr ähnlich. Wenn ich Interlaken nehme, mit Eiger und Jungfrau, diese Destination ist sehr stark vom Individualtourismus geprägt, die Leute kommen, nur um einmal den *Top of Europe* zu sehen, zu besteigen. Das ist ähnlich wie bei uns am Königssee. Der Königssee ist auch ein Tagesausflugsziel, das von Gästen besucht wird, die einzig den See, nicht die umliegende Region besuchen wollen. Damit kann man sich sehr gut ergänzen. Tourismusorte wie zum Beispiel Nizza haben völlig andere Problemstellungen, die haben eigene Probleme [...]. Komplett andere Aufgabenstellungen, andere Problemfelder. Das ist sehr, sehr speziell in den Alpen. (Hans, Bürgermeister)

I think that the same way of living and the touristic arrangements can be extended to the Alpine arc, to the entire Alpine arc. When I go to Val d'Isère, the difference is not that big, concerning the touristic offer. And the problem that one can find there, concerning the streets, sewage water, water management, are very, very similar, with the seasonal use. [...] The type of problems and the way we resolve them are very similar. If I take Interlaken, with the Eiger and the Jungfrau, with very strong individual tourism, where tourists go just to say that they were once on the Top of Europe... This is similar to what happens here at the Königssee. The Königssee is also a one-day excursion for visitors. For this reason, we are complementary. Tourist places like Nice have entirely different types of problems; they have their own problems [...]. A completely different way of resolving the problems, and other problem areas. It is very, very particular in the Alps. (Hans, mayor)

Niklas, member of Alliance in the Alps, admits that the awareness of sharing the same problems lets him feel closer (*näher* in German) to the municipalities situated in the Alpine arc, than is the case for Hans:

Ich glaube, rein vom Gefühl her sind uns, sind mir persönlich, schon die Alpengemeinden näher, ob die in Österreich sind oder nicht, weil die Probleme [gleich] sind. Ob es im Tessin [Schweiz] ist oder in Frankreich in den Alpen, oder in Österreich in den Alpen, die werden wahrscheinlich überall gleich sein. Die Grundthemen: Abwanderung, Hangsicherung und im Winter Lawinen, im Sommer Hochwasser, also Hochwasser nicht aber Erosion und so weiter. Ich glaube, dass die Themen mit unseren verwandt sind, und darum glaube ich, dass der Bezug zu diesen Gemeinden eher vorhanden ist als zu solchen im Flachland mit 20'000 Einwohnern. Also ich glaube, dort habe ich überhaupt keinen Bezug, überhaupt keinen! (Niklas, Bürgermeister)

I think that I feel emotionally closer to the municipalities in the Alps, whether in Austria or not, because the problems are [the same], whether in Ticino [Switzerland] or in the French Alps, or in the Austrian Alps. They are probably the same everywhere. The basic themes: emigration, slope reinforcements, avalanches during winter, floods in the summer, well, not floods, but erosion and so on. I think that our problems are related and that is why, I think, these municipalities are closer to us than those in the plain, with 20000 inhabitants. So, I think that there I do not have any bond, none at all! (Niklas, mayor)

This shared feeling of community building is taken into account by the networks of local political actors. They seem to be a good tool in order to, on the one hand, let the common problems arise and be shared (through the possibilities of dialogue offered in the meetings, for example). On the other hand, they have a facilitating role in providing information about the possibilities of collective resolution:

> [Alparc] connects protected areas all over the Alps, and all over the Alps they can cooperate about certain problems, about certain points that are common to all of them. (Ivica, employee of the Alpine Convention)

According to the mayors and managers of protected areas, these common problems must be shared and faced together on a larger scale in order to explore possible local solutions. This is made possible, as Claudio says, notably by individual entities joining an international network:

Il vantaggio di un'alleanza è che ti permette di riscoprire e poi di presentare dei territori che erano partiti con gli stessi problemi e che li hanno risolti, questo è sicuramente interessante. (Claudio, segretario regionale)

The advantage of an alliance is that it enables you to rediscover and present territories that started with the same problems and have solved them. (Claudio, regional secretary)

To conclude, feeling they are part of a "community of problems" makes local leaders act at a transnational level and thereby contributes to what Häkli calls a "politics of bridge", where "one seeks to appropriate the unifying rather than the separating aspect of the mountains" (Häkli, 2004: 65). As Crivelli writes, the specificity of a region is no longer affirmed by its difference with the other but through a sort of collective contract that creates a system of relations between different subjects with the aim of handling the relations with the "outside world" (Crivelli in Gillet *et al.*, 1996: 18).

5.2.5 *Differences between rangers and managers of parks in the identification process*

Following the analysis of how leaders and members of the networks identify with the Alps, this section introduces a new type of actor: rangers of protected areas. As mentioned earlier (see chapter 3), the questionnaire sent to the managers of protected areas showed that they were not actively involving the rangers in pan-Alpine activities. As a result, I judged it important to hear their voices so as to discern whether a lack of interest (and thus a lack of identification with the Alpine arc) on the part of the rangers dictated this attitude, or whether there were other reasons. The opportunity to meet a number of rangers came up in 2007, when Petra Arnuš and I joined the rangers' meeting for the Danilo Re Trophy.

The question of the rangers' uniform

Danilo Re was born in Boves (Piedmont, Italy) in 1956. In 1981, he achieved his dream of becoming a ranger and started working for the Piedmont Alta Valle Pesio Regional Park (extended 10 years later to include the Val Tanaro).[102] He died November 23, 1995 while serving his park, and since then a sports tournament is held every year in his honor. Every protected Alpine area is invited to participate in the event.

From the March 15 to 18, 2007, the 12th tournament was held by the Mercantour National Park, in France (see official logo, fig. 14). Petra and I followed the competitions, enjoyed the "Alpine buffet" with the guests, distributed a questionnaire and interviewed some rangers.

Fig. 14: Official logo of the 12th Danilo Re Trophy.

One point that is worth underlining here and will be analysed along with the interviews that Bernard Debarbieux and Petra Arnuš conducted later with some French managers of protected areas. In my field notes book, I recorded a discussion that I had with an Italian and a French ranger: "The Italian ranger said that there is a will to create a common uniform for all the Italian rangers. He added that he personally would prefer a common uniform for all the Alpine rangers, but for that to happen would probably take at least 100 years."

After this first discussion with the Italian ranger, we decided to ask further questions on this point, since we believed it was an important assertion concerning the identification of rangers with the Al-

102 <www.alparc.org/danilore/index.php/en/>, consulted on the 16.03.2015.

pine region. In fact, the uniform has a symbolic value for those wearing it. For that reason, we wanted to know whether other rangers shared this idea or if it was just the personal feeling of that particular ranger. During the Trophy, we had informal and more formal discussions on this issue with some rangers, especially with the Italian rangers, who were the ones showing the highest level of frustration concerning their professional activities. The informal discussions I had with rangers employed at the Swiss National Park, for example, did not give me such an impression of dissatisfaction concerning the issue of the uniform in particular or their professional activities in general.

Bernard Debarbieux and Petra Arnuš also discussed these themes with French managers of protected areas, whom they interviewed some weeks later. It would probably have been worth delving deeply into the question to render a much broader picture by rangers and managers from all over the Alps, but unfortunately time and resources did not allow for this to happen. Still, the impressions we collected are already interesting and deserve to be presented here.

The discussions I had with the rangers show that there are two factors with which the Italian rangers do not identify: The work they are doing does not correspond to the profession they have chosen; and the uniform they are wearing is considered to be too "narrow", as I will explain later.

The first point can be evidenced through Dario's words, who was present at the Danilo Re Trophy, representing the Parco delle Orobie Valtellinesi.[103] In the following extract, Dario explains his professional status as a policeman. In fact, even for someone like me who has no expertise in protected areas, the uniform of the Parco delle Orobie Valtellinese employees did not go unnoticed amongst the other participants in the Trophy (see fig. 15).

103 The information collected in Allos has been completed by an e-mail exchange between Dario and me in April 2011.

Fig. 15: Dario (on the left) at work with his uniform
© Massimiliano Pasini.

While discussing the different tasks of the rangers with participants, Orobie was always mentioned as an emblematic example of the variety of rangers' statuses existing in the different Alpine protected areas. In fact, as corroborated by Dario in an e-mail of April 4, 2011:

L'attività nel Parco delle Orobie la svolgiamo sempre perché il parco ha una sola guardia, ma sulla divisa non abbiamo nulla che ci identifichi come guardiaparco. [...] Il termine di guardacaccia è andato un po' in disuso (mi vien da dire, per certi versi, purtroppo). (Dario, guardiaparco)	We still work for the Parco delle Orobie, because the park has only one ranger, but concerning the uniform, we do not have anything that identifies us as rangers. [...] The term gamekeeper is no longer in use (in some ways, I would say this is unfortunate). (Dario, ranger)

Concerning the historical evolution of his tasks, Dario explains that until 1992 he and his colleagues were overseeing hunting, fishing and the protection of the environment. However, with their former status, they could not arrest people. With the new law, the wardens joined the police force and have been armed. In the interview with Dario, he recounts the chronological path of his career. He first passed the competitive examination to become gamekeeper. When the Lombardy Region decided to transform the local authority structure (the province, the municipality), they created a local police force, different from the *carabinieri* and the police. The local police would take care of the city centre, the roads, etc. At the same time, they transformed the provincial force, composed of former gamekeepers. If once the main task Dario and his colleagues had was to oversee hunting, they now have the same functions as policemen and *carabinieri*.

This reform of the administration, which also had the effect of reorganizing the work task of people like Dario, caused frustration with the rangers, now policemen, and misunderstanding between employees of the park and inhabitants of the region:

Dario (guardiaparco): A me che identifica tanto è il ruolo della guardia e come noi siamo conosciuti all'interno delle nostre realtà locali.	Dario (ranger): I identify myself with the ranger role and the way we are known inside our local reality.
Cristina: E come siete conosciuti?	Cristina: And how are you known?
Dario: Siamo conosciuti come guardia caccia ancora, come andare a spiegargli che siamo polizia locale? Come andare a fare un verbale ad un cacciatore perché non aveva allacciato la cintura? Ti dice: "Ma cosa sei qua a rompermi le scatole?". "Non l'allaccio perché non l'allacci neanche tu tante volte". Capito cosa succede?	Dario: We are still known as gamekeepers. How can you explain to people that we are now local police? How can you impose a fine on a citizen because he did not fasten his belt? He will answer you: "But why do you bother me with that?", "I do not fasten it, because you do not fasten it either". Do you understand what happens?

Giovanni, a ranger in another protected area, does not feel comfortable with the diversity of uniforms used for Alpine rangers either, even though he and his colleagues do not wear a police uniform. When I asked him to tell me more about the uniform, he readily voiced his preference for a common uniform, and then he expressed his wish for a harmonization of the tasks at the regional level. Later, when I openly questioned him about the possibility of wearing an Alpine uniform, he declared he could easily identify with it:

Cristina: Qualcuno mi parlava di un progetto di unificare le uniformi...

Giovanni: La Regione è un po' che ne parla. Non è riuscita neanche a far questo! Secondo me sarebbe il primo passo. Se tu vedi, ci sono dei guardia parco, quelli vestiti di verde, poi ci sono quelli vestiti di grigio-nero, poi ci sono quelli vestiti di grigio. Noi siamo più verdi, quindi c'è tutto un caos.

Cristina: Ti vedresti con un'uniforme unica alpina?

Giovanni: Sì, senz'altro!

Cristina: Preferiresti?

Giovanni: Sì, perché secondo me è molto importante che il cittadino riconosca... Cioè, dato che le Alpi in realtà per i cittadini non hanno confine, perché io quando vado a fare sci alpinismo sconfino dalla Svizzera, all'Italia, alla Francia, al Tirolo, ecc. Avere questa figura uguale dappertutto vorrebbe dire che il cittadino immediatamente identifica la persone che ha davanti.

Cristina: Somebody told me about a project to harmonize the uniforms...

Giovanni: The region has been talking about it for a while. They could not even manage to do this. It would be, in my opinion, the first step. If you have a look, there are rangers who wear green uniforms, then the ones with black and grey, then the grey. We are greener, at the end it is just chaos.

Cristina: Would you see yourself in an Alpine uniform?

Giovanni: Yes, certainly!

Cristina: Would you prefer it?

Giovanni: Yes, because I think that it is really important that citizens recognize... I mean, since the Alps, for citizens, do not have any borders, because if I go skiing, I cross the Swiss, Italian, French, Tyrol borders, etc. If there were the same figure, everywhere, citizens would immediately identify the person he or she has in front of them.

From this statement, it is clear that, although I was the first to introduce the topic of an Alpine uniform, Giovanni undoubtedly supports the idea. Interestingly, the reason he considers it to be positive is determined by external factors ("It is practical for citizens") and not by his identification with the Alps. Giovanni did not say that he would like to wear an Alpine uniform because he identifies with the job that the other Alpine rangers are doing or with the Alps as such, but because sightseers would immediately recognize a ranger, in whatever Alpine country. The question of the identification of rangers with the Alps was clearly introduced by asking the rangers about the feeling of belonging to different territorial scales. In the following paragraphs, I analyse the answers to this issue.

Identification of the rangers with the Alps

In the interview with Dario, I asked him to tell me more about a question that Petra and I had put in the questionnaire. We explicitly asked which territorial entity (the municipality, region or country where one lives, the protected areas where one works, the Alps or the mountains in general) is the most important for the interviewee. Dario clearly answers "the Alps" in the questionnaire, and he explained why in the interview:

Dario (guardiaparco): Perché io vedrei la regione alpina come un'unica entità. Perché, prendiamo il discorso faunistico, la popolazione di stambecco non conosce "di là di qua" [della frontiera]. Io vedrei dunque meglio un discorso più largo.	Dario (ranger): Because I see the Alpine region as a unique entity. Because, looking at it from the point of view of the fauna, the ibex population does not know a "here and there" [of the border]. So, I would better see a more global discourse.
Cristina: Anche a livello personale ti senti attaccato a...	Cristina: And at a personal level, you feel attached to...
Dario: Alle Alpi, sì, molto!	Dario: The Alps, yes, very much!

Although Dario has strong bonds to the place where he lives and works, he obviously chose his profession because it allowed him to stay in contact with the mountains and the Alps, a place with which he strongly identifies. The same is true for Laura, whose choice to work in a protected area was dictated by a vocation and a strong identification with the Alps. In the following extract, Laura, an Italian ranger even though she decided to speak French with me, explains, as Dario did, the answer to the question:

Laura: [Dans le questionnaire j'ai dit que] je suis assez attachée à mon pays... assez...	Laura: [In the questionnaire] I said that I was fairly attached to my country... fairly...
Cristina: Donc l'Italie?	Cristina: Italy?
Laura: L'Italie, oui. Assez attachée à la région. Très attachée aux Alpes. Très attachée à la montagne en général. Peu attachée à l'Espace Protégé ou peut-être attachée à l'Espace protégé... Et pas du tout à la commune.	Laura: Italy, yes. Fairly attached to the region. Very attached to the Alps. Very attached to the mountains in general. Little attached to the protected area, or maybe just attached to the protected area... And not at all attached to the municipality.
[...]	[...]

Cristina: Vous avez dit avant que c'est vraiment les Alpes et la montagne en général qui…	Cristina: You said that the Alps and the mountains in general…
Laura: Ouais, ouais, certainement! Parce que j'aime toute la montagne de tous les pays. Il suffit que ça soit au-dessus de 300 mètres de dénivelé et c'est déjà de la montagne. Si c'est 600 mètres c'est mieux. Si c'est plus que 1000 c'est encore mieux. Plus que 2000 c'est magnifique. 3000 c'est le paradis. 4000 c'est le rêve.	Laura: Yes, yes, of course. Because I love all the mountains all over the world. It is enough for it to be above 300 m. For me, that is already a mountain. If it is over 600 m., it is better. If it is more than 1000 m., even better. More than 2000 m. is splendid, 3000 m. is heaven, 4000 m. is a dream.

Laura first of all expresses her strong attachment to the mountains in general and the Alps in particular. I had the impression that this is true for all the rangers that participated in the Trophy.

The discussion about the uniform can be taken as a symbolic epiphenomenon of a general impression that I could feel during the Trophy but is difficult to express in words. The popular saying that mountaineers, just like seamen, understand each other, even without speaking, possibly best describes this great complicity that I witnessed between the rangers from all over the Alps. An anecdote that happened during the Danilo Re Trophy may help to convey that atmosphere.

Thursday, March 15, 2007. Everybody was waiting for the Alpine buffet to be served. Two rangers, a Swiss and a Slovenian, and I were standing outside the building trying to communicate. Normally, in the Alpine meetings, there is always at least one person that can speak the language of the other and ensure the interpretation in order to aid the communication. In that case it was impossible. We had no common language. I was feeling really sad, because the Swiss ranger told me, in German, that he would have liked very much to be able to speak with the Slovenian colleague. He told me that they would have much to talk about.

Some minutes later, Petra came, so that the two rangers were able to communicate. I could see in their eyes that they were really happy that, finally, they could exchange some words. But in the meantime, while they could just exchange looks, it was impossible not to sense the strong feeling that those two persons shared beyond words. A strong commitment to their profession and the conviction of being part of the same reality: the Alps. Paolo, the person who as a boy had a strong passion for taxonomy, justifies his "Alpine species" with few words that reflect what I tried to express above. He uses an Italian colloquial expression, "*fare comunella*", which literally means "gang up on". The word *comunella* best synthesizes the idea that he develops below:

C'è sicuramente qualcosa che unisce. Io vedo quando ci sono delle riunioni e arrivano rappresentanti degli Stati vicini, degli Stati alpini, c'è subito una grande intesa, alle volte al di là delle lingue. Effettivamente la gente di montagna si capisce. (Paolo, impiegato del Centro di Ecologia Alpina)	There is certainly something that unites us. I see it during meetings when there are representatives of the neighbouring states, Alpine states. There is, from the beginning, a great understanding, sometimes beyond language. Really, mountain people understand each other. (Paolo, employee of the Centre of Alpine Ecology)

Rangers, probably given that they have chosen their profession because of their love of mountains and seeing that they are constantly in touch with them and with their inhabitants during their daily activities, seem to identify with them more strongly than their bosses do, as I explain in the next section.

Identification of the managers of protected areas with the Alps: the French example

The anecdote reported in the previous section illustrates the rangers' strong identification with the Alps and their intense feeling of "being mountaineers". This point contrasts with the impressions collected by Bernard Debarbieux and Petra Arnuš in the interviews with the French directors of protected areas. The example of Jean, at the time of the interview the director of a protected Alpine area and member of Alparc's steering committee, is representative of the situation of the French system. Although this experience cannot be generalized to the entire Alpine arc, it highlights how the attitudes towards the Alps can differ from that of the rangers.

Like his colleagues, Jean arrived in the Alps from another park in France and has since left that protected Alpine area to work for another park in south-western France. This particular system results in a regular movement of government officials (*hauts fonctionnaires*) as directors of parks within France. Thus, for Jean, as for all the other directors of French parks, the Alpine experience is just a moment in a long career where the function – and not the type of environment in which they work – is important. This constant change of locality makes it impossible for French officials to identify themselves with a particular place. Jean, in his interview, argues that he has been freshly nominated as director of the park as justification for not having any in-depth knowledge of the Alpine Convention. As he says, "I know Alparc a bit, but I feel more like somebody discovering all of this".

Discovering a new place, Jean was from the beginning immersed in the pan-Alpine projects. His predecessor was a member of Alparc's steering committee, so it seemed "natural" that Jean would replace him in this task as well. Here, Jean explains why he accepted this task:

Jean (directeur d'un espace protégé): Moi, le coté international me plaît bien, et donc voilà j'ai découvert cela. Au bout de deux ans, moi j'y trouve personnellement mon compte et je pense que c'est vraiment important pour la structure d'avoir ce travail à l'échelle des Alpes. Sur la notion de 'international' cela me semble évident: à partir du moment où on est au bord d'un autre pays, cela me semble évident qu'il faut passer la frontière.	Jean (director of a protected area): I like the international side of my job, so I said 'yes' and discovered all that. After two years, I feel comfortable, and I think that it is really important for our structure to have this work at the Alpine scale. On the point concerning the international work, this seems evident to me: Since we are at the edge of another country, it seems evident that we must cross the border.
Bernard: Alors, montagne ou pas montagne, cela serait pareil en Belgique, en France...	Bernard: Which means, mountain or no mountain, it would have been the same in Belgium, in France...
Jean: Exactement. Je serais au bord de la Belgique, j'aurais ce même besoin, envie, ou même enjeu stratégique de discuter avec les voisins, parce qu'il n'y a pas de frontières entre les thématiques sur lesquelles on bosse. Après, le Réseau Alpin c'est encore autre chose. Les autrichiens ne sont pas à côté de chez moi, et pourtant cela me semble important de travailler avec eux, et les slovènes sont	Jean: Exactly. If I were close to Belgium, I would have the same need, or desire, or even strategic issue of discussing with my neighbours, because there are no borders for the themes on which we work. Then, the Alpine Network is something else. Austrians are not close to me, and yet I think it is important to work with them. Slovenians are even farther, and yet I think it is interesting

encore plus loin de chez moi et cela me semble intéressant de bosser avec eux. Là je crois qu'il y a vraiment cette notion d'entité géographique, naturelle, historique, je ne sais pas quoi, des Alpes, qui fait que c'est intéressant et important de partager des pratiques, de partager des préoccupations. Voilà, parce qu'il y a cette entité-là.

to work with them. In this case, there is really a notion of a geographical, natural, historical entity, call it what you want, namely the Alps. This makes it interesting and important to share practices, to share worries. Because there is this entity.

One can feel Jean's willingness to spend time international networks. Yet, this does not arise from his identification with the Alps but from a personal preference and vision of his work. The way Jean expresses the idea of the Alps as one entity is clearly "distant". He understands that, given the Alpine configuration, people share their worries and know-how, but he formulates his idea as if he were outside "their" world. Later in the interview, Jean admits that it is a priority to invest in the local level. Involvement at the pan-Alpine level is thus the result of a personal skill, the "network culture", as he says, and not due to a strong commitment to the pan-Alpine project: "There are people that have the 'network culture,' and there are others who do not have it." Obviously, he does.

Of course, this is happening in France, which is a very centralized state. In Austria and Germany, for example, this would be impossible, first of all because the protected areas are placed under the regional administration *(Länder)*, and therefore their directors have more leeway to impose their decisions. Local inhabitants are welcome to work for the parks and become directors. They can stay as long as they want: Vogel, for example, is a long-time director of the Berchtesgaden National Park and president of Alparc. Still, the French case, and Jean as an example, demonstrates well that the commitment to the pan-Alpine project through the pan-Alpine networks does not depend on a strong identification with the Alpine region alone. For Jean, Alparc is an opportunity that can satisfy his desire for international exchanges.

5.2.6 *The case of WWF: pan-Alpine involvement without identification with the Alps*

Another example, institutional this time, should also be noted. Nicolò, from the WWF, while describing the environmental organization's engagement in the Alps, thinks rationally in terms of efficiency of nature conservation. The Alps are one region amongst others within the organization's strategic principles, i.e. the safeguard of biodiversity. Non-identification with the Alpine region is evident in Nicolò's speech; he shows more concern that biodiversity be maintained, no matter whether in the Alps or in other regions of the world:

Abbiamo iniziato a lavorare sul concetto di ecoregione in seguito ad una constatazione che è stata fatta a partire dal 1992 con la convenzione sulla biodiversità, dove si è capito che, nonostante i successi di conservazione che ci sono stati a livelli di sito, la tendenza a livello globale è che la perdita di biodiversità si è accentuata. [...]	We started working on the concept of ecoregion after an evaluation that was done in 1992 with the convention on biodiversity, when it became clear that, despite successes in conservation at the site level, the global tendency was more marked by a loss of biodiversity. [...]
Allora ad un certo punto ci si è detti: "Come si può massimizzare il proprio impatto a livello globale, [pur non avendo risorse illimitate]?". E si è detto: "Dobbiamo agire su quelle regioni nel pianeta che sono più importanti per la biodiversità, secondo il principio che salvando i punti giusti, quindi vincendo le battaglie giuste, vinci la guerra". Quindi cosa è stato fatto? Sono state identificate 238 ecoregioni, che sono regioni importanti, cruciali o per il mantenimento della biodiversità. (Nicolò, impiegato presso il WWF)	So, at a certain point, we said: "How can we maximize our global impact, [even if we do not have endless resources]?" We said: "We have to act on those regions on the planet that are the most important for biodiversity, following the principle that if one safeguards the right places, i.e. by winning the right battles, one can win the war." So, what did we do? We identified 238 ecoregions that are important regions, crucial for maintaining biodiversity. (Nicolò, employee of WWF)

The Alps are one of these 238 regions in which the WWF decided to invest. What clearly emerges from Nicolò's account is that what encourages him and WWF to act at an ecoregional level is a question of efficiency, and not identification with the Alpine perimeter as such. Giulia, who at the time of the ecoregional strategy's implementation was working for the WWF, similarly retraces the advent of the WWF's new methodology in terms of conservation:

Allora [prima degli anni 1990], non si era ancora imparato a ragionare per grandi ecosistemi con un approccio integrato. Questo è cominciato negli anni 1990 nella grande famiglia delle ONG per la conservazione. E negli anni 1990, il WWF, Nature Conservation e Conservation International hanno lavorato insieme per identificare gli ecosistemi mondiali più importanti. Sono 238 regioni più importanti al mondo, e le Alpi sono una di queste. (Giulia, impiegata presso il WWF)	At that time [before the 1990s], conservation NGOs had not yet learned to reason in terms of large ecosystems with an integrative approach. This started in the 1990s in the big family of conservation NGOs. In the 1990s, the WWF, Nature Conservation and Conservation International worked together in order to identify the most important ecosystems in the world. These are 238 regions amongst the most important habitats in the world, and the Alps are one of those. (Giulia, WWF employee)

These testimonials, those of Nicolò and Giulia, but also of Jean, show that a strong personal (and institutional) identification with the Alps is not an indispensable prerequisite in order to invest the Alpine perimeter with transnational projects. Additionally, there are also various registers of identification and involvement, as the two foremost personalities of Alliance in the Alps, Antonio and Rainer, demonstrate. This conclusion supports the doubts that Debarbieux expresses in an editorial in the *Journal of Alpine Research* in reply to Paasi's statement, "It is helpful to think analytically that 'identity' is part of the institutionalization of regions, the process through which regions come into being" (Paasi, 2002: 140).

Debarbieux expresses his doubts by firstly arguing that "numerous examples show that the absence or weakness of a shared feeling of belonging does not necessarily burden a strictly regional process" (Debarbieux, 2009: 14), as the WWF Alpine Programme illustrates. Secondly, the geographer criticizes the exclusivist and utilitarian thesis concerning regional identity, which reduces the range of identities to a single territorial (regional) identity (Debarbieux, 2009: 15), as Rainer's example emblematically demonstrates. In fact, the example of Rainer, who is probably one of the most active people on the pan-Alpine scene, shows that Alpine identification does not rule out possessing other social and political identifiers. Identities, and this becomes obvious from Rainer's example, change and evolve while being performed and, therefore, actuated (Jenkins, in Häkli, 2000: 15).

Poche provides an interesting way to think of identities as "performed". In an article written by the sociologist in 2000, he affirms that identity phenomena make sense only if they are experienced (*vécus*), and experienced as needed. Within these contexts, the group adjusts its responses (Poche, 2000: 213). But this need ("Alpine problems *cannot* be solved alone") is not territorially fixed: It is felt that some things need to be resolved at the pan-Alpine level, while others can be resolved at different levels. This non-exclusive engagement for the Alps by the Alpine actors will be illustrated in the following section by showing in which ways the latter identify with other scales, in parallel to the Alpine one.

5.2.7 Identification with other scales than the Alps

Although Rainer combines different scales of involvement in order to implement the principles of sustainable development, it is possible to conceive of different alliances that, distinct from the Alpine one, will result in other territorial configurations and in different "communities of problems". I asked Niklas whether it made sense, in his eyes, to build a network of small towns at the Austrian or the European level, regardless of whether they are situated in the mountains or in the plain. He answered as follows:

Es würde Sinn machen, glaube ich schon. Es wird wieder ein Netzwerk mehr sein innerhalb vieler Netzwerke, aber das würde sicher Sinn machen. Was die Kleingemeinden angeht, gibt es schon Themen, ob die in den Alpen sind oder im Flachland, es ist egal, es gibt auch dort Themen, die verwandt sind mit unseren. Also, es würde nicht schaden. (Niklas, Bürgermeister)	It would make sense, I think. The problem is that it would be an extra network amongst a lot of networks, but it would certainly make sense, because simply the themes are the same where small towns are concerned. There are common themes, in the Alps or in the plain, it is the same; there are also themes that are familiar to us. It would do no harm. (Niklas, mayor)

Although I could find only this example in the interviews I collected, I am quite sure that, had I asked the question directly, interviewees would have seen the advantage of creating additional alliances, and thereby different territorial configurations, for the different problems they encounter. What is universally mentioned is the reference to a more local scale, the area of daily life that can lead to parochialism, as in Dario's case, who stresses his local roots by using dialect words. Dario is always ready to joke, and this comment he made at the very beginning of the interview should be interpreted as a provocation, although the "local competition" he mentions could probably be confirmed in real life:

Autoctono della val Chiavenna, che è un'appendice della Valtellina. E noi ci teniamo a fare questa distinzione: *la Valtellina l'è un meschté, la Valciavenna l'è n'altro*[104] (Dario, guardiaparco)	I am a native of the Chiavenna Valley, an appendix of the Valtellina. And this distinction is important to us: The Valtellina "is something, the Chiavenna Valley is something else" (Dario, ranger)

In most cases, this bond with local reality is expressed in other terms, simply by defining the space where interviewees go on a daily basis. This can be the valley where they live, or the regional transnational space they frequent, in the case of people living close to borders. In his answer to my question, "At which territorial unit do you feel closer?", Giovanni, for example, admits that he identifies with the area he frequents:

Giovanni (guardiaparco): Io sono un dipendente della Provincia da tanti anni, per cui mi sento legato alla Provincia come istituzione, però è una questione di dipendenza funzionale.	Giovanni (ranger): I've been working for the province for so many years that I feel a close bond to the province as an institution, but this is a matter of work.
Cristina: Abbandona la divisa un attimo!	Cristina: Forget your uniform for a moment!
Giovanni: Ah, come persona! Forse alla Regione in cui abito, perché è uno spazio che io frequento. Io anche quando vado a fare una gita, magari vado in Provincia di Cuneo, vado magari in Valle Ossola, però comunque sempre più o meno all'interno della mia Regione.	Giovanni: Oh, you mean as a person! Maybe to the region where I reside then, because it is a space where I normally hang out. If I go on a trip, I may go to the province of Cuneo, maybe the Ossola Valley, but in any case always inside my region.

104 Expression in the local dialect.

In the case of the Great Walser Valley, Ulrike, who is the regional manager and manager of the Biosphere Park, affirms that the inhabitants of the valley "identify themselves as Walser or as *Grosseswalsertalern*. And in their daily life, they identify even more with their municipality".

Nevertheless, the interviews with mayors from the valley reveal that this is not true for everybody. Mayor Henrik, for example, identifies with a wider area (the *Land* of Vorarlberg) than what the region manager identifies with the inhabitants of the valley (Great Walser Valley):

Ich fühle mich im Land Vorarlberg sehr wohl, ich finde wir haben ein sehr schönes Land. Wenn ich hierher komme, denke ich immer: "Wir haben es auch schön!". Es gibt viele schöne Gegenden. In Europa gibt es viele schöne Gegenden, aber ich fühle mich eigentlich im Land Vorarlberg wohl. Und ich möchte es nicht nur aufs Walsertal beschränke. Ich fühle mich im Vorarlberg wohl. (Henrik, Bürgermeister)	I feel really good in the *Land* of Vorarlberg, I think we have a really nice *Land*. When I come here, I always think: "It is really beautiful here!" There are a lot of nice places. In Europe there are a lot of beautiful places as well, but I really feel good in the *Land* of Vorarlberg. And I would not restrict it to the Great Walser Valley. I feel good in Vorarlberg. (Henrik, mayor)

So, Henrik recognizes the perimeter of an administrative unit (the *Land* where he lives and works) as the one with which he identifies the strongest. This perimeter is bigger than the valley's. Generally, the impression I got from the mayors of the valley is that there is indeed a common identification with the valley. People are usually very proud to be *Grosseswalsertalern* (i.e. from the Great Walser Valley) and very connected to their local reality.

Ulrike partly confirms Maximilian's declarations. When asked how important the Alps are for him, he mentions something that is shared by numerous residents of the Great Walser Valley, which is the belonging to their ethnic group, the Walser. As he says: "My heart beats for the Walser. I am a Walser and proud to be [one]; I am active in the Walser association, too." This link to the Walser community brings a double association: For one, it facilitates relationships everywhere that Walsers are living; this in turn creates a space of action complementing that of the Alps.

For others, especially for those living close to borders, their space of daily life is a regional transnational space, as the interview with Sabine shows:

Cristina: In welcher geographischen Einheit fühlst du dich am wohlsten?	Cristina: In which geographic unit do you feel the best?
Sabine (Betreuerin): Für mich ist es nicht an Grenzen fest zu machen, für mich ist es so dieser Raum Vorarlberg-Schweiz, der ganz leicht in den Bodensee hinaus geht. Ich würde sagen, so 50 bis 100 km im Umkreis. Das ist mein Raum, weil ich hier in mehreren Orten schon gewohnt habe. Für mich ist es die kulturelle Bindung auch an diesen alemannischen Sprachraum, der die Schweiz mit Süddeutschland verbindet, dieser Dialekt hat eine relativ starke Bedeutung. Wobei der Alpenraum an sich mir sehr wichtig ist.	Sabine (animator): For me this space is not attached to borders; for me it is this Vorarlberg-Switzerland space, a bit around the Bodensee. I would say more or less 50–100 km around Mäder. This is my space, because I have already lived in many of the localities there. For me, it is this cultural linkage with the Alemannic German area, which connects Switzerland and southern Germany, as this dialect is quite strong. Even if the Alpine space is also very important for me.

If for Sabine the transnational area to which she is referring corresponds to a perimeter included in the Alps as defined by the Alpine Convention, for Manuele, his transnational space of reference goes beyond the Alpine region to include the Italian plain:

Mi sento molto più vicino alla zona dell'Italia probabilmente come pensiero, come persona e come stila di vita, non alpino, ma più da pianura padana. (Manuele, impiegato comunale)	I feel closer to the Italian area, as a way of thinking, as a person and as a lifestyle, not Alpine, but more of the Po Valley. (Manuele, municipality employee)

The added "not Alpine", expressed by Manuele, reflects Leonardo's observations when he tried (with difficulty) to find new members in the Ticino Canton for Alliance in the Alps:

In effetti secondo me il Ticino fa un po' fatica a capire che fa parte delle Alpi. (Leonardo, stagista in una regione)	In fact, I think that Ticino has trouble understanding that it is part of the Alps. (Leonardo, trainee in a region)

This observation about the people's non-identification with the Alpine region (inhabitants, but also members of the network) must be borne in mind for the following chapters, when the functioning and challenges of the networks will be discussed. For the moment, going back to the issue of the identification's geographical units, we need to underline that this transnational space makes sense not only for individuals, but at the institutional level, too.

Fabio, who is from the same region as Manuele, also refers (although not exclusively) to the same geographical perimeter, including the southern part of Switzerland and northern Italy when describing the projects carried out by the Malcantone Region where he works, for example, a project on the Lema mountain called "*Lema, la montagna che unisce*" ("Lema, the mountain that unites") and a project concerning chestnuts trees, a historically important plant in the area:

Hai, in parallelo, altre reti. Abbiamo fatto dei progetti Interreg A, quelli transfrontalieri. Abbiamo lavorato bene con le comunità montane in Italia, con i parchi naturali, il focus è il castagno, perché noi abbiamo comunque un'identità molto forte legata al castagno, e quindi abbiamo altri reti più bilaterali, però con questa arriviamo fino a Varese, Milano. (Fabio, segretario regionale)	There are other networks. We participated in some Interreg A projects, those that concerned cross-border areas. We worked really well with the *comunità montane* in Italy, with the protected areas. The focus is the chestnut tree, because we have a really strong identity linked to the chestnut tree, and so we have other networks, more bilateral, and with them we reach all the way to Varese, Milan. (Fabio, regional secretary)

These institutional links, based on common projects, in a regional transboundary perimeter, are common even for collaborations between protected areas. Julien even admits that these can be stronger than the ones that protected areas built with the Alpine perimeter:

De longue date ici il y avait ce partenariat avec le Parc du Grand Paradis, qui est évidemment continu, transfrontalier, même si le relief se prête moins que dans d'autres espaces transfrontaliers à des échanges très réguliers. Mais c'était quand même le point fort de l'action au niveau international pour le parc, plus que par rapport au réseau Alparc. (Julien, Directeur d'un espace protégé)	For a long time now, we have had a partnership with the Gran Paradiso Park, which is obviously contiguous, transboundary, even if the mountainous landscape makes it difficult to have very regular exchanges. But this was the main focus of the international action of the park, more than the link to the Alparc network. (Julien, director of a protected area)

There is another degree of identification that should be mentioned here, which is the one relating to the mountains in general, not only those within the Alpine perimeter but spreading all over the world. Andrej is probably the clearest on this point, when he proclaims his local attachment, coupled with a strong bond with the mountain:

> Petra: Affiliation to Slovenia and to the Alps?
>
> Andrej (ranger): Yes. To the Triglav National Park.
>
> Petra: Not to the Alps in general?
>
> Andrej: To the mountains. Not to the Alps. We have hills, mountains, Himalayas, Peru, Chile, Kyrgyzstan. Mountains.
>
> Petra: All this is important?
>
> Andrej: It's the same. I respect Kyrgyz, the ones who come from the mountains, and behave like normal "mountaineers", and it's not important whether they come from the Alps or from the Caucasus. They are mountain dwellers, *montagnards* for me.

Mathieu speaks of "a common mountain identity" that "builds character". Andrej would probably agree with this declaration:

On a une identité commune qui est la montagne, parce que cela forge les caractères et les âmes. On a une approche à la nature qui n'est pas la même qu'une personne qui vient de la plaine. Ce n'est pas que les autres n'ont pas une approche avec l'environnement, mais ils n'ont pas la même que nous. Sinon, effectivement, chacun a une culture nationale qui s'ajoute à cela. Je pense qu'il y a ces deux niveaux. (Mathieu, employé d'une commune)	We have a common identity, which is the mountain, because it builds character and souls. We have an approach to nature that is not the same as that of someone from the plain. It is not that the others do not have an approach with the environment, but it is not the same as ours. And then, on the identity aspect, everybody has a national culture that may be added to this. I think there are these two levels. (Mathieu, municipality employee)

The aspect concerning the different "national cultures" present in the Alps will be discussed in chapter 7. For the moment, it must be said that, regarding the regional transboundary area, there is not only an individual link to the mountains but also institutional attempts to instigate collaborations with other mountain ranges in the world. One of the most symbolic collaborations is probably the foundation of the Alliance of Central Asian Mountain Communities (AGOCA), the Central Asia municipality network, created on the model of Alliance in the Alps and firmly desired by the two main players of Alliance in the Alps: Rainer and Antonio.

To finish this section, I would like to present two extracts of interviews that illustrate how some people, when asked about their affiliation to geographical perimeters, show a complicated geometry of identification spaces, spreading to the entire world. First, the example of Niklas:

Das ist klar, das ist für mich die Region Großes Walsertal. Für mich, vielleicht ist es blöd wenn ich es sage, aber ich denke; für mich ist das wirklich ganz Europa! Eigentlich die ganze Welt, das ist einfach wichtig, dass es funktioniert. Wenn es ihnen gut geht, dann ist die Chance, dass es mir auch gut geht, viel größer. Geht es den Amerikanern gut, den Chinesen gut, dann geht's uns einfach wahrscheinlich auch viel besser. (Niklas, Bürgermeister)	It is clear, for me, it is the Great Walser Valley Region. For me, it is a bit stupid when I say it, but actually to me it is all of Europe! Actually, the entire world. It is important that it works well. If the others feel good, there are chances that I will feel good as well. If the Americans feel good, the Chinese feel good, then we will probably also feel good. (Niklas, mayor)

The last quotation is taken from the interview with Jakob and shows this capacity of a single individual to feel linked to different perimeters, from the town to Europe. It is interesting to note that this capacity has one exception, the Austrian national space symbolized by its capital, Vienna:

Wien ist für mich sehr weit weg, Wien ist praktisch genauso weit weg wie Paris. Ich war ein paar Mal in Wien und zweimal in Paris und für mich rückt der Nationalstaat ein bisschen in den Hintergrund und die Region kommt eher zum Vorschein. Also für mich sage ich jetzt Vorarlberg. Und meine enge Identität ist meine Gemeinde und der Große Walsertal Biosphärenpark. Aber sonst bin ich im gewissen Sinne ein stolzer Vorarlberger, darf ich doch so sagen,	Vienna is for me really far away; Vienna is actually as far away as Paris. I was in Vienna a couple of times and twice in Paris, and for me the national state recedes a bit into the background, and the region comes to the surface. For me, I would say Vorarlberg. And my narrow identity is my municipality and the Biosphere Park of the Great Walser Valley, but I am in a certain way a proud inhabitant of the Vorarlberg, I have to admit, not presumptuously, but aware of all the

nicht überheblich, sondern im Bewusstsein aller Stärken und Schwächen. Und ich bin wirtschaftlich sehr viel unterwegs. Und ich bin ein Anhänger der europäischen Idee. (Jakob, Bürgermeister)

strength and all the weaknesses. And I often travel professionally. And I am a supporter of the European idea. (Jakob, mayor)

5.3 Conclusion

All these examples show the theoretical debate on identity, which considers identity as something plural and changing over time, creating intricate territorial assemblages, as Allen and Cochrane point out (Allen & Cochrane, 2007). I finish this section with the words of Jakob, who perfectly captures this idea:

Im Prinzip hat jeder Mensch mehrere Identitäten. Heute ist der Mensch von seiner Identität her vielschichtiger. Früher, also vor zwei Generationen, war er sicherlich wesentlich nationalstaatlicher geprägt, aber in unserer Welt verschwindet das eher. Man baut neue Identitäten auf, verschiedene Identitäten, und das hilft natürlich eine neue Identität aufzubauen. (Jakob, Bürgermeister)

In principle, every person has more identities. Nowadays, individuals have multi-layered identities. Before, I would say two generations ago, individuals were more affected by the national state, but in our world it is disappearing. One builds new identities, different identities, and this obviously helps to build a new identity. (Jakob, mayor)

What Jakob says can be illustrated by the fact that the same persons are quoted in this chapter for different purposes. From time to time, the same individuals are quoted for their commitment to the Alps or for their commitment to other scales than the Alps. Although this could seem quite contradictory at first glance, it actually highlights these multi-layered, sometimes identified as post-modern, identities, which can be flexibly combined by a same individual. It justifies Di Méo's answer: "Today, do you belong only to one social group, to one *territoire*? Certainly not! I would like to say less than ever, seeing the increased mobility of our contemporary people and their always broader social and spatial experiences" (Di Méo, 2002: 177).

Additionally, and as was shown, there can be different reasons why local political actors use the Alps as their reference area. Emblematically, the example of Antonio and Rainer showed that both have some interest in empowering the Alps and in identifying with them, while, at the same time, especially for Rainer, the Alps are just one identification perimeter amongst many others. By contrast, the WWF or the ministries of the environment that signed the Alpine Convention do not strongly identify with the Alps. Yet, they invest in this perimeter in order to implement projects and therefore, by those actions, contribute to the construction of the pan-Alpine region.

6. Pan-Alpine networking: performing the regional assemblage

In the previous chapter, I analysed how the Alps are imagined and objectivized by the local political actors involved in the networking process. By creating a common and shared image of the Alps, local political actors are simultaneously internalizing and externalizing the idea and representation they have of them. The interviews I conducted with local political actors, as well as the websites of the networks in the study, show that this image is not homogeneous.

Nevertheless, common characteristics can be drawn, for example the Alpine socio-cultural and natural diversity as a foundation of its unity. According to the local political actors, the beauty of the Alpine nature and heritage should, on the one hand, be protected and, on the other, promoted and managed in a sustainable way. In a second instance, the concept of identification was introduced in order to demonstrate the capacity some local political actors have to identify themselves inside the perimeter they objectivized as the Alps. With local political actors' increased awareness that mountains inherently entail something positive, some of them seize the opportunity to invest in that perimeter by proposing innovative initiatives and to implement projects in their local reality by creating institutions, such as the pan-Alpine networks, capable of supporting those projects.

In so doing, as Debarbieux contends in an article published in the *Journal of Alpine Research*, mountain regions might become a referent for collective action (Debarbieux, 2009). However, the geographer posits the following question: When and how does a mountain region become such a referent? He answers with two possible outcomes: firstly, when a region perceived as a mountain entity is used to delimit and problematize governance issues (taking into account the region's mountainous character) that are sufficiently important and recurrent to give rise to an institutional reconfiguration at this scale; secondly, when, at the scale of pre-existing entities, and possibly within the population directly concerned, there are resources and room for *manoeuvring* that enable local political actors to promote this institutional reconfiguration, and, if need be, one based on identity, which is complementary to existing configurations (Debarbieux, 2009: 16).

In this chapter, I discuss Debarbieux's statements and analyse, on the one hand, the institutional reconfiguration of the Alpine region and, on the other, the room for manoeuvring that pan-Alpine actors have in this process. In other words, the focus will be on the networks of local political actors as the primary institutions reconfiguring the regional Alpine governance by transcending national borders by means of their activities.

6.1 Introducing the pan-Alpine assemblage

Networks of local political actors draw a new map of the Alps by transcending national boundaries and building an Alpine region delimited by the Alpine Convention's boundaries. The perimeter of the Alpine Convention thus limits the localities that can claim to be part of the Alps. Networks can only acquire members inside this perimeter; therefore, the Alpine Convention's framework decides the geography of the networks. Nevertheless, the pan-Alpine region cannot be only topographically understood and conceptualized, since the actors/institutions conceiving and acting *in* the Alpine region are not necessarily *of* the region. This means they cannot be localized on the same map, as Allen and Cochrane's concept of "assemblage" correctly underlines (see chapter 2). Two examples taken from my case study clearly illustrate how the actors operating for the region are not necessarily of the region.

Firstly, the Alpine case shows that it is not because Brussels or Rome are geographically distant from the Alps that the discussions or decisions of the European Union or the national representatives carry less weight than the words of Alpine social and political actors. In fact, the reaction of some scholars and some local politicians who criticized the process during the establishment and the writing of the Alpine Convention clearly illustrates the contrary. Alpine actors openly criticized national authorities for the way in which the Alpine Convention had been initiated and conducted, qualifying it as a "colonial manner of acting" (Bätzing, 1994: 186).

This shows that, during the Alpine regionalization process, persons and/or institutions acting for the region are not necessarily localized inside the region: The Austrian Minister of the Environment, a signatory to the convention, is not working in the Alps but in Vienna. Still, that signature has a critical incidence on the Alpine destiny.

Secondly, it is not because the Italian Minister of the Environment is working in Rome, which is about 800 km away from Innsbruck, where the Permanent Secretariat of the Alpine Convention has its seat, that his words have less weight than those of the president of CIPRA, who works in Schaan (Liechtenstein), situated inside the Alpine Convention area and slightly less than 200 km away from Innsbruck. Power is not proportional to the geometric distance, as Allen and Cochrane point out (Allen & Cochrane, 2007).

So, the Alpine case shows exactly what the concept of regional assemblage implies, i.e. that the commitment of the actors for the region is defined by their practices and not by some predefined scalar arrangement (Allen & Cochrane, 2007: 1171). The practices in the Alps are, on the one hand, the projects that local political actors implement inside the Alpine Convention's area and on the other, the institutional arrangements that they are building in order to regulate and govern their actions. In that sense, they contribute to new forms of governance inside the European territory. Balsiger points out underlying three features of the Alpine cross-border area that may lead to governance reforms:

> The first of these relates to its eco-regional but comprehensive sustainable development scope, the second to its process-based approach to the framework protocol model of international agreement, and the third to its permeability to non-state actor influence (Balsiger, 2008: 18).

The important deduction to be made from this statement is that the combination of these three features enables the Alpine initiatives to contribute to a process of institutionalization. Debarbieux confirms that the different Alpine initiatives bring about a "process of political institutionalization" (Debarbieux, 2008a: 101). The latter corresponds to one of the four shapes, i.e. "development of institutions", identified by Paasi in his regional institutionalization theory. In line with Paasi's theory, the establishment of a "territorial shape" can be identified in the Alps by the establishment of the Alpine Convention's boundaries (see chapter 2). The second shape is the constitution of a "symbolic and conceptual shape" through a recognition of the importance given to the value of sustainable development (see chapter 5). In parallel, as will be explained in this chapter, the Alpine region is developing an important structure of institutions, as evidenced by the establishment of institutionalized networks of local political actors, which corresponds to the third shape defined by Paasi. The fourth shape, the "establishment of the region" will be analysed in the next chapter (see chapter 7).

As discussed in the theoretical framework, the institutionalization of the Alps is influenced by two processes: the horizontal initiatives undertaken by local leaders and some Alpine institutions (for example the CIPRA), and the important catalyst initiatives of the European Union, which promotes transnational projects. The former, which includes the establishment of reticular initiatives, are particularly relevant to the regionalization process. Almost all the existing Alpine networks have been created by CIPRA with the aim of finding specific and appropriate tools to implement the Alpine Convention. The main pan-Alpine actors investigated here have been presented in chapter 4. These networks, which can be considered institutions created *ad hoc* for the ambitious pan-Alpine project, represent the main success of the Alpine regionalization process. Research allowed me to look deeply inside these "ambassadors of the Alpine Convention", as Götz defines them (Götz, 2011: 10), and the main results of the analysis of the networks' functioning will be presented in this chapter.

One of the major results of this horizontal mobilization is the materialization of these initiatives in regional institutions. Goran, the director of a Slovene protected area, reveals this process in the interview conducted by Petra Arnuš:

> Petra: Because of the existence of ever more organisations, partnerships, networks and projects in the Alps, do you think this political area will be more alive in the future?
>
> Goran: In what sense?
>
> Petra: As an economic, social and ecological region?
>
> Goran: I think it will be. It would be clever to think about it in terms of European regionalization, because of the comparable possibilities for development and because of the comparable problems. In those coincidences the connections are good, and also some political organizations are created.

Goran, in view of the comparable problems that people of the Alps are facing, welcomes the new connections that are built within the Alpine region and sees how new political organizations follow this evolution. But how is it possible to define this "regionalization", this "creation of organizations" and this "regional institutionalization"?

In this chapter, I answer the latter as well as the following questions: How do these new territorial and institutional entities, namely the networks, really work? What are the challenges they face? These questions will be complemented in the following chapter by the following: What can we expect from these new territorial and institutional reconfigurations? And what is the social and political significance of the Alpine process? The pertinence of those topics is given by MacLeod, who comments that, in the contemporary world, the theory of the institutionalization of regions may provide useful elements to understand the emergence of particular socially cohesive territories, their institutionalization and their establishment as specific sites for political governance (MacLeod, 1998: 840).

Thus, in the following sections, I uncover the contents and challenges of this pan-Alpine assemblage. In so doing, I will pay attention to the networks of local political actors. Before entering this debate, I would like to report a passage from the interview I conducted with Peter Brandauer, president of Alpine Pearls. He hopes that new areas, supported by systems more flexible than those of the national states, will be key to further cooperation inside the Alps. This cooperation, which in his view goes beyond the national frames, should focus on precise themes, contents and shared problems:

Ich denke, dass sich einiges ändern wird in der Form der Kooperation, der öffentlichen Verwaltung. Ich glaube man wird sich jetzt immer mehr von diesem relativen starren System verabschieden und ich hoffe, dass man hier im Laufe der Zeit einfach ein flexibleres System entwickelt. Dass es einfach verschiedene Ebenen gibt, wo man mehr oder weniger kooperiert und zusammen arbeitet. Und dass es nicht einfach so ein starres Gefüge gibt, sondern dass es ein offeneres Gefüge gibt. Und da denke ich, dass es nicht so entscheidend ist, ob man jetzt Staatsgebiete verändert, sondern dass es vielmehr darum geht Kooperationen aufzubauen, die immer ein spezielles Thema, einen speziellen Inhalt haben, der Sinn macht. Das ist auch meine Vision, dass es in der Zukunft hier offener wird, dass man nicht so an starren Strukturen festklebt. (Peter, Präsident der Alpine Pearls)

I think that quite a lot of things will change in terms of cooperation and management. I think we will abandon this stringent system, and I hope that in the future a new and more flexible system will be developed – a system with different degrees of cooperation, in which one can get more or less involved and find different ways of working together. I hope there will not be such rigid structures but rather more open ones. And, from this point of view, I do not think it is crucial to change the national borders; instead, it is important to develop cooperation, which must have a special theme, a special content, which makes sense. This is my vision; I hope in the future we can become more open, and that we no longer stick to rigid structures. (Peter, president of Alpine Pearls)

The establishment of networks of local political actors can already be considered a step in the direction Brandauer wishes: a more flexible, less nationally dependent, system of cooperation. This is what gives meaning to an institutionalized region, in which the barriers are falling and cooperation within the regional perimeter is increasing. In the following sections, I will analyse how this system works and what the challenges are of this new *modus operandi*.

6.2 The networks of local political actors: builders of the Alpine region

In the first part of this section, I explain the interviewees' justification for the Alpine networks of local political actors having been created and developed since the 1990s. Then I will analyse the reasons why local political actors decide to join the network's initiatives. Those will be put into perspective and compared with the interviews I conducted with non-members (but potential members) of the Alliance in the Alps network that share their reasons for not joining a network.

6.2.1 Birth of networks in the 1990s: members' reasons

I tried to systematically ask the persons I interviewed how they explain that Alpine networking started in the 1990s and has developed really fast since then. Some of the most interesting answers are reported here and structured according to three groups of reasons: the new possibilities of communication and mobility; the change in the individual and institutional mentalities and the transformation of the political context.

One of the most important reasons given by the interviewees to explain the increased popularity of reticular and networked cooperation is, on the one side, the intensification of individual mobility and, on the other, the advent of the Internet. Both multiplied the opportunities for people and institutions to meet and to build connections, something that would have been hard or even impossible to do in this way as recently as 20 years ago.

Niklas, for example, highlights the possibility of increased exchanges of information, which appeared with the spread of the Internet from the 1990s. Although he seems quite sure that the Internet is one of the sources of new connections, he does not explain why this translated into the creation of networks of local political actors. In other words, how the virtual connections were transformed into factual and institutional links:

Eine Ursache sind einfach die digitalen Kommunikationsmöglichkeiten. Die sind in den 90er Jahren nicht einfach geboren worden, die hat's immer schon gegeben. Der Unterschied ist, dass die dann für jedermann zugänglich geworden sind. Und dies auch zu einem Preis, der nicht so sehr wehgetan hat. Also es ist einfach, dass sich tausende Menschen ohne große finanzielle Investitionen un-	One reason is certainly the advent of digital communication. It was, of course, not invented in the 1990s, it existed already, but at that moment it became available to everybody, and at an affordable price. This means, simply, that thousands of people could be linked to each other without big financial investments; and we observed that through the World Wide Web it was re-

tereinander vernetzen können. Man hat es durch das World Wide Web auch gemerkt, dass es sehr einfach ist sich auszutauschen und dass man sich dort sehr viel Input holen kann. Und ich glaube schon, dass es eine Ursache sein könnte, warum [die bessere Vernetzung] genau in den 90er Jahren stattgefunden hat. (Niklas, Bürgermeister)

ally easy to exchange, and it was easy to get a lot of information. And I think this could be one of the reasons why exactly everything started in the 1990s. (Niklas, mayor)

Jakob, without explicitly mentioning the Internet, as Niklas did, gives some more details on how new contacts were established. He uses, quite emblematically, the concept of "global village". We will recall that this is a concept coined in the 1960s by the media theorist Marshall McLuhan, who used it to describe the potentialities of new technologies to contract the world into a "village", since information from all across the globe was starting to become instantly available.[105] Today, the concept of "global village" is part of everyday vocabulary, and it is generally used to signify the potentialities of the Web. So, indirectly, Jakob probably intended to use this concept to underline that the Internet is a powerful tool that helps people to know more from each other. The possibility of getting more information about events in other regions, and thereby ensuring that similar problems are brought to light, is part of the reason why, in the view of Jakob, people are keener to establish contacts:

Das weiß ich nicht, ich kann mir vorstellen, wir sind ein "globales Dorf", wir wissen viel mehr voneinander, als vor hundert Jahren. [...] Ich denke, dass hier einfach die Probleme aufgetaucht sind, die die Initiatoren im bestmöglich lösen bzw. bewältigen wollten. Die sind in den anderen Regionen auch aufgetaucht, aber und hier hat man über die Kontakte verfügt oder die Kontakte sind zu Stande gekommen. (Jakob, Bürgermeister)

I don't know, but I can imagine that we are a "global village", we know much more about each other now than in the last 100 years. [...] I think contacts were established when we realized that the problems that have occurred here (and that the initiators have solved) were also an issue in other regions. This is how the contacts were established. (Jakob, mayor)

105 See McLuhan, 1964.

And once the contacts are established, they can be more easily maintained thanks to the Internet. In fact, as Henrik (mayor) points out, the Internet "closes the ranks" and makes it easier to build connections between municipalities. Not only are the virtual contacts made possible by the Internet seen as one of the major reasons why networks were established, but the increased mobility of people is also considered a factor that opened up the horizons of the inhabitants and local authorities, as Maximilian indicates:

Genau weiss ich es auch nicht, aber ich glaube schon, dass es einerseits mit Mobilität etwas zu tun hat. Wir sind heutzutage wesentlich mobiler. [Damals war es so, wir hatten immer gesagt:] "Wir sind wir" und "ihr seid ihr". [...] Es war immer das Bewusstsein: "Wir müssen besser sein als die da". Jetzt [...] hat sich einfach eine Wanderung ergeben und man ist eigentlich einfach mehr bestrebt zusammen zu arbeiten, auch bei uns in der Region. Früher, vor 20 Jahren, hatte man den Bürgermeister in Raggal vielleicht ein Mal im Jahr getroffen, heute sehe ich ihn fast zwei Mal im Monat. (Maximilian, Bürgermeister)

I don't know exactly, but I think that this, in part, has to do with mobility, with the fact that we are much more mobile now. [Once upon a time it was not like that, we always said:] "We are we", and "you are you". [...] There was always the idea: "We have to be better than the other ones there." Now [...] change has happened, and more efforts are made to cooperate, even here in our region. Twenty years ago, one might have met the mayor of Raggal once a year; today, I see him twice a month. (Maximilian, mayor)

This quote by Maximilian clearly suggests that a new way of governing small Alpine municipalities appeared in the past two decades. If, 20 years ago, rivalry between neighboring municipalities was the usual way of (non-)communicating, now cooperation is the new way of working. Maximilian stresses this point:

Die Überlegung ist sicher, dass man sagt: "Wo gibt es Möglichkeiten? Wo gibt es die Synergien?". Und die Möglichkeit zur Zusammenarbeit wird heute überall gepredigt. Die Leute haben sich ein bisschen verändert, durch welche Einflüsse auch immer! Ich denke schon, dass Mobilität da einen wesentlichen Einfluss gehabt hat. Sicherlich, früher sind die Leute

We now think: "Where can we find new possibilities? Where can we find synergies?" And opportunities for cooperation are welcomed today. People changed a bit, although I don't know the reasons for that. Mobility, I think, played a role. For sure, people living in small places were much closer before and had contacts amongst themselves

in den kleineren Orten wesentlich mehr auf sich oder auf den Nachbarn angewiesen gewesen, das ist heute eigentlich nicht mehr so. Heute [...] versucht man mehr in größeren Kreisen zu denken. (Maximilian, Bürgermeister)

or with their neighbours; today it is not like that anymore, today [...] one tries to have contacts in bigger circles. (Maximilian, mayor)

The new opportunities offered by new technologies occurred while new social challenges, described by Leonardo as "identity crises", appeared in Alpine municipalities:

C'è stata un'evoluzione più legata allo sviluppo del territorio. E c'era anche questa crisi d'identità che i comuni sentivano, le persone più sensibili sentivano questi problemi d'identità, di perdita di valori e anche di perdita di territorio e hanno cercato di farsi forza uno con l'altro, quindi di mettersi anche in rete, di vedere cosa stavano facendo gli altri grazie a questi interscambi e da lì sono nate abbastanza spontaneamente queste reti. (Leonardo, ex stagista in una regione)

There has been an evolution linked to the development of the territory. There was also this "identity crisis" experienced by the municipalities. The most sensitive persons were feeling these problems of identity, of a loss of values and of territory. So, the municipalities tried to muster up the courage, sometimes by building networks, by looking at what other municipalities were doing. From there, these networks were born almost spontaneously. (Leonardo, former trainee in a region)

The networks thus offer the most active municipalities an opportunity to intensify links with other municipalities and, through that strategy, to face the identity crises brought on by the loss of values and of linkage with the territory, as Leonardo says. Leonardo and Maximilian see networks as a political solution to social problems that municipalities have had to tackle in the past decades.

Henrik has another perspective and stresses that networks are a social answer in a changing political context. In fact, as he underscores, thanks to the European Union's project, it became necessary to be more open, to see beyond the boundaries of the town or the valley. He admits as much in the interview when he simply says that "the European Union made people become more open". Thus, for Henrik, the new socio-political context prepared the ground for networking and can be regarded as one reason for the boom of networks. At the same time, networks can be considered political solutions and answers in a changing political context, as Niklas affirms. In his view, the European project made it necessary for small municipalities to find new strategies in order to be heard. Networking is thus one strategy that helps to increase the political weight of small municipalities:

Vielleicht durch diese Öffnung in Europa, die eigentlich in den 90er Jahren schon lange in vielen Köpfen war, hat man die Notwendigkeit gesehen [sich ebenfalls zu öffnen] weil ansonsten die Kleingemeinden eventuell stärker unter die Räder gekommen wären. Ich glaube, dass ist dieses politische Gewicht [von Netzwerken] im Hinterkopf da. (Niklas, Bürgermeister)

Maybe it has to do with this "opening" in Europe that was already in the mind of a lot of people in the 1990s and then became necessary. If not, the small municipalities would have gone to rack and ruin. I think this political weight [of networks] was already in the back of their minds. (Niklas, mayor)

The European Union did not just "socially and politically prepare the ground" for more cooperation in the form of networking, but it allocated specific funding instruments in order to reach this goal. This point is highlighted by Peter as follows:

Es hängt zusammen mit diesen Förderinstrumenten. Da hat man natürlich einfach Unterstützung bekommen und das war auch der politische Wille einer Kooperation, ein Netzwerk aufzubauen. Das waren am Anfang ganz stark nur auf die Laufzeit des Projektes bezogene Kooperationen. Aber es wird immer aktueller, dass man versucht Kooperationen zu entwickeln, die auch über das Ende der Projektlaufzeit hinaus wirken. Das ist hier [mit Alpine Pearls] ein typisches Beispiel. Man hat sich vorgenommen, eine Kooperation zu gründen, die eben über die Laufzeit des Projektes Alps Mobility II hinaus wirkt und aktiv ist, auch wenn das EU-Projekt längst abgeschlossen ist. (Peter, Präsident von Alpine Pearls)

It has to do with the funding instruments. We received support, and there was a political will for cooperation and for the establishment of networks. At the beginning, this cooperation was just limited to the duration of a project. But it became always more fashionable to develop partnerships, which have continued even after the project ended. Alpine Pearls is a typical example. We thought about establishing a partnership, which would be active and would last even after the Alps Mobility II project, even when the EU project has long been finished. (Peter, president of Alpine Pearls)

As Peter highlights, if funding instruments were mainly thought to allow partnerships in short-term projects, the actors involved in those projects tried to extend their existence by building institutions. The case of Alpine Pearls is emblematic of this evolution, since the now established network was born as two short-term projects called Alps Mobility. Many actors also emphasize this idea of development from projects to institutions as one of the main reasons for municipalities to join an already established network, as will be discussed in the next section, which will focus on local authorities' motivations for being part of a network.

6.2.2 *Reasons to join the network*

In the first phase of this research, a survey by questionnaires for the three networks (Alparc, Alliance in the Alps and Alpine Pearls) was carried out.[106] The question "What were the reasons for your municipality joining the network?" was asked only to the members of Alliance in the Alps and Alpine Pearls, but not to the members of Alparc, whose membership is compulsory and therefore not the result of a choice.[107]

For Alliance in the Alps, the answers show that the great majority of the members joined the network to promote sustainable development and to profit from the exchange of experience between Alpine municipalities (see fig. 16).[108]

106 The methodology and main results for the Alliance in the Alps case study can be found in a summary document called "Pan-alpine networks: The case of Alliance in the Alps" (Del Biaggio, 2007).

107 The questionnaire to Alpine Pearls was distributed and analysed by Nicolas De Groote. Since he received only nine answers, the results cannot be used in order to generalize tendencies.

108 Replies to the question "For what reasons did your commune join Alliance in the Alps?" The possible replies and answers' modalities ("important", "quite important", "not very important", "unimportant") were suggested by the authors; respondents had to cross the most appropriate answer. N = 89.

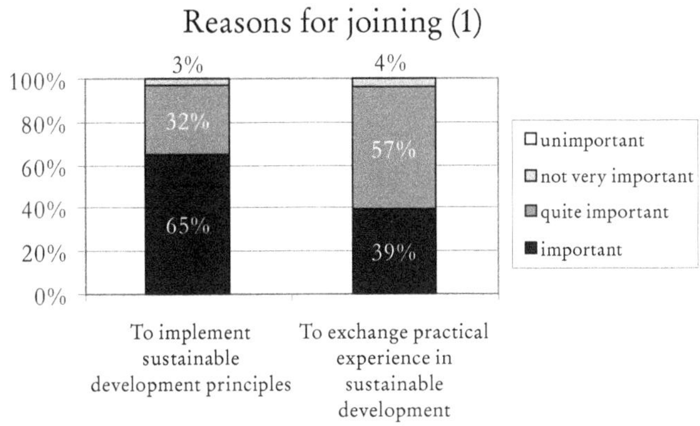

Fig. 16: Reasons linked to sustainable development for joining Alliance in the Alps.

These two kinds of motivations fit the original objectives of the association; those of Alliance in the Alps are therefore clearly identified and supported by the members. In comparison with the two main reasons that represent the spirit underlying the network, the more practical reasons for involvement are relatively less important (see fig. 17).

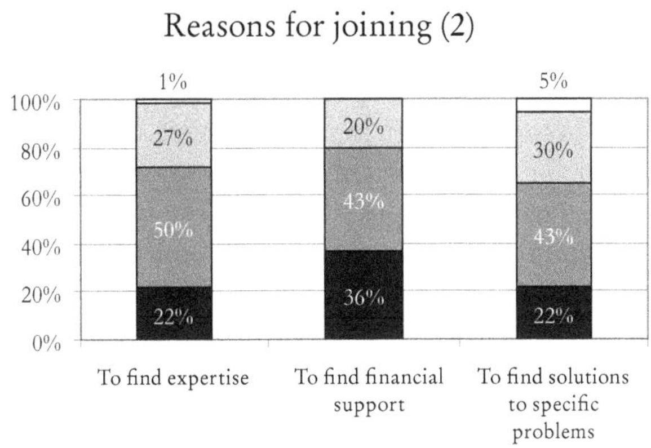

Fig. 17: Practical reasons for joining Alliance in the Alps.

It can therefore be said that Alliance in the Alps is not perceived mainly as a provider of practical assistance, e.g. in technical consulting or through financial support. Additionally, the members of Alliance in the Alps do not all feel equally concerned with the issue of building a pressure group directed at the Alpine and European states to promote sustainable development in the Alps. These statistical figures are also supported by the statements of interviewees whom Nicolas De Groote and I met during our meetings with representatives of Alliance in the Alps and Alpine Pearls.

In line with what was shown in the previous paragraph, for some interviewees the idea to adhere to a network constitutes a "natural path" for municipalities that were already active with projects of sustainable development. This is the case, for example, for the municipality of Berchtesgaden (Germany), which, before the decision to adhere to Alpine Pearls, had already developed a concept of soft mobility, as De Groote reports in his master's dissertation (De Groote, 2009). The choice to join a network committed to sustainable development was therefore seen as an opportunity to further develop a policy that was already anchored in the municipality. The network gives the municipality the possibility to commit to sustainable development in a more institutionalized context by adhering to an association expressly devoted to the spread and implementation of the idea of sustainable development.

Claudio provides more specific elements for joining a transnational network. As is explicitly written in the guiding principles of the network, one of Alliance in the Alps' main aims is the exchange of experiences between municipalities located inside the Alpine arc. It is what I call the "pragmatic dimension of the network". But, interestingly, the interview with Claudio (regional secretary) shows how this pragmatic dimension can be presented as a political argument in order to justify membership. The possibility not only of exchanging experiences but, thanks to the experiences collected via the network, of initiating new dynamics and new projects inside the region, has been a decisive argument for Claudio to join it. In fact, the information that the region can get from other members facing similar problems can be used to propose innovative ideas for mayors, who are, like Claudio, preoccupied with the day-to-day life of their municipalities:

Abbiamo subito visto che era interessante, anche perché erano dei progetti che si sviluppavano nell'arco alpino. Siccome noi avevamo un po' questa esigenza, soprattutto perché siamo una regione pedemontana, che è un po' tipica anche della zona del Friuli, dovevamo vedere che tipo di progetti si poteva eventualmente proporre, in forma nuova e un po' coinvolgente, per i politici locali. Oramai, soprattutto nelle zone a basso potenziale di sviluppo e finanziariamente deboli, per i politici la preoccupazione era quella di assicurare il quotidiano, senza una visione di gestione sia del patrimonio forestale, che da noi è importante, sia del sostegno all'attività del settore primario. E questo anche in un'ottica di cercare di sviluppare il turismo, il turismo di qualità. Da lì abbiamo fatto delle riflessioni, abbiamo ritenuto che era importante come regione partecipare ad Alleanza nelle Alpi. (Claudio, segretario regionale)

We saw from the beginning that it was interesting, also because the network develops projects concerning the entire Alpine arc. Since we had this need, especially because we are a piedmont region, which is a bit typical as well of the Friuli region, we had to see which kinds of projects we could propose, in a new and involving way, to our local politicians. Subsequently, especially in financially weak areas with low development potential, politicians' major concern was to ensure the daily running – with neither a vision for the management of the forest heritage, which is important here, nor the support of the activity of the primary sector. And this was with the idea to develop the tourist sector, quality tourism. So, those were our reflections, and we considered that it was important, as a region, to participate in Alliance in the Alps. (Claudio, regional secretary)

One of the most complete answers to the question "Why do municipalities join a network?" was given by an animator of Alliance in the Alps, Robert, to Marie-Anne Guérin. In his long answer to this question, Robert provides a detailed list of benefits that a municipality receives upon entering a transnational network. These points will be analysed in the following section as part of a discussion of the different dimensions of the network. Robert's response will therefore be split into different parts, which will be explained in different sections.

For the moment, it is interesting to note the main reasons he cites while answering the specific question of the reasons members of Alliance in the Alps mention for joining the network. Robert highlights that he cannot give a single answer for the entire Alpine arc, since the reasons to join the network differ from country to country. Nevertheless, he underscores the importance of the presence of some "dynamic persons" in the municipalities, as he calls them; people who are curious and willing to search for new possibilities. Robert seems to suggest that without the presence of these people, it would have been impossible for some municipalities to come up with the idea of joining a transnational network:

Je pense que c'est très variable, ça dépend des pays, je ne peux pas juger à large échelle, mais je dirais pour les communes que je connais, c'est d'abord dépendant d'une ou deux personnes dynamiques qui ce sont dit: "Là il y a un truc qui m'intéresse", qui sont curieuses. (Robert, animateur)	I think this really varies; it depends on the country. I cannot judge on a large scale, but I would say that, for the municipalities I know, it first of all depends on one or two dynamic persons who say: "There is something interesting there." They are curious. (Robert, animator)

In his study of Alpine Pearls, De Groote also highlights the importance of the presence of particularly committed individuals when it comes to joining the network. In his master's dissertation (De Groote, 2009), he states that Neukirchen am Grossvenediger's (Germany) membership in Alpine Pearls was made possible because its mayor strongly believes and acts in favour of sustainable development and therefore did not hesitate to join the network when the possibility arose.

Alliance in the Alps' structure is designed to "capture" such particularly dynamic individuals. Indeed, the animator's role is to search for new members and convince potential interested mayors to join the network. This system apparently worked quite well for the Lugano Valley Region (Switzerland), which became a member of Alliance in the Alps thanks to its contact with the animator of the network. The task of the animator in detecting, contacting and convincing municipalities to join the network seems quite central. Olivier, who at the time of the interview with Marie-Anne Guérin was the animator for Alliance in the Alps, explains the strategy and arguments he uses to convince local authorities to join Alliance in the Alps. This gives us an idea of the reasons why it is worthwhile for a municipality to become a member. In the interview, he lists three of the reasons: the possibility to get useful information from the network; the opportunity to benefit from the counsel of a coach (the animator); the prospects of receiving funding for projects:

Cela leur apporte une base d'informations et de données qui leur permet, dans leurs initiatives, d'avoir des supports d'analyse, une méthodologie, des "astuces". Par exemple, un maire qui a envie de lancer un projet de mobilité douce dans sa commune, dans tout l'arc alpin sur les communes [membres d'Alliance dans les Alpes], tu en as peut-être une dizaine qui ont lancé ce type de projets. [Le maire] aura à sa disposition une dizaine de fiches synthétiques sur ces projets qui ont été menés dans les Alpes, sur les méthodes qui ont été employées, les résultats qu'il y a eus, les déboires aussi, les problèmes que les communes ont rencontrés. Disons tout un corpus d'expériences qui permettra à ce maire qui a envie de faire quelque chose de se dire: "Tiens, ça c'est bien". Et en fonction des critères politiques, environnementaux, sociaux, économiques de sa commune, il pourra jongler avec tous ces éléments et développer au mieux son projet. Je leur dis: "Qu'est-ce que cela vous apporte?", je leur dis: "Vous avez envie d'engager un projet de développement durable dans votre commune? Vous ne partez pas de zéro, vous n'allez pas réinventer la roue, comme on dit, et ça ira beaucoup plus vite, ça sera beaucoup plus facile, vous aurez à votre disposition des personnes ressources, vous pourrez appeler les maires". Cela c'est une première chose.

This brings them a base of data and information that allows them to have, for their initiatives, support for analysis, a methodology, some "tricks". For example, a mayor who would like to launch a project of slow mobility in his municipality knows that in the entire Alpine arc, among the municipalities [members of Alliance in the Alps], probably around 10 of them have already done this kind of project. [The mayor] will therefore have at his disposal about 10 fact sheets concerning those projects conducted in the Alps, in which the methods employed, results and problems encountered are explained. There is the possibility to consult a knowledge database, which will allow the mayor to say: "Look, this is good." And contingent on the political, environmental, social and economic criteria of his municipality, he will be able to juggle all these elements and develop his project in the best way. I ask them: "What does it bring to you?" I answer: "Do you want to start a sustainable development project in your municipality? You don't start from nothing, you do not have to reinvent the wheel, as we say, and it will be much faster, it will be much easier, you will have at your disposal resource people, you will be able to call the mayors." This is the first thing.

Une seconde chose: on leur explique que je suis là pour les supporter dans leur projet. Nous sommes censés répondre aux besoins de ces communes, en terme de méthodologie de projet, de réflexion. [...]

A second thing: We explain that I am here for them, to support their projects. We are supposed to answer their needs, in terms of a project's methodology, in terms of reflection. [...]

Troisième point, cette année ils peuvent bénéficier d'un financement suite à l'appel à projet Dynalp2 à hauteur de 20'000 euros. (Olivier, animateur)

Thirdly, this year the municipalities can benefit from a funding of up to 20000 euros for projects inside the Dynalp2 programme. (Olivier, animator)

This last point that Olivier raised, namely the possibility of obtaining funding to implement projects, has to be analysed in more detail. A few paragraphs further up, I highlighted the fact that Claudio appreciates Alliance in the Alps for the possibility it gives, among others, to stimulate local politicians with new impulses and ideas coming from all over the Alps. However, later in the interview, Claudio admits that actually they, as a region, decided to join the network primarily because of the possibility of getting funds from the Dynalp project. This admission shows the importance of "money" in the decision to join the network. This is confirmed by statistical features. Let us review the graphic (see fig. 18) that I produced for the seminar "Mountain networks", held in Geneva on the 28th of November 2007 and organized by our department's research group.

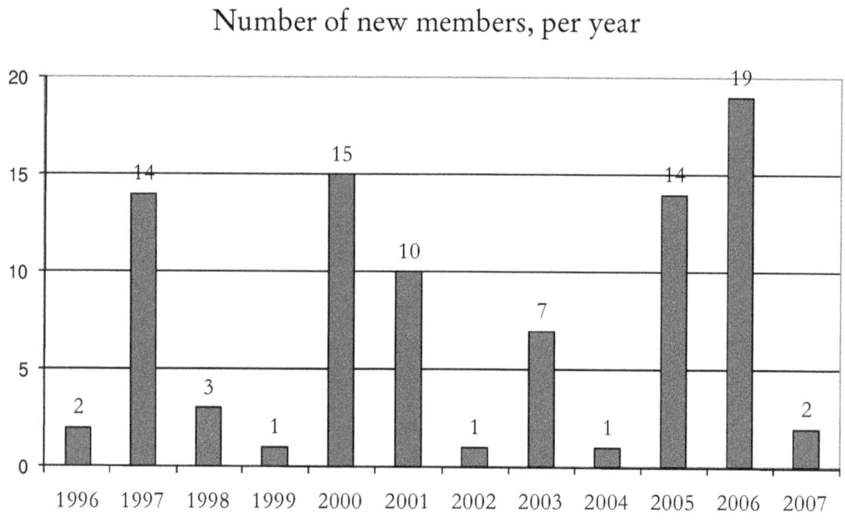

Fig. 18: Year of membership, Alliance in the Alps.

Fig. 18 illustrates the year of membership of the municipalities of Alliance of the Alps that answered our questionnaire. It shows that the municipalities have joined the network in successive waves, in particular after the creation of the association (1997) and in the years 2000–2001 and 2005–2006. The reasons for the two peaks were discussed by Claire Simon, who at the time of the meeting was assistant director of CIPRA International and coordinator for Alliance in the Alps,[109] during the roundtable organized in the research seminar.

During her intervention, Claire first underscored the fact that from the statistical features I highlighted it appears that the practical aspects (as well as the financial aspects) are secondary reasons to join the network (see fig. 16 and fig. 17). Yet, as Claire remarked, not without a bit of regret, this statement seems to be contradictory, since the two waves correspond exactly to the years when Alliance in the Alps was preparing a call for allocating money for a programme called Dynalp. When this occurred with Dynalp 1 (2003–2006) and Dynalp 2 (2006–2009) the network gained new members that could benefit from the allocated funds. Indeed, it seems that "money" is one of the major factors that attract municipalities to join the network. It is also one of the most important elements for refusing membership, if those financial elements are not clearly explained to potential members, as will be discussed later.

Nevertheless, an interesting point is raised by Sabine, animator for Alliance in the Alps, who observes a recent evolution that is worthy of note. The role of animators, as already explained, is also to meet new potential members and to discuss with them the aims and conditions for being members of the association. During one of the interviews she conducted with them, Sabine could experience new arguments put forward by local authorities for joining the network. On that occasion, she realized that for the first time the arguments were not linked to funds they could receive by being members but to the possibility of belonging to a "club of sustainable municipalities" and raising the profile of the image those municipalities want to show the exterior world:

Sabine (Betreuerin): Was hat sich geändert? Es ist ganz interessant, dass ich vor ein paar Tagen von der Region Bregenzerwald [Österreich] erfahren habe, wo ein paar Gemeinden schon Mitglied sind. Aber jetzt überlegt ist die Region von sich aus, ob sie der Allianz beitreten

Sabine (animator): What changed? It is quite interesting, what I experienced a few days ago with the Bregenzerwald Region [Austria], where some municipalities are already members. Now the entire region is thinking of joining Alliance, which means some 15–16 additional municipal-

109 Claire Simon is now director of CIPRA International.

möchte. Das wären dann zusätzlich um die 15–16 Gemeinden, ohne dass sie sich vorgestellt hätten Dass immer mehr Gemeinden von sich aus an das Netzwerk herantreten, das heisst, dass das langsam greift: "Da gibt es etwas, was uns unterstützen würde", unabhängig von Dynalp II.

Cristina: Unabhängig vom Geld?

Sabine: Unabhängig vom Geld, sondern wirklich: "Könnte uns das was bringen?". [...] Wegen des Austauschs, wegen des Images, dazu zu gehören in diesen Kreis der nachhaltigen Gemeinden, auch genommen zu werden.

ities, and this without anybody meeting them to ask whether they want to become members. The fact that more and more municipalities are deciding spontaneously to join the network means that slowly they grasp the idea that "there is something that could help us", and this independently from Dynalp 2.

Cristina: So, independently from money.

Sabine: Independently from money, but really: "Can this bring us something?" [...] Because of the exchanges and the image and in order to belong to this circle of sustainable municipalities.

I found the same idea in only one other interview. Antonio explains how excursions organized by the network have been, for him, *stimuli* not to think only in terms of funding but also in terms of an ideal, i. e. the implementation of the objectives of the Alpine Convention:

Quando uno tornava [da escursioni organizzate dalla rete] con delle idee positive si è cominciato proprio a lavorare per cercare di costruire dei progetti nel comune che fossero non tanto legati ad un finanziamento, ma ad un obiettivo. L'obiettivo era quello di costruire le cose secondo le finalità della Convenzione delle Alpi. (Antonio, sindaco)

When one returns [from excursions organized by the network] with positive ideas, one starts working towards building some projects in the municipality. Those are not linked to funding but to an objective. The objective is to build "things" in accordance with the aims of the Alpine Convention. (Antonio, mayor)

If what Sabine and Antonio have experienced is a phenomenon that certainly cannot be generalized for the entire Alpine arc, it remains an interesting evolution that will maybe grow in the next few years. Sabine's observation, along with the words of Antonio (who underlines that the network allows local authorities to reach an ideal, namely the implementation of the Alpine Convention, and not only to get funds for concrete projects), is probably to be seen as a further step, more political and ideological, in the networking process. However, the capacity of members to take this further step still seems to be a prerogative of only a few members. It is probably not a coincidence that Antonio is not a "simple mayor" but one of the initiators and leaders of the network. Antonio was already convinced, from the beginning, that the network represented a great opportunity for his municipality. When I met him after 10 years of networking experience, his municipality had already benefited from the economic support for several projects, so that his aims evolved in a more political direction, i.e. the implementation of the Alpine Convention. It will be interesting to research further in order to find out whether this trend is unique to a few members or denotes a larger phenomenon.

6.2.3 Reasons not to join the network

During the research process, I decided that it would be interesting to expand my interviews and include mayors or regional secretaries who knew of the existence of Alliance in the Alps via the animator but decided not to join it. I chose to interview some municipalities and regions from the Ticino Canton (Switzerland), since I know that Fabio, the animator of Alliance in the Alps for the Canton, has regular contacts with them and tried to convince them to enter the network.

On the 6th of June 2007 I interviewed Claudio (regional secretary). After the meeting, he asked me whether I wanted to accompany him to a local event he was invited to that could be interesting for me. A family in the region presented a local liqueur made of walnuts (*nocino*) to representatives from Bern (Switzerland) in order to try to get the "Geographical indication of origin" label. By sheer coincidence, at that occasion I met Marta and Giorgia, both from a region that could potentially join Alliance in the Alps. However, the region had never taken the step, even though it had received a Dynalp grant for a project it subsequently implemented, along with another region that was a member of the network. Actually, from the words of Fabio, by applying together for the Dynalp funds the "tacit agreement" between the two regions was: "If we obtain the project, then you join the network." The project was funded, but the region never joined Alliance in the Alps. Fabio recalls this "tacit contract" as follows:

Il mio obiettivo principale, poi fallito, di Dynalp era che [la Regione con cui avevamo fatto il progetto] entrasse almeno nella rete [Alleanza nelle Alpi]. Abbiamo fatto il Dynalp assieme, convinti, non convinti, perché "costa". E gli ho detto "ve lo paghiamo noi per i primi anni. Quello che prendo dal coordinamento ve lo do per entrare", tanto sono pochi soldi. E dopo lì si sono persi un po' loro, con i loro interessi. (Fabio, animatore)

The first goal of Dynalp, which failed, was for [the region with which we did the project] to join Alliance in the Alps. We did Dynalp together; at first, they were convinced, then they were not convinced anymore, because "it costs". I told them "I will pay your fees for the first years; the money I receive for the animation, I will give it to you", which is not much money in any case. And then they lost themselves with their own interests. (Fabio, animator)

Marta, the coordinator of the region mentioned by Fabio, whom I first met separately, told me that she would have agreed to join Alliance, but in order to succeed, it would have been necessary to give a document to Giorgia, the president of the region, that would list, in writing, what the region would gain, in terms of economic advantages, from a membership. Later, when I was able to meet the two representatives of the region together in a local bar, Giorgia confirmed what Marta had already told me: The membership fee is, in her view, too expensive, and she does not know what the region gets in return.

In my field notes, I wrote: "The financial reason seems to be really the only criterion taken into consideration by the region for not joining." Nevertheless, not truly convinced by this first response, as Fabio's statement already suggests in the last sentence ("they lost themselves with their own interests"), I decided to ask them for an interview, which took place two years later in November 2009. The interview brought out a much more nuanced answer that went beyond the simple economic factor. This interview constitutes the central thread of this section, which will be enriched by other interviews.

The amount of the fee does not constitute the only reason for not joining the network. This question was first mentioned by Marta and Giorgia at the beginning of the interview in 2009. In fact, Marta stresses that the yearly fee of 4000 CHF that the region would spend to become member of Alliance of the Alps is a great amount of money, considering their total budget of 200000 CHF. Given this situation, the question they asked themselves before taking the decision to join was: "What do I get out of that?" This worry about the "return on investment" seems to be shared by other people. Maximilian, now a member of Alliance in the Alps via a regional membership, admits that, at the beginning, he was not persuaded by the decision to join Alliance in the Alps:

Ich habe für mich gedacht, "Irgendwo zahlt man einen Mitgliedsbeitrag und weiß eigentlich auch nicht genau was das bringt und was das nützt". (Maximilian, Bürgermeister)

I thought to myself: "We pay fees without actually knowing what this will bring and whether it is somehow useful" (Maximilian, mayor)

In order to avoid just talking about the cost of membership and to hear some other arguments, I started the discussion with Marta and Giorgia by asking them: "Why would it be interesting to join the network, if it were free?" Giorgia answered my question as follows:

Il lato positivo è la conoscenza di altre persone che hanno problemi simili. La possibilità di scambiarsi le esperienze. Questo sicuramente. Ed effettivamente *al dì d'incö nüm sempru tütt quel che em fai l'è propi sempru da cercàà da mia guardà dumà l'internu, ma da cercàà anca da vedéé un pu quel che gheva in giir. Questo chi senz'altru, la [entrare nell'Alleanza] pudeva vess un'ottima roba.*[110] (Giorgia, presidente di una regione)

The positive side would be the possibility to meet other persons who have similar problems. The possibility to exchange experiences. This, of course. And actually everything we did in the region we did to try to look at not only what is going on inside the region, but also to see what others were doing. And for that, of course, [joining Alliance] would have been an excellent idea. (Giorgia, president of a region)

Marta agrees on the opportunities that Alliance would provide them in terms of exchange, but she points out that she still does not understand what the region would gain from it, besides the chance to "exchange experiences, go somewhere and see what they do". Although there is an awareness that being in an Alpine alliance would increase the possibilities of exchanges over a bigger territory, the region is already engaged in such activities, even outside the network.

110 During the interview, Giorgia often spoke the local dialect. As with every dialect, it is a spoken language and not a written one. Therefore I tried to transcribe it in an intuitive and phonetic way, without any fixed rules.

The main point is thus the lack of understanding of the direct utility of membership. This concern seems to be shared by other potential members, which decided not to join the network, as Enrico underscores:

Probabilment nüm a sem mia denta perché chéla sensazion lì che la poda vess ütil par nüm la vegn mia föra Probabilment ma schbai, ma se al Fabio al ma diseresa: "Per nüm l'è schtai molto util par cheschto" magari podresi pensac sü. Lü l'a sempro pasa al mesagg da cosa i fa, ma [mai] "l'è ütil per...". (Enrico, segretario regionale)

We are probably not in the network because we could not grasp the idea of how a membership would be useful for us. Most probably, I am wrong, but if Fabio had told us: "For my region it has been very useful for that", maybe we could have thought about it. He always told us what they are doing, but [never] "it is useful for..." (Enrico, regional secretary)

Indirectly, there is an equal measure of criticism directed at the regional animator, who failed to "sell his product Alliance in the Alps" by concretely demonstrating the usefulness of a transnational network. Thus, there is a problem of measuring the advantages that a transnational alliance can bring, but also a more technical problem, namely the approach to and arguments put forward to potential members in order to convince them. Additionally, while probing more in depth during the interview, it became clear that non-membership is often related to internal political problems. As Marta avows:

Noi, davanti ai nostri [sindaci], non avremmo avuto nessuna possibilità di giustificarci, perché i nostri comuni non hanno nessuna capacità di guardare oltre il loro confine comunale. [...] I nostri comuni, che non capiscono cos'è la regione, capirebbero ancora meno cos'è un'Alleanza nelle Alpi. (Marta, segretaria regionale)

We, in front of our [mayors], would not have had the means to justify the membership, because our municipalities do not have any capacity to look beyond their own boundaries. [...] Our local authorities do not even understand what the region is; they would understand even less what Alliance in the Alps is. (Marta, regional secretary)

Although with a less critical tone, Enrico, regional secretary, also underlines the lack of openness of the municipalities composing the region. The latter, he admits, already have difficulties understanding the utility of working with some national institutions created to support the Swiss regions, such as the Swiss Association for the support of regions and municipalities (SEREC) or the Swiss group for mountain regions (SAB), so that it would be even more difficult to convince them to join a transnational alliance. This "closed vision" makes Marta's work quite difficult; she admits the only way to convince their mayors is by citing the economic factor:

Io faccio passare i progetti in Consiglio regionale perché posso dire: *A schpendum pocc e ga guadagnum.* Non posso fare altro, non riesco a fare altro. (Marta, segretaria regionale)	I can make the Regional Council accept some projects only if I tell them: "We spend a bit of money, and we will earn a lot." I cannot do anything else; I cannot manage to do more. (Marta, regional secretary)

Furthermore, for some persons, not even the financial incentives are a strong enough motivation to join the network. During the interview with Claudio, it became clear that even the possibility of receiving money to implement what he considers to be "small" projects would not be enough to convince his sceptical audience. I mentioned to him the example of the Malcantone region (Switzerland), which received 10000 euros for a project called "Sustainable use of woods – Study on the ecological value of managed and abandoned chestnut woods", which focuses on the ecology of a species of bats living in the chestnut woods of the region. In reaction to my statement, Enrico first of all replied by saying this is uncommon ("The fact that a member receives funds probably happens just once"). What is likely more important here, however, is that Enrico's words reveal the argument that mayors would not consider a "small" project on "insignificant" animals such as bats a serious motivation to join the network. His words seem to suggest that the "real" worries for mayors lay somewhere else:

Però che ricevi indrè di soldi, l'è una roba che süced una volta, pensi. [...] Mi mal vedi, se disi [ai sindaci]: "Devon metic 2000 franc perché doman a riesum a veg di finanziament per i pipistrei"... A ghè mia in cheschto mument la capacità da ciapà sul serio una roba dal gener. (Enrico, segretario regionale)	But I think that it will happen just once, that I will get money back. [...] I can see it already, if I say [to the mayors]: "We have to spend 2000 francs, because in the future we will get funds for studying the bats"... There is not, at the moment, the capacity to take this kind of thing seriously. (Enrico, regional secretary)

Those quotes clearly reveal a disparity of vision between the local and the regional administrators. The latter, somewhat jokingly, but not without some factual basis, answer as follows to my question regarding their will to profit from experiences that other municipalities and regions developed: "How far would you have to look for other experiences?":

Giorgia (presidente di una regione): *Nüm gà ném mia da cunfin!*	Giorgia (president of a region): We do not have boundaries!
Marta (segretaria regionale): Fino al Giappone...	Marta (regional secretary): All the way to Japan...

While they are ready to get experience "all the way to Japan", the municipalities composing the region seem to be unable to see what is going on just outside their towns. This gap between the local politicians and the promoters of the region is equally present when Marta speaks about the (closed) mentality of the inhabitants. Here, she opposes the differing visions of the mountain between what she calls the Germanic and Lombard cultures:

Perché la visione della montagna di quelli "d'in dentro", dei tedeschi, germanofoni è molto più bella, è molto più piena d'identità che non la nostra lombarda. La gente non ha un'identità che crede della montagna. La nostra valle si è ripopolata perché costa meno mettere a posto la casa del papà che pagare un affitto a Chiasso, e la qualità di vita è diversa, ma il ripopolamento è un dato di questo tipo. Non perché ci credono. E non c'è più nessuna relazione con il territorio. [...] C'è un problema di identità, c'è un problema ad andare in avanti. (Marta, segretaria regionale)	Because "on the other side of the Gotthard Pass", the vision of the mountain of the German-speaking is much nicer, much fuller of identity, than ours, the Lombard. People here do not have an identity that grows from the mountain. People come to our valley, because it costs less to renovate Daddy's house than to pay rent in Chiasso, and the quality of life is different, but this trend is due to those kinds of factors, not because they believe in it. And there is no relation to the territory. [...] There is a problem of identity, a problem of moving forwards. (Marta, regional secretary)

Marta suggests that this lack of "real contact" with the territory makes it impossible for the local population to understand the goodness of projects connected to sustainable development. This means, implicitly, that they would not understand the choice of their municipality to eventually join a network whose main objective is precisely the implementation of sustainable development. She uses the concept of "identity" to justify their understanding. It is what I called "identification": The people in her valley do not identify with the mountain, and so they would not support the region joining a network that clearly identifies with the Alps.

Leonardo presents another problem of "mentality" or "identity", one that ultimately leads to the same result of non-involvement in the network that Marta highlights. What Leonardo portrays is more a problem of "negative parochialism", which he identifies as being typical of the people living in the Ticino Canton. In this extract, Leonardo responds to my astonishment about the fact that the members of Alliance in the Alps in Ticino are situated in the southern part of the canton, considered to be pre-Alpine, while no members can be found in the northern part of Ticino, located in a high mountain setting:

Secondo me in Ticino siamo molto ignoranti in queste cose, e secondo me se le regioni, appunto, quelle prettamente alpine, non sono membri di Alleanza è proprio perché si è fatto un discorso che è venuto forse dal Sottoceneri.[111] Secondo me se veniva un discorso più dagli altri cantoni limitrofi, dal Canton Uri per esempio, a presentare con dei comuni della Svizzera interna forse era più facile. (Leonardo, stagista in una regione, membro di Alleanza nelle Alpi)	In my view, in Ticino we are really ignorant of these kinds of things, and I think that if the "real" Alpine regions are not members of Alliance it is really because the idea came from the Sottoceneri.[112] In my opinion, if the idea had come from the other neighbouring cantons, from the Uri Canton, for example, or from other German-speaking cantons, maybe it would have been easier. (Leonardo, trainee in a region member of Alliance in the Alps)

In addition to this parochialism bred by a distinction between "Sopra-" and "Sotto-" Ceneri, which is the same as the distinction between "mountaineers" and "city dwellers", or between "rich" and "poor", another peculiarity that emerged from the interviews and was well expressed by Leonardo and Enrico is the non-identification by the inhabitants and local political actors from Ticino with the

111 Il Sottoceneri è la parte sud del Ticino, che si trova a Sud del Monte Ceneri.

112 This is how the southern part of Ticino, south from the Monte Ceneri, is usually called.

Alps, as well as an obvious lack overall of what they call an "Alpine culture". This recalls what Marta said about the "identity problems" she observes in the valley where she lives and works. Enrico explicitly advances this argument to suggest that without a strong identification with the Alps, there is not the capacity to "look elsewhere", i.e. to look at and work with other Alpine municipalities, since local political actors and the broader population do not consider the "Alps" argument to be a strong identifier:

In effetti secondo me il Ticino fa un po' fatica a capire che fa parte delle Alpi, a differenza di altri cantoni come il Vallese o Uri o Grigioni, che hanno un'identità chiara alpina. (Leonardo, stagista in una regione, membro di Alleanza nelle Alpi).

In fact, in my opinion, the Ticino Canton has difficulties understanding that it is part of the Alps, unlike other cantons such as the Valais or Uri or the Grisons, which have a clear Alpine identity. (Leonardo, trainee in a region, member of Alliance in the Alps).

A ghem di gross diferenz tra chel che l'è una nazion dove ghè amò una cültüra alpina e dove invece ghè mia. Se mi pensi a Trento, Innsbruck, Grenoble, ma vegn sübit l'idea da città che vivono nelle Alpi, fiere di essere nelle Alpi, con una cultura delle Alpi, perché i ga una massa critica. I nos città (Locarno, Bellinzona, Lügan) i sa sent pü fieri da vess nelle Alpi o il centro di una cultura alpina. Per cui una specie da sub-cultura [alpina] la reschta, però l'è sconfitta, la ga mia un futuro. [...] E dove una schtrütura dal gener l'è debola, è sconfitta, è ovvio che la ga mia i forz e la capacità da guardaa altrove. (Enrico, segretario regionale)

There are big differences between nations where there is still an Alpine culture and where there is none. If I think about Trento, Innsbruck, Grenoble, I think about cities that live in the Alps, proud to be in the Alps, with a culture of the Alps, because they have a critical mass. Our cities (Locarno, Bellinzona, Lugano) are no longer proud to be part of the Alps or in the centre of an Alpine culture. There is still a sort of [Alpine] sub-culture, but it is defeated; it does not have any future. [...] And where such a structure is weak, defeated, obviously it does not have the strength and the capacity to look elsewhere. (Enrico, regional secretary)

Silvia, an employee of a municipality that could potentially become a member of Alliance in the Alps and/or Alpine Pearls, expresses in concrete terms how this general attitude of non-identification with the Alps works when a municipality considers the value or not of joining an Alpine network. Silvia also explains what she thought when she heard of Alpine Pearls for the first time. Since the image of what an "Alpine town" does not fit her own perception of her municipality, in her eyes it would make no sense to join the network:

Lei ha posto questa domanda: "Perché non aderiamo?". Io ho l'impressione che sono delle realtà diverse. L'ho visto per caso, non lo so se c'è una specie di associazione delle Perle Alpine, e io ho visto che erano dei centri, dove io ho avuto l'impressione che fossero prettamente legati all'alpinismo, con un'economia proprio legata al territorio alpino. E' per quello che mi sono detta, cioè, forse come idea è sbagliata, però, proprio territorialmente parlando mi sembrava qualcosa di veramente molto diversa. Per me le Alpi [sono] l'immagine della cartolina della Svizzera, del turismo. (Silvia, impiegata in un comune non membro di Alleanza nelle Alpi)

You asked this question: "Why do we not join the network?" I have the impression that these are different realities. I saw by chance that there is a sort of association called Alpine Pearls, and I had the impression that the members are towns that are really linked to alpinism; they have an economy that is strongly linked to the Alpine territory. For that reason I thought, maybe it was the wrong idea, but territorially speaking I had the feeling that it was something really different. For me, the Alps [are] the image of the Swiss postcard, of tourism. (Silvia, employee in a municipality that is not a member of Alliance in the Alps)

Obviously, the image she has of the Alps does not correspond to the one she has of her own municipality. Yet, even when local political actors identify with the Alpine territory, it is not enough of a condition to want to join a pan-Alpine network, as Giorgia points out by saying that the "Alpine characteristic of the network" is certainly interesting, but not a sufficient argument:

La caratteristica di "Alpi" [di Alleanza] la ma va bén, però l'è mia asé par mia pensà da guardà anca mò oltre le Alpi. Però al podares vess un bel tassell, anca parché nüm sa ritegnum centrai in di Alp, anca se semm in bass al Tisin, quasi incüneà nala Pianura Padana, però sem sempru denta in di Alp. Quant che andava a schöla i ma disevan "nelle Prealpi". (Giorgia, Presidente di una regione)

The "Alps" characteristic [of Alliance in the Alps] is fine for me, but this is not enough. We should look even beyond the Alps. But this could already be a good start, because we think we are central in the Alps, even though we are situated in the southern part of Ticino, almost inside the Po Valley, but we are inside the Alpine region. When I went to school, I was told [we were] "in the pre-Alps". (Giorgia, president of a region)

In that sense, and this reflects the discussion in chapter 5 about identification with the Alpine region, the Alps as such are not a sufficient argument to decide to join the network, nor is non-identification a deterrent to making that choice. This was largely explained in the discussion about the municipalities of Mäder and Mendrisio. In addition, further practical arguments are provided by Marta (regional secretary), as well as by Enrico, in their decision not to join Alliance in the Alps. Obviously, there is a problem in terms of the time and resources that are available. In order to show me this is a real problem, Marta explained that she is working 40 % for the region, while Giorgia is doing voluntary work. So, if Marta decides to participate in a two-day seminar, not including two days of travelling (one for going and one for coming back), she would "lose" two weeks of work that she is supposed to do for the region. Marta asserts that if she worked 100 %, things could be different. She would even take Giorgia on the excursions. In addition to the "pure" calculation of time, Marta says there is also a question of the themes discussed during the meetings. In fact, as she points out: "If they discuss an argument that really interests me, I will go; if not, I will not. The problem is that you have to choose amongst 50000 offers. So, what should I do?" She solved this dilemma by deciding not to join the network, because she found the membership too demanding for both her and the region.

That all agree that, if time were an extendable variable, it would be worthwhile to do the tours in the Alps, as it would enable them to meet new people and develop new ideas. Since time is a scarce resource, regional secretaries have to choose between different offers, as Marta indicates. Moreover, and specifically in Marta and Giorgia's case, there is the memory of a complicated contact experienced with Alliance in the Alps:

Noi abbiamo fatto il Dynalp, che è stato estremamente contorto. Fortunatamente non l'abbiamo gestito noi, l'ha gestito Fabio. Però il problema è che Alleanza nelle Alpi è una struttura molto macchinosa secondo me. E lontana da noi, perciò ancora meno facile avere i contatti. (Marta, segretaria regionale)	We did the Dynalp project, which ended up extremely twisted. Thankfully we were not responsible for its management. Fabio managed it. But the problem is that Alliance in the Alps is a complex structure, in my opinion. And it is far away from us, so that it is even more difficult to maintain contacts. (Marta, regional secretary)

This difficult first and last encounter with the network likely influenced her opinion about its general way of working. More generally, and this is probably valid for the entire Ticino Canton, Marta underscores Alliance in the Alps' lack of visibility and of political weight. This argument leads the potential members to wonder why they should invest so much money and time in something that does not produce many benefits. Thus, Alliance in the Alps does not represent a politically strong entity that would justify their region's adherence.

[Il problema è anche] la visibilità di Alleanza nelle Alpi in Ticino. Non ha potere contrattuale. Se Alleanza fosse qualcosa che ci servirebbe per andare in avanti, lo prenderemmo, però non ha potere in Ticino. (Marta, segretaria regionale)

[The problem is also] the visibility of Alliance in the Alps in Ticino. It does not have any contractual power. If Alliance were useful in propelling us forwards, we would take it, but it has no power in Ticino. (Marta, regional secretary)

In order to understand the network's political weight and visibility at a higher administrative level, I called the Ticino cantonal authorities and tried to get in touch with people likely to tell me something about Alliance in the Alps. The attempts ended up in a cul-de-sac: Every time I tried to speak about "Alliance in the Alps", I was wrongly corrected by being told that I did not mean to say Alliance in the Alps but rather the Alpine Initiative.[113] This shows the lack of understanding and knowledge about the network. Nevertheless, I decided to meet somebody from the cantonal office for economic development to see whether I could get more information.

This interview took place on the 29th of June 2007, when I interviewed Tommaso. During the discussion, it was clear to me that there existed a possibly insurmountable divergence not only between his vision and that of local political actors such Marta, but also between his and that of Fabio and Claudio. Two examples should be sufficient to illustrate this divergence. Tommaso related to me a couple of "good examples to follow" in the field of regional development; one of them is the decision by Hugo Boss to establish a factory in Coldrerio, which, to quote Tommaso, "brings added value and working places". On the other hand, he criticized the way local newspapers glorified the inauguration of small infrastructure that was built in a remote valley. During the entire interview, Tommaso had only words of praise for the cities and accusatory comments for the peripheries

113 <www.alpeninitiative.ch>, consulted on the 19.04.2015.

and valleys while avoiding the feeling of frustration resulting from his office wanting to promote dynamic projects, but the context in which he has to practice is generally static.

This story is important because Marta and Giorgia openly addressed their criticisms to the canton, and specifically to Tommaso. They criticize the canton because its administrators do not prioritize "small" territorial projects and are not willing to support (financially as well as morally) their sustainable development projects. In that sense, and given this context, Marta says it would not be of any strategic use to join the network in order to influence the cantonal territorial politics. Although others, such as Antonio, used the "Alliance of the Alps strategy" to counter the non-recognition of their local politics by the administrative units, as explained in Section 5.2, Marta and Giorgia, who are aware of the little influence the network has in their local reality, had to play another card and did not participate in an international network:

Allora io ho messo in piedi un discorso di promozione turistica [sostenibile] di un certo tipo e sono andata a prendermi un professore dell'Istituto del Turismo dell'Università di Losanna,[114] perché ha un Curriculum Vitae lungo così. Così quando dovrò andare a discutere con i diversi Direttori dell'Ente Turistico, che non hanno nessuna capacità professionale di fare queste cose, perché loro al massimo distribuiscono i volantini, io sono andata a prendere il Big Boss e presento il mio concetto di promozione turistica con lui. (Marta, segretaria regionale)

So, I tried to hold a discourse of [sustainable] tourist promotion, and I decided to collaborate with a professor from the Tourism Institute at the University of Lausanne, because he has such a long Curriculum Vitae. In that way, when I have to discuss with the directors of the Tourism offices, who do not have any professional capacity for doing this kind of thing, because at most they will distribute flyers, I instead decided to go to the Big Boss to present my concept of tourist promotion to him. (Marta, regional secretary)

The legitimacy of their choices was therefore supported by the strength of their contacts with this eminent professor, well known by local politicians, instead of concluding alliances with other municipalities all over the Alps.

114 The Tourism Institute of the University of Lausanne no longer exists under this name.

The interviewees exposed different reasons both to join and not to join the network. The network is seen as a good opportunity for mayors and regional administrators to exchange information. This opportunity can be used in a more pragmatic way, i.e. as a source of information on how to implement projects of sustainable development in a practical way or in a more political way, namely as input for showing local authorities some innovative ways of dealing with regional development.

The network also offers the welcome possibility of funding for local projects and provides personal animators who give advice on their implementation. All these arguments have been presented as important reasons to join the network. Nevertheless, these arguments are not strong enough, unless there is at least one committed person in the municipality. On the other hand, the arguments for not joining the network stem from a non-understanding of the benefits (primarily economic) that membership can yield for the municipality or the region. In a static political context where mayors have to ensure the day-to-day running of the municipality, energy, time and resources are not available to invest in such a demanding network. If, additionally, there is no identification of the municipality with the Alps, the thought of joining a pan-Alpine network becomes even more remote.

6.3 The dimensions of networking

In the previous sections, I analysed the arguments that the interviewees advanced for (not) joining the networks. However, once the decision is made to join the networks, members discover the "world of networking" with its potentialities and limitations. These are presented and discussed in the following paragraphs.

When I started my fieldwork in 2007, Alliance in the Alps celebrated its 10th anniversary. A press release from the 22nd of March 2007 portrays the aims of the network and presents the programme of the birthday's events. A birthday, of people or of institutions, is always a good occasion to assess the goals achieved and to express wishes for the future. In the press release, the main successes are presented and cover different dimensions of networking. Here, the emphasis is clearly on the multifaceted political dimensions of the network. In the very first paragraph, Alliance in the Alps announces that its objective is to implement the principles of the Alpine Convention at the local level, which can be viewed as the political will of overcoming the "traditional" national boundaries and collaborating at the horizontal level in the name of a transnational project. Thanks to this new partnership, as the press release suggests, the network not only gained growing political influence inside the Alpine region but also beyond its borders. In order to show the growing weight of the network within the Alps, the authors publish some quantitative figures proudly disclosing that 230 local authorities, representing 4.5 % of the Alpine population and 5.2 % of the area, joined the network. To illustrate the growing significance of the network outside the Alps, another argument is put forward, namely the establishment in 2003 of a "sister organization" (Alliance of Central Asian Mountain Communities, AGOCA) created in Central Asia on the model of Alliance in the Alps.

The socio-cultural component represents another dimension communicated in the press release. This is highlighted by the claim that, thanks to the network, members developed a common Alpine identity and helped to improve the quality of the Alpine inhabitants' lives. Of course, the more "pragmatic" aspect of the network, namely its capacity to stimulate exchanges of information and experience, is also underlined. Finally, the ecological factor is presented, albeit in rather a veiled way, as it is advertised in the text commemorating the 10th anniversary of the network: "The network has also made progress on the long road to sustainable development in the Alps."

The press release that Alliance in the Alps published emphasizes four positive dimensions of networking (political, socio-cultural, pragmatic and ecological). These are also present in the interviews I conducted, as I show in the following paragraphs. In addition to these four elements, some interviewees also added the economic advantages of networking. In total, then, five dimensions will be considered.

6.3.1 Political dimensions of the network

In a short article that appeared in the *Jahrbuch des Vereins zum Schutz der Bergwelt*, Andreas Götz, director of CIPRA International, applauds the increased political strength of the local authorities that decided to join Alliance in the Alps. Götz concludes his article by declaring that "of course, thanks to the collaboration with other municipalities, the members could improve their image, both towards their inhabitants and outside their boundaries". Götz elaborates on the impact of this better image by saying that "The ones who have a strong voice will be heard". It is possible to speculate whether what Götz declares is shared by the members of the network as well. Using the interviews with members and representatives of the Alpine networks, I not only answer the question whether the concerned municipalities recognize this political strength of the network but also analyse what the form is in which they acknowledge it.

The network used for lobbying work

Mathieu, an employee in a municipality member of Alpine Pearls, distinguishes three levels at which the network's political strength is used in their local reality. Firstly, the network proposes concrete arguments to the population by showing that "another way of living is possible", if I can translate thus the example that Mathieu brings up. Secondly, the network gives elements to contest the politics proposed by the traditional administrative authorities. Finally, it facilitates the access to funds at the national level. Mathieu develops his arguments as follows:

On va distinguer trois échelles. Première échelle, c'était de convaincre la population locale qu'on peut se balader autrement qu'avec son gros "quatre-quatre". Je crois qu'on n'y est pas forcément toujours très bien arrivé, parce que [la population] c'est les plus difficile à convaincre. Ensuite, on avait une deuxième cible qui était le	I will distinguish three levels. Firstly, we wanted to convince the local population that it is possible to move around otherwise than with a big SUV. I think we did not completely succeed, because [the population] is the most difficult to convince. Then, we had a second target, the *Conseil général*.[115]

115 The *Conseils Généraux* are the assemblies of the French Departments.

Conseil général. Avec celui-là on s'est battu parfois méchamment, on a même parfois levé le compte à une réunion en claquant la porte et en se barrant pour leur montrer qu'à un moment donné ils étaient complètement à la rue d'un point de vue de définition de la politique. Donc là, on a fait bouger des choses. Pas suffisamment à notre goût, mais on a enclenché un mouvement. Et le troisième point, c'était également apporter une réflexion au niveau national pour obtenir des financements. (Mathieu, employé dans une commune)	With them we fight, from time to time, really hard. Once, we even slammed the door during a meeting to show that they were completely wrong in their definition of the policies. There, we got things moving. Not enough, to our liking, but we launched a movement. The third point was to bring new thoughts at the national level in order to get funds. (Mathieu, employee of a municipality)

Mathieu was the only interviewee to distinguish different levels by defining the increased power of intervention that the municipality could acquire thanks to the network. Nevertheless, two of the three levels mentioned by Mathieu (the regional and the national) are mentioned, albeit in a different manner, by other interviewees. The third level, the broader population, however, was not brought up by others in answer to my question concerning the political weight of the network. Later in the interview, Mathieu explains that, since Alpine Pearls is considered a European label by others, it legitimizes the action of the municipality towards the Assembly of the Department. In fact, he reasons that the legitimacy of the network helped them "knock on the right door", in order to be heard. The strategy was worthwhile since the municipality did indeed receive some funds from the department, and its political strategy is now better accepted.

During the same interview, conducted by Nicolas De Groote, a second person, an employee of another municipality that is a member of the Alpine Pearls, summarizes the ideas developed by his colleague with the word "lobby". Luc clearly states that the network helps to constitute a lobby that can be used towards the administrative authorities:

L'idée du réseau c'est vraiment de faire un lobbying, de dire que le fait de s'être rassemblés, de mettre en avant les points identiques que l'on rencontre, c'est vraiment le poids politique vis-à-vis des autorités. Luc (employé d'une commune)	The idea of networking is really to create a lobby. By being together and putting forward the same themes that we are facing, a political weight *vis-à-vis* the authorities is created. (Luc, employee of a municipality)

Normann, a tourist office employee in charge of the relations between his local authority and the Alpine Pearls network, explains in a concrete manner how the network was used by the municipality to politically justify their project of increasing the offer of public local transport:

Wir haben in diesen zweieinhalb Jahren sehr viel erreicht. Wenn ich daran denke, wie oft jetzt in den Tälern der Bus fährt, da war vor zwei Jahren noch keine Rede davon. Das ist auch heutzutage nicht mehr ganz so einfach, mit den finanziellen Mitteln einen Linienbus einzusetzen, der alle zwei Stunden ins Tal hinein fährt. Also das haben wir geschafft. Das sind so Aktionen, die wir auch durch die Mithilfe von Perlen der Alpen geschafft haben. Jetzt nicht unbedingt die finanzielle oder die persönliche Mithilfe, aber einfach durch die Kooperation, dass die Politiker verstehen, was wir hier wollen, beziehungsweise dass die Mobilität generell für die Touristen immer wichtiger wird. (Normann, Mitarbeiter im Fremdenverkehrsbüro)

We achieved a lot in these two and a half years. If I think how often buses go up and down in our valley now, this was not the case two years ago. And nowadays, with the financial means given to us, it is not easy to have buses running every two hours in the valley. So, we managed. These are actions that we achieved thanks to Alpine Pearls as well. Not necessarily through its personal or financial help but simply through the cooperation. Now the politicians understand what we want and that mobility for tourists will become more and more important. (Normann, employee of a tourist office)

As Normann shows, the strength of the network is not only the possibility to get money or to have at the disposal of the municipality persons that can provide technical support. This strength can also be seen in the political support the municipality receives when trying to implement projects that do not match the priority of local authorities. As Normann says later in the interview, it is a matter of being able, thanks to the network, to draw attention (*Aufmerksamkeit erregen*) and in so doing to make it possible for small municipalities to be heard by "higher" politicians. In that sense, as Robert underlines, it is in the networks' interest to increase their visibility, since this also increases their political strength:

Ce réseau il a une certaine force; plus il sera visible, plus il aura de force. Il a une force qui est bénéfique pour

This network has a certain power; the more it is visible, the more powerful it will be. Its strength is good for its

les communes membres. La force par exemple en Suisse, c'est que c'est une démarche qui est reconnue comme étant très positive par les Cantons et au niveau de la Confédération. Les communes qui sont membres du réseau sont des communes qui sont considérées comme des communes qui ont une volonté d'aller de l'avant, des communes, courageuses, organisées, bref qui ont beaucoup de qualités, qui sortent du lot et lorsqu'elles vont demander une aide, demander de participer à un projet-pilote, elles vont tout de suite avoir les portes qui s'ouvrent. [...] Et à la Confédération ils obtiennent une oreille bienveillante. Lorsque par exemple l'Office fédéral de l'agriculture va lancer des projets-pilote avec financement à l'appui, comme par hasard on retrouve les communes qui sont membres du réseau Alliance dans les Alpes. Elles ont tiré profit de cela. (Robert, animateur)

members. For example, in Switzerland, strength is about being part of a network whose approach the cantons and the Confederation recognize as positive. The municipalities that are members of the network are seen as willing to look forward [and be] fearless, organized municipalities. In short, municipalities with a lot of qualities, above average, so that when they ask for help, when they ask to be part of a pilot project, they are heard. [...] And these municipalities are heard by the Confederation. When, for example, the Federal Office for Agriculture launches pilot projects, with supporting funds, "surprise surprise", those municipalities that are members of Alliance in the Alps are chosen. They profit from that. (Robert, animator)

This way of using the network in order to form a lobby at the higher administrative instances seems to work not only in Switzerland, but in the other Alpine countries as well. Antonio tells a similar story concerning Italy. He relates how, following the example of other member municipalities, he could use strong arguments to block a project of enlargement of the ski area in his municipality, a project that was strongly wanted by the Friuli-Venezia Giulia Region:

Il Col Colmier l'abbiamo chiamato "Montagna dei ragazzi". Questo c'è servito anche per difendere la nostra montagna dall'ampliamento delle piste da sci. Se la montagna fosse passata nell'indifferenza, sicuramente avrebbero già fatto gli impianti sciistici. Non sono ancora esclusi questi pericoli, perché i tentativi di farli ci sono sempre, però abbiamo avuto la forza di fare uno scontro con la Regione e per il momento la cosa è

We called the Col Colmier the "Mountain of the boys and girls". This was useful in order to defend our mountain against the enlargement of the ski area. If the people had become unconcerned by the mountain, for sure the ski resort would have been constructed. These threats are not excluded yet, because they always try to do it, but we had the strength of contesting the region, and for the moment the project is blocked.

stata bloccata. [...] Quindi le problematiche dell'arco alpino, messe sulle piccole problematiche di Budoia, hanno dato il coraggio di prendere una posizione. [...] Questo è stato possibile [...] perché si può dire: "Guardate che non è vero che l'economia è solo collegata agli impianti da sci, guardate questo paese della Francia, piuttosto che quell'altro, cosa hanno fatto". Sono esempi concreti. E noi facendo parte di Alleanza possiamo dire che queste cose da qualche parte le hanno risolte in un altro modo. (Antonio, sindaco)

[...] So, the problems of the Alpine arc, if one transposes them onto the small reality of Budoia, gave courage to stand up against the project. [...] And this was possible [...] because one can say: "Have a look, it is not true that the economy is only linked to ski resorts; have a look at this village in France, or that other, have a look at what they did." The are concrete examples. And since we are part of Alliance, we can say that these things, in other contexts, were solved in a different way. (Antonio, mayor)

As the rest of the interview also shows, the municipality could, thanks to Alliance in the Alps, increase the sense of belonging of the broader population to the upstanding mountain territory. Antonio therefore sees a correlation between this awareness towards the mountain and the fact that the population stood up against the project of new ski resorts.

The network used as source of inspiration

As the story told by Antonio suggests, the possibility, presented by the network, to exchange experiences amongst Alpine municipalities reinforced the political arguments of the opponents of the project, since they acknowledged that other municipalities reoriented their winter tourism without the development of more ski areas. This second argument has been used by Claudio as well, who says that the examples taken by other members of the network can be used in arguments with politicians:

E' sempre interessante, quando si va da un politico, poter dire: "In Alleanza nelle Alpi c'è stato un comune che l'ha fatto, ha avuto dei vantaggi, c'è stata una ricaduta positiva". [Anche se è qualcosa] difficile da monetizzare. (Claudio, segretario regionale)

It is always interesting, when one meets a politician, to be able to say: "In Alliance in the Alps a municipality did it, they could take advantage of it, they had a positive outcome." [Even if it is something] that is difficult to translate into money. (Claudio, regional secretary)

And, even if the financial benefits are difficult to measure, as Claudio says, Alliance in the Alps helps, thanks to facility of exchanging experiences, to morally support local authorities in their choices:

Più che coraggio, sicuramente [...], Alleanza nelle Alpi deve essere qualcosa che collega. E' presente a livello internazionale con un progetto positivo, mostra che si può andare avanti migliorandosi e confrontandosi. E' un segnale che siamo nella direzione giusta. (Claudio, segretario regionale)

More than courage [...], Alliance in the Alps should be something that creates links. It is present at the international level with a positive project; it shows that one can go ahead by exchanging experiences and by always doing better. It is a sign that we are going in the right direction. (Claudio, regional secretary)

Thus, the large amount of experiences within the network is used as convincing examples to persuade local politicians that "another way of doing" is possible; it is, as Claudio says, a sign that the local authorities are "going in the right direction". It becomes, in that sense, a strong argument because it is difficult for local authorities to find counter-arguments, since concrete implemented examples can be shown and their own project defended on this basis.

On the other hand, as Claudio adds at the end, it is difficult to estimate the economic value of this advantage offered by the network. Know-how and information are indeed hardly calculable goods. This recalls the difficulty people like Marta and Giorgia face in trying to convince their sceptical mayors to join the network. In fact, if those mayors only want to know how much money they can receive from the network, it is nearly impossible, because many of the advantages of belonging to a network, as Claudio suggests, are not calculable in economic terms.

However, some interviewees still seem to be more aware of the direct economic advantages that the network can offer. For example, Olivier, an animator of Alliance in the Alps, relates how he uses the network to receive funds from the regional authorities:

On a sollicité un financement à la région, notamment dans le cadre de la Convention Interrégionale pour le Massif des Alpes (CIMA), qui sera un peu le relais des financements régionaux des projets

We asked for funding from the region, notably via the *Convention Interrégionale pour le Massif des Alpes* (CIMA), which will be the local intermediary of the regional funding for mountain projects.

en montagne. Les instances régionales sont tout à fait au courant de l'existence de ce réseau. Et donc à son président, à son vice-président, on a adressé un courrier de lobbying rappelant l'existence de ce réseau et de communes engagées pour l'application des protocoles de la Convention alpine. (Olivier, animateur)

The regional authorities know about this network. So we sent a "lobbying" letter to the president and the vice president of CIMA in which we reminded them of the existence of the network and of the municipalities that are involved in the implementation of the Alpine Convention. (Olivier, animator)

This occurred when, at the time of the interview, in October 2006, only four French members belonged to the Alliance in the Alps network. Still, as Olivier later admits, even if there are only a few French members, the municipalities can rely on almost 300 Alpine local authorities to create, as he calls it, an "Alpine municipalities' logic" capable of providing examples of new ways to conceive mountain economy:

A l'échelle de l'arc alpin, ils sont presque 300 et cela prouve qu'il y a une "logique communes alpines", qui supporte toutes les politiques touristiques qu'on [nous, dans le réseau] connaît bien et qu'il y a réellement besoin de développer une autre approche de l'économie montagnarde. Et à travers ce réseau il y a beaucoup de maires qui souhaitent engager des politiques alternatives, qui seraient susceptibles de rejoindre le réseau et trouver une certaine force de lobbying. [...] Les communes ne sentent pas du tout l'Etat impulser ce genre d'initiatives, et du coup un maire qui a envie de rallier ce réseau il y trouve une certaine assurance, une certaine légitimité. Il s'assure de la valeur de ses convictions, de ses démarches, alors qu'effectivement s'il n'a pas ça à sa disposition, il est tout seul face à lui-même et puis il se dit "Est-ce que vraiment ce que j'ai engagé ça peut se faire?", on peut vite renoncer. Alors que le réseau apporte un contexte dynamisant. (Olivier, animateur)

At the Alpine arc scale, there are almost 300 members, and this proves that there is an "Alpine municipalities' logic" that supports the tourism policies that we [in the network] know well and that there is a need to develop another approach of mountain economy. Thanks to this network there are many mayors who want to develop alternative policies and who are ready to join the network in order to attain a sort of lobbying power. [...] The municipalities do not feel that the state encourages this kind of initiatives, so that a mayor who wants to join the network finds within it a certain confidence, a certain legitimacy. He finds reassurance of the value of his convictions, of his approach. Without this tool, he would be alone and say to himself: "Is it really possible to achieve what I have begun?", and one can easily give up, whereas the network brings a stimulating context. (Olivier, animator)

So, the network seems to bring a "stimulating context", as Olivier suggests, capable of catalyzing new concepts and ideas that are circulating in the Alps and that attentive mayors can use as concrete arguments to convince their population and their administrations of the legitimacy of their choices.

Only one interviewee, Peter, president of Alpine Pearls, suggests another possible target: the private economy. According to him, the critical mass of a network makes it possible to create new collaborations with some private firms, for example with railways companies. Peter further underlines the importance of creating this kind of cooperation because of a specific project that he wants to develop that would offer the possibility of providing guests of the Pearls with a transport ticket valid throughout the entire Alpine territory for a number of consecutive days. In a report by the network (Alpine Pearls, 2006), the authors take the Swiss Pass ticket[116] as an example that could be generalized to the entire Alpine arc. In order to implement this ambitious project, of course, the network requires strong collaboration with regional and national railway companies. Peter also admits that a large number of members would help render the network more credible and make it a stronger partner:

> Für uns ist es da sicher besser, eine kritische Größe zu haben, damit wir auch gehört werden, damit wir auch die Chance haben, zum Beispiel mit Bahnunternehmen oder auch mit Wirtschaftsunternehmen eine Kooperation einzugehen. Da müssen wir natürlich eine gewisse Stärke und Größe haben. (Peter, Alpine Pearls Präsident)

> For us, it is better to have a critical mass, so that we can be heard, so that we have the chance to create new collaborations with railway companies or with private firms. To do that, of course, we need a certain strength and size. (Peter, president of Alpine Pearls)

In this section, I described how local authorities felt about pan-Alpine networks being used as a political lobby and gave some insights into its different dimensions. As I explained, the majority of the interviewees conceive the network as an important tool to get funds from their administrations (the regional or the national authorities). That being said, they also see it as an instrument to form a group of com-

116 <www.swiss-pass.ch/en/>, consulted on the 19.04.2015.

mitted municipalities that fight together to raise political awareness concerning the importance of implementing sustainable development at the local and pan-Alpine levels. Although the concept of sustainable development was not mentioned in the interviews, it is clearly what was meant when people spoke about an "alternative economy" or when they mentioned "the aims of the Alpine Convention". Although these different political dimensions seem to be clear for the networks' members, not everybody agrees on their current political weight, as I will discuss in the following sections.

The current political weight of networks

When I asked my interlocutors whether they felt that the pan-Alpine networks are able to wield political power, the answers were discordant; in fact, the interviewees tended to say "no". Still, some of them recognize the networks' important political power. This is the case, for example, of Erik, an employee of Alparc, who recounted his experiences with the Econnect project,[117] which was carried out by Alparc:

Je pense qu'à travers les réseaux il y a un certain poids [politique] qui se met en place. Pour les réseaux écologiques, d'une part on est reconnu par le Réseau écologique pan-européen (REP) en tant qu'entité alpine. [...] Et puis, d'autre part une certaine reconnaissance, très au début, de la part de l'Europe aussi, parce que les projets menés au niveau alpin concernant la politique des réseaux écologiques se sont arrêtés à la Direction Générale – Environnement de l'Union Européenne, ils nous ont entendus. Ils ont trouvé cela intéressant et ils nous ont proposé de collaborer. Donc il y a des premiers pas qui sont engagés dans cette direction. (Erik, employé d'Alparc)

I think that through the networks there is a certain political weight that is established. For the ecological networks, we are, on the one hand, recognized by the Pan-European Ecological Network (PEEN) as an Alpine entity. [...] And then, on the other hand, we gained a certain recognition from Europe, since the projects carried out at the Alpine level concerning the policies of the ecological networks found a receptive audience at the Directorate-General for the Environment of the European Union. They listened to us, found what they heard interesting and proposed a collaboration. So, there are some first steps in this direction. (Erik, employee of Alparc)

117 For more information, see: <www.econnectproject.eu/cms>, consulted on the 20.04.2015.

In contrast to what Erik says, Jakob says he does not believe in the strength of the network if confronted with the national or European authorities. At least, as he admits, not yet, i.e. not until the number of members increases:

Ich glaube nicht, dass wenn die Allianz Themen aufgreift wie Verkehr, dass sie besonders großes Gehör findet in Bern, in Wien oder in Brüssel. Aber die Bedeutung nimmt zu, davon bin ich überzeugt. Die Allianz ist als politisch unabhängige Organisation gedacht, die Fehlentwicklungen oder Lösungsmöglichkeiten aufzeigen soll und so weiter... Tatsache ist, dass sich mit der wachsenden Zahl an Mitgliedern natürlich schon ein gewisses politisches Gewicht ergibt. Und wenn die Mitglieder alle bei einem Problem aufstehen, dann besitzt das natürlich schon eine gewisse Schlagkraft. Und dann würde man international marschieren und nicht national. (Jakob, Bürgermeister)

I do not think that when Alliance takes up themes such as traffic, it is really heard in Bern or Vienna or Brussels. But its importance is increasing, I am convinced of that. Alliance is thought of as a politically independent organization that should bring to light flawed developments and search for new solutions. The fact is that, if the number of members increases, there is a growing resulting political weight. And when all the members react together on a problem, there is an impact. And by doing so, we stand up internationally and not nationally. (Jakob, mayor)

For Jakob, the fact that a growing compact and international "voice", speaking in the name of all Alpine local authorities, is emerging is a sign that networks can exert a certain political influence. However, in his view, this influence is still small, unless the number of members grows. As experienced a person as Antonio replies as follows to my question about the importance of having such an organization as Alliance in the Alps in order to give more political weight to the politics of sustainable development in the Alps:

Antonio: Sì, in questo credo che Alleanza nelle Alpi in qualche modo abbia dato un piccolo contributo. Non solo lei, però penso che Alleanza abbia dato un segnale di impostazione. Perché quello che mi è rimasto, che mi ha sorpreso in diverse occasioni è che noi [di Alleanza nelle Alpi], che non ci vediamo tanto, non siamo dei professionisti, non

Antonio (mayor): Yes, I believe that Alliance in the Alps contributed in that sense. Not only Alliance, but I think that Alliance gave the direction. What I experienced, and was often surprised by, is that in different instances we [from Alliance in the Alps] – who do not meet regularly, are not professionals and do not do only that, because we

facciamo solo questo, perché lavoriamo, siamo piccoli comuni, però, quando ci presentiamo, veniamo tenuti in grande considerazione, rispetto magari ad organizzazioni che hanno più una valenza politica, come le associazioni di montagna, tipo l'Unione Nazionale Comuni Comunità Enti Montani (UNCEM). Tutte quelle associazioni in cui i rappresentanti sono nominati secondo le suddivisioni di tipo politico.

work, because we are small municipalities... nevertheless, when we introduce ourselves, we are highly considered, for example compared with other organizations that have greater political worth, such as the *Unione Nazionale Comuni Comunità Enti Montani* (UNCEM). All those associations in which the representatives are appointed following the political administrative subdivisions.

Cristina: Quindi voi in un certo senso, se posso dire così, avete oltrepassato le frontiere amministrative per creare una nuova configurazione di lavoro?

Cristina: So, in a certain way, if I can put it that way, you, with Alliance, have overcome the administrative boundaries in order to create a new configuration of work?

Antonio: Sì, probabilmente sì, siamo usciti un po' da quegli schemi, quindi abbiamo sostenuto che si può anche fare delle cose senza questi ruoli che sono legati alle [suddivisioni] politiche [tradizionali]. Presentiamo delle cose pratiche, siamo rappresentanti dei cittadini delle Alpi, quindi questo forse ci dà anche considerazione. Non è che vogliamo distruggere le altre organizzazioni, però noi ci presentiamo come rappresentanti di comunità, che hanno queste problematiche e che in modo concreto le portano avanti. Qui io ho visto che Alleanza nelle Alpi è sempre vista con un certo interesse.

Antonio: Yes, probably yes, we managed to get out of those patterns. We showed that it is possible to do some things even without those roles linked to the [traditional] political [subdivisions]. We present some practical things. We are representative of the inhabitants of the Alps; that is probably why we are taken into account. We do not want to destroy the other organizations, but we introduce ourselves as representatives of local authorities and have those issues to carry on. On those occasions, I have noticed that Alliance in the Alps is viewed with interest.

Although Antonio does not mention by whom the network is considered as an interesting organization, knowing Antonio's wide range of activities inside the network, this could be, for example, during meetings he and the head of the network had with national or European representatives. Nevertheless, and despite Antonio's positive feeling, later in the interview he admits he has "the impression that the Alps are not considered by the European Union as an important region". This is because, according to him, the "political power is still linked to the most inhabited cities".

Niklas's opinion is equally interesting, since, unlike Antonio, he is not a particularly involved mayor in the activities of the network. Niklas recognizes the political weight of Alliance because, in his view, the network gives strength to small municipalities, a strength comparable to that of workers organized in a union:

Niklas (Bürgermeister): Ich glaube, [dass Allianz in den Alpen] einfach eine Stärkung der Kleingemeinden bringt. [...] Ich sehe es wie eine Gewerkschaft für die Arbeiternehmer, so sehe ich die Allianz in den Alpen für Kleingemeinden.	Niklas (mayor): I think that Alliance in the Alps brings strength to the small municipalities. [...] I see it as a workers' union. This is the way I see Alliance in the Alps for small municipalities.
Cristina: Glauben Sie denn, dass Allianz in den Alpen ein politisches Gewicht bringt?	Cristina: So, do you think that Alliance in the Alps brings a political weight to municipalities?
Niklas: Ich glaube schon, ja! Man könnte es noch mehr ausnützten, noch mehr machen europaweit. [...] Alle zusammen verfolgen sie ein Ziel, das ist einfach die Stärkung und die Sicherung der Alpgemeinden.	Niklas: I think so, yes! It would be possible to use it even more, to do more at the European level. [...] All together, they pursue an aim, and simply put this is the strengthening and the safeguard of the Alpine municipalities.

Later in the interview, the certainty expressed by Niklas at the beginning slowly gives way to a more mitigated opinion on the real political power of the network, already mentioned when he said it "would be possible to use more political weight at the European level":

Ich habe nicht gemeint, dass es jetzt schon soweit ist. Dazu praktiziert man das [das politische Gewicht] im Moment auch noch zu wenig. Aber ich glaube es nicht, dass wir so in dieser Form wie es jetzt besteht überhaupt kein politisches Gewicht haben. [...] Wir, mit der wenigen Erfahrung [der sechs Gemeinden, die eine Region aufgebaut haben], haben auch unser politisches Gewicht in die Waagschale werfen können. Darum wäre es nicht richtig zu sagen, dass eine Allianz von mehr als 200 Gemeinden kein politisches Gewicht hätten, das glaube ich nicht. (Niklas, Bürgermeister)	I did not say that it is already like that. At the moment, we still exert [this political weight] too little, but I do not think that in this form as it is now we do not have any political weight. [...] Our small experience [of six municipalities that joined into a regional administration] shows that we could already increase our political weight thanks to that. For that reason, I do not think that an alliance comprising more than 200 municipalities does not have any political weight. (Niklas, mayor)

Reading the interview with Niklas, one can clearly see his statements stem from a conclusion that he has drawn from his own experience. As he says, if a municipality gains political weight by joining five other municipalities, it is simply illogical that an alliance of more than 200 local authorities does not increase the political weight of its members. In reality, however, Niklas is critical of the network's actual political influence towards higher authorities, such as the European Union. This feeling of the network's (in)capacity to influence the regional, national and international administrations reminds us of what Marta and Giorgia (administrators of a region) said about the network's recognition as an important partner for the Ticino Canton, where they work. Their feelings are shared by Leonardo, who sees Alliance in the Alps as not having any political weight in the Ticino Canton.

The interview that Nicolas De Groote conducted with Antje, a member of the Alpine Pearls network, also points to doubts about the network's political influence:

Nicolas: Wissen Sie, ob Perlen der Alpen ein gewisses politisches Gewicht ausüben will?	Nicolas: Do you know whether Alpine Pearls would like to exercise political weight?
Antje (Mitarbeiterin im Fremdenverkehrsbüro): Das wünscht sich jede Vereinigung. Wie stark das sein kann, weiß ich nicht. Da habe ich eher meine Zweifel, dass das sehr gewichtig ist.	Antje (employee of a tourist office): Every association wishes to. But how strong this can be, I do not know. I have some doubts that this counts for much.

The examples here do not mean these impressions can be applied to the entire Alpine arc. Nevertheless, it shows that the question of the political influence of such reticular organizations is open, and the answers sometimes conflicting. In the next section, I present an interesting case of how Alliance in the Alps was used, ultimately without success, to influence national (in this case, Swiss) decisions. Furthermore, this example will not provide a unique answer to the issue debated here but will probably raise even more questions about the actual political weight of the pan-Alpine networks. Future developments of the networks will tell us whether criticism towards the capacity of the networks of influencing national and European politics was justified.

The "special Swiss case"

Before discussing the Swiss case, it is important to recognize the general context. As reported in the presentation of the Alpine Convention (see chapter 4), two countries did not ratify the protocols of the Convention: Italy and Switzerland. This fact, combined with other elements, is a clear indication for Sandro (employee of the Alpine Convention) that the macro-Alpine scale, as he calls it, is facing a period of crisis:

Non sono stati fatti nuovi protocolli, sono state fatte soltanto delle dichiarazioni sul clima e poi il protocollo popolazione e cultura. Due paesi non hanno ancora ratificato i protocolli, la Svizzera e l'Italia. L'Unione Europea non si interessa più alla Convenzione delle Alpi. Quindi c'è un po' una crisi del livello macro. [...] La Convenzione delle Alpi ha avuto gradualmente una diminuzione di importanza politica, oggi stiamo toccando uno dei minimi storici, perché effettivamente noi lo vediamo, la Convenzione delle Alpi, per tutte le parti contraenti, ha molto poca priorità politica. Però per una gran parte delle parti contraenti la priorità politica è molto bassa. (Sandro, impiegato alla Convenzione delle Alpi)	No new protocols were elaborated; only two declarations (one on climate and one on population and culture) were delivered. Two countries did not ratify the protocols, Switzerland and Italy. The European Union is not interested in the Alpine Convention anymore. So, at the macro level there is a bit of a crisis. [...] The Alpine Convention is experiencing a decline in its political importance; we are now at an all-time low, because in effect we can see that, for the contracting parties, the Alpine Convention is not a political priority. But for a majority of the contracting parties, the political priority is low. (Sandro, employee of the Alpine Convention)

In this general context, portrayed by Sandro as a "crisis context" in which the Alpine Convention is not taken into account, especially by the national authorities, the pan-Alpine networks could play an important role in pushing the national states to give the Alpine Convention greater importance. In the Swiss case, they are seen by some as a prominent actor that could provide a positive impulsion to induce Bern into ratifying the protocols. Robert, animator of Alliance in the Alps, explains how Alliance in the Alps tried to put pressure on the Swiss authorities by means of a press conference. The story is worth telling:

Il y a eu une conférence de presse pour activer le traitement des protocoles à Berne, le 3 mars 2003. [...] C'est la "direction du projet" qui a tenu cette conférence avec le Groupement suisse pour les régions de montagne (SAB). Et puis il y a eu des politiciens communaux qui ont pris la parole, qui ont rapporté justement leur expérience positive, concrète, en relation avec la Convention alpine. Il y a eu un large écho dans les media, ce qui a conduit, contrairement aux attentes, au fait [que] le Conseil des Etats n'a pas refusé de ratifier des protocoles, mais ils ont demandé une nouvelle évaluation. Je crois qu'on en est toujours là, mais ça a eu comme effet de bloquer... Ils [le Conseil] étaient partis pour dire carrément "non" à la ratification, et les différentes actions de lobbying exercées, dont celui d'Alliance dans les Alpes. [...] a eu comme effet qu'ils ont reporté leur décision. (Robert, animateur)

On the 3rd of March 2003, a press conference was held to try to activate the treatment of the protocols in Bern. [...] The "direction of the project" and the Swiss group for mountain regions (SAB) organized the conference. Some local politicians spoke; they reported their positive and concrete experiences related to the Alpine Convention. There was a large echo in the media, and contrary to expectations, this led to the Swiss Council of States not refusing to ratify the protocols but instead asking for a new evaluation of the question. I think the situation did not evolve since then, but it has had the effect of blocking... They [the Council] were ready to bluntly say "no" to the ratification, but the actions carried out by the lobbies, including that of Alliance in the Alps, resulted in the postponement of their decision. (Robert, animator)

In the press release, members of Alliance in the Alps demanded the protocols be ratified, which underlined the fact that the Alpine Convention did not need to be seen as an obstacle for the development of mountain local authorities. Knowing that the 11th of March 2003 (in other words, a few days following the publication of the press release) the Swiss Council of States would have debated the ratification question and knowing that the Commission of the Environment of the Council had recommended in December 2002 that the Council not ratify the protocols, the members of Alliance wanted to show the importance of the ratification via this lobbying effort.

This strong lobbying drive resulted, however, in a defeat for the supporters of the Alpine Convention. In fact, some years later, in September 2010, the lower chamber, the National Council of Switzerland, definitively "buried the ratification", as the title of an article published by the *Courrier* clearly informs.[118] The result of the vote

118 "La ratification est enterrée", *Le Courrier*, 30.09.2010.

of the Council was quite clear: Some 102 National Councillors voted against (and only 76 for) the ratification of the nine protocols of the Alpine Convention.

This example shows that, on the one hand, the members of the network are aware of the political influence they can exert. If they had not been convinced of that, they would not have mobilized in the capital city to try to convince the senators and the national councillors to give the Alpine Convention a chance. The possibility that the network provided them to speak in the name of a group of municipalities gave them the strength to fight. On the other hand, the result of this approach confirms the doubts expressed by some members about the network's actual political power. Even if the lobbying work managed to postpone the national authorities' decision, the result, in the end, was a defeat for the defenders of the Alpine Convention, including Alliance in the Alps.

6.3.2 Socio-cultural dimensions of the network

While the political impact of the pan-Alpine networks, especially towards the national states and the European Union, is debatable, everybody seems to agree on networks' ability to provide opportunities to meet new people and new cultures. It is what I call the "socio-cultural dimensions of the network".

The importance to meet personally

The heads of pan-Alpine networks always advocate for personal meetings and not only for the "virtual" contact that the Internet makes possible today. All through the year, networks organize a number of thematic and research workshops, official meetings and excursions. These are all occasions for the members to meet, exchange experiences and see how others are dealing with some particular issues. I had the possibility to participate in some of the events organized by Alliance in the Alps, Alparc and the Alpine Pearls and could observe the relaxed and pleasant atmosphere that the members and managers of the networks create. I always had the impression of being amongst

a group of good friends and, after a couple of days spent with them, the departure was always a bit sad.

This picture, taken on the occasion of the 10th anniversary of Alliance in the Alps, along with the smiles of Rainer Siegele (left), president, and Claire Simon, manager of the network, is probably more illustrative than any number of words (see fig. 19).

Fig. 19: Birthday cake for the 10th anniversary of Alliance in the Alps © Alpenallianz.

I discussed with Andreas Götz, former director of CIPRA International the issue of the social capacity of the head of the network to create a pleasant working atmosphere. Here are his reactions to my observation:

Es wird sehr zelebriert von Rainer Siegele [Präsident von Allianz in den Alpen], auch das man gemeinsam Essen und Wein trinken muss. Aber das ist so. Die besten Kooperationen entstehen so, weil die Leute sich anfangen zu mögen. [...] Wir haben das bemerkt. Wir haben wirklich aus fachlichen Gründen angefangen, unsere *Workshops* mit Übernachtung zu machen. Das ist natürlich

Rainer Siegele [president of Alliance in the Alps] celebrates this well. For him, we have to eat together, drink wine. ... But it is like that, because the best cooperation develops when people start liking each other. We noticed that, so for technical reasons, we began doing our workshops with overnight stays. It is of course practical, but we do not say: "Everybody can sleep wherever, and we

praktisch, aber wir sagen nicht: "Jeder kann irgendwo übernachten und wir treffen uns dann am Morgen", sondern wir machen einen Programmpunkt schon am Tag vorher. Sie sollen zusammen Abendessen, ein Glas Wein miteinander trinken. Und dann werden sie sich kennen und am nächsten Tag läuft es ganz anders. Das der eine sagt: "Wir machen das so", und der andere sagt: "Wow, das ist der Wahnsinn". "Ja, kommt dann ja vorbei". Und dann fahren sie mit deren Gemeinderäten dahin und schauen sich das an und trinken noch mal ein Glas Wein. So sind Netzwerke. Man kann das schon versuchen zu versachlichen und sagen: "Es geht hier um die Arbeit und nicht um das Vergnügen". So ist das nicht, wenn du Leuten vertraust, wenn du denkst: "Das war angenehm", dann glaubst du ihnen, was sie erzählen, und dann glaubst du, dass du das selber auch machen könntest. So sind Netzwerke, das ist ein Erfolgsrezept von Rainer. Es hat viel damit zu tun, mit diesen Sozialkompetenzen, mit dieser Festfreude und auch das du weisst, dass nachhaltige Entwicklung nicht eine bittere und knallharte Knochenarbeit ist, sondern eine Freude. [...] Man hat nicht viele solche, die das so können wie Rainer, aber das ist ein Erfolgsrezept, das ist es für mich ganz eindeutig. (Andreas, Direktor von CIPRA International)

meet the day after"; we start with something from the programme the day before. They eat together and drink a glass of wine together. They get to know each other, and the following day it works completely differently. One says: "We do like that." And the other says: "Wow, this is incredible!" "Yes, come and visit us!" And then they go there with their municipal council, and again they drink a glass of wine. Networks are like that. We tried to make it more technical by saying: "It is about work and not about amusement." But it does not work, if you trust people, then you think: "It was pleasant", and then you believe what they say, and you think that you can do the same. That is what networks are like, and it is Rainer's recipe for success. It has a lot to do with social capacities, with this capacity to party, so that everybody knows that sustainable development is not painful, hard labour but rather a delight. [...] There are not many people that can do it as Rainer does, but this is a recipe for success; it is absolutely clear to me. (Andreas, director of CIPRA International)

Andreas' words clearly show that this pleasant atmosphere does not simply "happen" but is an explicit strategy by some leaders to create a favourable climate for working on "serious" issues like sustainable development. It is, moreover, thanks to the visits organized in the framework of the networks' workshops that the self-esteem and the confidence on what its own territory can offer can increase, as Antonio highlights:

Molti hanno visitato Budoia, quindi anche questo è stato visto come un aspetto positivo. Io stesso ho avuto all'inizio il timore che Budoia non avesse niente di interessante per gli altri. Molte volte si è molto legati al proprio territorio, però si ha sempre un timore quando si è visitati, si vede sempre gli altri [come se hanno qualcosa in più da offrire]. Poi invece si scopre che anche per gli altri c'è qualcosa di interessante, e quindi questo rafforza anche l'identità. (Antonio, sindaco)

A lot of people have visited Budoia, so that this has been seen as a positive aspect. Even I, at the beginning, was afraid that Budoia did not have anything interesting to offer. Often we are really attached to our territory, but we are afraid when somebody pays a visit, because we always have the impression that others [have more to offer]. Then we discover that for the others as well there is something interesting, and this reinforces our identity, too. (Antonio, mayor)

However, what Antonio says about the possibility of increasing self-confidence through an exterior look did not appear in any other interview. What I found more often are the attempts to describe the pleasant atmosphere experienced during the meetings organized by the networks. For example, Robert, an animator of Alliance in the Alps, while remembering when he once organized a meeting with potential members of the network, speaks of the meeting using the familiar French word "*bonnard*", which could be translated as "cool". Robert observed how grateful the mayors attending the meeting were to have the opportunity to share some common worries and not only amongst French elected representatives but also people from the entire Alpine arc. During those meetings, some particularly interesting characters emerged. Andreas Götz takes Rainer Siegele as an example; Jakob, an Austrian mayor, remembers another person:

Jakob (Bürgermeister): Ich kenne auch einen Italiener...

Cristina: Zambon?

Jakob: Zambon, genau! Das Schöne ist, ich kenne ihn auch, aber er wird mich jetzt nicht kennen, aber er ist mir in Erinnerung geblieben. Wenn man hier [mit Allianz in den Alpen] Leute trifft, trifft man die, die Engagement zeigen und das ist auch noch was Besonderes. Also nicht die langweiligen Typen, Bürgertypen, oder denen das egal ist. Das ist auch was

Jakob (mayor): I also know an Italian...

Cristina: Zambon?

Jakob: Zambon, exactly! The nice thing is that I know him, but he will not recognize me, but I still remember him. When we meet people [with Alliance in the Alps], we meet those who show commitment, and this is the special thing. Not the boring "bourgeois" guys or the ones for whom it does not matter. It is special when you have the chance to

Besonderes, wenn man sich in einem solchen Kreise trifft und noch dazu in einem internationalen Kreis. Das ist also das Reizvolle.

be part of such a circle and, in addition, an international circle. This is exciting.

Social function of gastronomic rituals

It is probably not a coincidence that those two persons, Rainer Siegele and Antonio Zambon, were, at the time of the interview, president and vice president of the network, respectively. However, their commitment is obviously not only formal, because of their official function, but equally personal. People who met them appreciate their way of creating a relaxed working atmosphere. Meeting with people from different countries around a table loaded with good food and drinks helps to overcome cultural and linguistic difficulties. The ritual of organizing an "Alpine buffet", where every Alpine protected area brings some culinary speciality, even goes a step further (see fig. 20).

Fig. 20: Alpine buffet during the 16th Memorial Danilo Re in the Triglav National Park © Alparc, 2011.

The Alpine buffet was established around the beginning of Alparc, 15 years ago, and adopted by the Danilo Re competition, too. On that occasion, it is not only a matter of having good food together, but also a way of letting the participants show their own products that make them proud, as Erik suggests:

Le buffet alpin qu'on vient d'avoir c'est pour moi un bon exemple de montrer que les gens sont fiers de ce que leur région produit et ils sont fiers de le montrer. (Erik, employé d'Alparc)	The Alpine buffet that we just saw is for me a good example to show that people are proud of what their region produces, and they are proud of showing it. (Erik, employee of Alparc)

Some other gastronomic rituals take place during the meetings organized by the networks and prove that "taste" plays a crucial role in building social contacts. During the 2nd "Alpine Week", which took place in L'Argentière-la-Bessée (France) in 2008,[119] I noticed two interesting details concerning the important social function of food. Firstly, and in the same spirit as the "Alpine buffet", I could witness an interesting "exchange of products" between members of the network. Apparently, at every meeting, some people bring specialities from their region, which they exchange for other specialities from different regions. And so I observed Antonio giving a case of sparkling white wine from Friuli to Rainer, who, in turn, gave something else to Antonio. Secondly, and more symbolically, after my presentation on the results of the survey conducted with the members of Alliance in the Alps, which I held in a parallel session of the general meeting, I had the honour of receiving directly from Rainer the "Spirit of the Alps", as he emblematically calls the bottle of schnapps with the Alliance in the Alps logo (see fig. 21). And the "Spirit of the Alps", as Rainer conceives it, is also this way in which they undertake their mission, i.e. implementing sustainable development in the Alps in a pleasant way. But this bottle is also an invitation to travel through the Alps and to see first-hand how sustainable development is put into practice at the local level. In fact, while giving the bottle, Rainer always says: "This bottle is special, because when you finish it, you can come to Mäder and we will refill it for you for free", which is a clear sign of the importance given by Alliance in the Alps to the personal exchanges *in situ*, a sort of "learning by doing", or, better said "learning by visiting".

119 <www.alpweek.org>, consulted on the 20.04.2015.

Fig. 21: Bottle of "Alliance in the Alps schnapps" © Alliance in the Alps.

The pleasant atmosphere as guarantor of durable contacts

The pleasant atmosphere of the meetings helps to form lasting relationships. As Andreas says, it is easier to work with people you like and with whom you spent an enjoyable evening. Those events probably also create one of the best environments to find potential partners. As Mirko, a Slovene mayor affirms, "many mayors know each other from those meetings. Moreover, through those personal contacts, we can run things further". This quotation seems to suggest that it is easier to build more durable contacts if you personally know the persons you would like to work with.

Additionally, thanks to those meetings, the members have the possibility of encountering different cultures, which can be enriching not only in a personal way but also for professional purposes. For example, Mathieu, an employee of a municipality member of Alpine Pearls, admits that those meetings "are good, because people can compare their different tourist cultures". He adds that "we do not perceive the clients and the services in the same way". Thus, to be part of a network, on the one hand, is a good opportunity to meet people from different national cultures and, on the other, as Olivier, animator of Alliance in the Alps, says, it shows members that the concept of sustainable development is universal:

[Cela] montre que ces convictions de développement durable n'ont pas de frontières. On retrouve cela partout dans les Alpes. J'ai un peu du mal à exprimer ma pensée, mais je crois qu'ils trouvent un certain universalisme, quelque chose qui est partagé par différents peuples alpins, qui ont les mêmes logiques. (Olivier, animateur)	[This] shows that the concept of sustainable development does not have boundaries. We see it all over the Alps. I have some difficulties expressing my thoughts, but I think that they find a kind of universalism, something that is shared by the different Alpine people, who share the same thought process. (Olivier, animator)

Once the social contacts are established and the link between members is reinforced, another important dimension of the network can be put into practice, namely the exchange of information and the search for funds to implement projects of sustainable development. It is what I call the "pragmatic dimensions of the network", which I shall explain in the following section.

6.3.3 Pragmatic dimensions of the network

As previously discussed, one of the animators' strongest arguments for convincing local authorities to join a network is the possibility that mayors have to share and solve problems by exchanging experiences with other colleagues in the Alpine arc. The focus is therefore on, as Petite says, a "pragmatic" dimension (Petite, 2009: 313). In the following paragraphs, I analyse this dimension in greater depth.

As Petite reported, one of the interesting aspects of these exchanges among members is the opportunity to go beyond the relationships with one's closest neighbours, towards whom one normally feels a measure of jealousy and envy. The networks offer the possibility to experience what people are doing "three valleys further" and to be inspired by it without experiencing the sentiment of distrust that is possible to feel for one's immediate neighbour (Petite, 2009: 314). In the interviews, a common expression appears regularly, namely the idea that, thanks to the network, "local authorities do not have to reinvent the wheel anymore". In that sense, networks are better than "written stuff" (brochures or Internet sites), because they allow one to have insights into the details of the project's implementation, as Dragan, working for CIPRA, highlights:

> Local authorities learn faster from other municipalities, and it is also better than hiring some consultants or waiting for an agency for regional development or for the state to help you. [...] If you can hear directly from somebody, what they did, what is good and what is bad, what succeeded and what didn't, and then you try to develop your own strategy on the basis of those experiences, it is easier than reading in some brochure or on the Internet a kind of "best practices" list. Usually, the "best practices" do not say anything about the problems they encounter. Usually they say only good things, and local authorities get lost. (Dragan, CIPRA Slovenia)

It is thanks to personal exchanges that one can really understand the process, comprising successes and failures, of the implementation of projects. In fact, this was the very first motivation to create the Alpine network of protected areas, as Markus, an employee of a national park, recalls:

Der Gedanke war eigentlich, dass sehr viele Projekte gemacht werden, die gleich sind, aber nicht koordiniert werden. Und meine Idee war, wenn ich etwas über Wasser mache und auf einer Tagung bin, dann erfahre ich, dass in Italien das gleiche gemacht wird und ich habe von dieser Arbeit bisher nichts erfahren. Und ich habe damals zu Guido [Direktor von Alparc] gesagt: "Es wäre gut, wenn man eine Verbindung hätte", ein Netzwerk, in dem alle diese Parks zusammen sind und man diese Erfahrungen austauschen kann. Das war eigentlich so der erste Gedanke. (Markus, Mitarbeiter im Nationalpark)

The idea was, actually, that a lot of similar projects were undertaken, but not coordinated. And my idea was: If I do something about water, and I am at a conference, then I can find out that in Italy they are doing the same, although I don't know about this work. Then I said to Guido [director of Alparc]: "It would be good if we had a link", a network where all the parks are together, and we could exchange experiences. This was the first thought. (Markus, employee of a national park)

Now that the networks have been implemented, members actually see the positive effect of exchange of information and know-how. Thus, for example, in an article published in the Austrian regional magazine *Vorum: Forum für Gemeindeentwicklung in Vorarblerg* (Vorum, 2005), Hilti, the mayor of Schaan (Liechtenstein), welcomes the possibility given by Alliance in the Alps to exchange information and experiences. However, he stresses that, thanks to the network, one does not meet "any old" municipality, but instead, as he calls them, "*zukunftsorientierte Gemeinden*" ("future oriented municipalities"), local authorities that "show the way". Robert, an animator, says it shows that "if others did it, it is not utopian", which, he adds, allows for the shaping of local policies based on the experiences of others. In fact, these innovative municipalities show the path, as Claudio, regional secretary, experienced by seeing what other Alpine municipalities did in the field of tourism:

Alleanza nelle Alpi era interessante perché evidentemente sono zone che hanno sviluppato il turismo rurale, come le esperienze dell'Austria che ha lavorato moltissimo sulle aziende agricole, e che ha creato 20'000 posti letto. Se guardiamo anche l'Italia, la Francia. Ci hanno indicato di poter dare un valore aggiunto all'agricoltura coinvolgendola nel turismo. (Claudio, segretario regionale)	Alliance in the Alps is interesting, because, obviously, they work in areas that have developed rural tourism, as in Austria, where they worked a lot with farms and created 20000 sleeping units. The same in Italy and France. They demonstrated that it is possible to create added value for agriculture by adding tourist activities. (Claudio, regional secretary)

This quotation suggests that networks increase awareness of one's own capabilities and possibilities. Instead of competing with others, the networked collaboration allows each municipality to take advantage of others' experiences and see, as Antonio highlights, that the small peculiarities present on a territory have a common purpose. Antonio refers to a particular example, the *Ciastélazz*, an abandoned archaeological site in Budoia that was covered in vegetation. The impetus to open the site to the public came whilst observing the experiences of others in the network:

L'abbiamo scoperto [la fortezza] in un modo diverso. Queste cose ci sono sempre state, però sono sempre state viste soprattutto da chi è appassionato di questo elemento, [...] mentre adesso abbiamo visto che le nostre cose hanno un senso comune. Quindi i processi di Alleanza nelle Alpi sono valsi la pena,	We discovered [the fortress] in a different way. These things have always been there, but only by those with a passion for them noticed them, while now we have seen that our things have a common purpose. So, the Alliance in the Alps process was valuable, because we saw that in other places small things

perché si è visto come anche da altre parti le piccole cose sono messe in rilievo, allora perché non riconoscerle anche noi? Questo processo di vedere come cose molte semplici hanno avuto una valorizzazione da altre parti è stato un incoraggiamento, una maturazione. Ha incoraggiato a farlo. (Antonio, sindaco)

were highlighted. So why not recognize them here? This process of seeing simple things exploited in other places has been encouragement, a maturity process. It encouraged us to do it. (Antonio, mayor)

The notion that Antonio expresses with the word "encouragement" finds its way into other interviews. In an article published in *Vorum*, Nussbaumer, the mayor of Langenegg (Austria), speaks about "receiving new impulses" (Vorum, 2005). Olivier, animator, summarizes this process in one word, "inspiration", while Claudio, regional secretary, speaks about "vision". All these concepts underline the spirit of the network, the very deep significance of horizontal exchange: It is not just a matter of exchanging technical information but also the possibility of exchanging ideas and ideals.

The following two excerpts are good examples of how this exchange of experiences works in concrete terms and the themes that are worth sharing. Firstly, Ivica, from the Alpine Convention, explains how, in the field of management of protected areas, members of Alparc could benefit from being part of a network:

> It connects protected areas all over the Alps, and they can cooperate on common points. For example, they prepare management plans for protected areas, which also means plans to guide the visitors in those protected areas, how to establish some balance in protected areas between use and protection, how to add value to the fact that they live near or in a protected area and how they can profit from it. This means underlining the link between protected areas, tourism, agriculture and forestry of the protected areas. These are concrete things that everyone living there also feels. (Ivica, Alpine Convention)

Secondly, Mirko, mayor of a Slovene municipality, gives a concrete example of how, thanks to a visit to a member of Alliance in the Alps, he could understand how mountain farms are managed in other Alpine countries:

> We went abroad to exchange experiences about high mountain farms. Although we are not an agricultural area, we wanted to know how they solved those kinds of problems. As a mayor, I saw for the first time a high mountainous well-ordered farm, with all the contemporary standards. Actually, I found out how those things work. I also found out [how] elsewhere they arranged things more simply and in an easier way than we do. (Mirko, mayor)

Mirko's example shows the importance of countries less experienced with sustainable development in mountain economies, such as Slovenia, being able to witness the strategies that others have already implemented. Sabine, animator, defines it as "*unterschiedliche Wissenständeaktivität*", which can be translated as "activities with a different level of knowledge", especially compared with the long experience of Austrian municipalities in the field of sustainable development. The challenge has always been, for the network, as Sabine highlights, to allow other municipalities outside Austria to take advantage of this knowledge. This was actually how the idea to build a network came about. But, in the end, a network is useful because all municipalities and all protected areas have to struggle with the same problems, no matter in which country they are, as Branko, former director of a protected area in Slovenia, declares:

> Cooperation! Those physical barriers have fallen: this idea that they are something more and we something less; like with the network, we are all equal, we all try hard, everywhere there's lack of money, and this is the same everywhere. And this common complaint connects us. (Branko, former director of a protected area)

Nevertheless, and although Branko puts all municipalities in the Alpine arc on the same level, the reality is that in some countries the idea of sustainable development has been established for longer and is stronger: Therefore, if it is easy to understand why and how Mirko's municipality can profit from a network, the question remains what other, more experienced municipalities, can gain by joining the network. Peter, president of Alpine Pearls, explains this issue by emphasizing that even the "strongest" municipalities have something to learn from others:

Es gibt natürlich ein sehr unterschiedliches Niveau hinsichtlich der Einstellung in puncto öffentlicher Mobilität. Die Schweiz ist natürlich das Musterbeispiel. Die müssten sich vielleicht fragen: "Warum arbeiten wir in so einem Projekt mit Partnern, die in diesem Punkt wesentlich schwächer sind?". Ich denke, dass auch die Schweizer Musterorte wie zum Beispiel Saas-Fee oder Zermatt es noch nicht so wirklich touristisch genutzt haben, diese Sanfte Mobilität als Tourismusangebot darzustellen, als Teil eines Produktes. Als wir in Werfenweng begonnen haben, uns mit dem Thema der Sanften Mobilität zu befassen, haben wir eine Exkursion gemacht. Wohin? In die Schweiz natürlich, um uns Wengen und Zermatt anzuschauen. Und im vorherigen Jahr ist die Delegation dieser Schweizer Musterorte nach Werfenweng gekommen. Nicht um sich das autofreie Dorf anzuschauen, denn das haben wir nicht, aber um sich anzuschauen, wie wir das Thema Sanfte Mobilität touristisch vermarkten. (Peter, Präsident von Alpine Pearls)

There is, of course, very different levels of concern towards public mobility. Of course, Switzerland is a model. They probably wonder: "Why do we work in a project with partners that are considerably weaker in this field?" I think that even places like Saas-Fee or Zermatt did not really use the possibility to present their soft mobility projects in a tourist offer, as part of a marketing product. When we, in Werfenweng, started to work in the field of soft mobility, we made an excursion. Where? To Switzerland, of course, in order to see Wengen and Zermatt. And two years ago a delegation from Switzerland came to visit us. Not to see how we deal with the "car-free" concept, because we do not have one, but to see how the soft mobility theme is used in a marketing concept. (Peter, president of Alpine Pearls)

The interesting point, which is the advantage of being part of a network, is that information circulates better, because those experiences are collected in an exhaustive database that the networks provide to the members.

As Peter suggests, local authorities can always learn from the experiences of other members, since nobody excels at everything. For some members, "receiving" (information, support, know-how) is only one side of the coin, and interviewees also stress the importance of "giving". As Normann, an employee of a tourist office, says, "co-operation with other partners allows us to learn a lot but also gives others the opportunity to learn from us".

In addition to this mutual learning process, being part of a network facilitates the process of finding members who are searching for partners to build new projects:

On a fait tout d'abord un voyage "initiatique" en Autriche afin de se rendre compte du retard que les destinations françaises avaient sur un certain nombre de services à la clientèle, notamment dans ce qui concerne le "dernier kilomètre". À cette occasion on est allés à Obersdorf, qui sortait de Alps Mobility I. Ils avaient un contact avec un réseau qui était en train de réfléchir à la mise en place de Alps Mobility II, et ils cherchaient tout simplement des partenaires français. La proposition allait dans le sens qu'on essayait d'insérer au niveau de nos communes. [...] Donc, tout naturellement, on a commencé à entrer en contact avec eux. (Mathieu, employé d'une mairie)

We first made an "initiation" trip to Austria during which we came to see that French destinations were lagging in certain client services, for example in what is called the "last kilometre". During that trip we went to Obersdorf, which had been part of the Alps Mobility I project. They were in contact with a network that was thinking about the next project, Alps Mobility II, and they were looking for French partners. The proposition was going in the same direction as what we were trying to implement in our municipality. So, naturally, we started having stronger contacts with them. (Mathieu, employee of a municipality)

For Mathieu, thus, it has been a matter of "being in the right place at the right time". It was a coincidence that the Austrian colleagues were searching for French partners. An opportunity, that, obviously, Mathieu's municipality did not miss. If his story is one of happy coincidence, it seems that members of the networks are very well aware of this potentiality, namely the possibility of being chosen as a partner for projects. This specific aspect of networks is used by their members and promoted by the networks themselves. Jean gives a concrete example of research of partners for a specific project:

[Nous avons] rencontré Guido [directeur d'Alparc] pour avoir son avis sur les possibilités de montage au niveau de la Commission Européenne. Aujourd'hui on a un avant-projet pour un futur LIFE, avec une liste de personnes d'espaces protégés à l'échelle alpine. Donc moi j'ai [conseillé à ma collaboratrice] d'aller le revoir, pour voir s'il connaît d'autres espaces protégés [avec qui collaborer], parce qu'on va peut-être élargir le projet et traiter de l'aigle royal. Comme cela nous pouvons voir quels espaces protégés au niveau de l'arc alpin ont travaillé sur l'aigle, des espaces qu'on n'aurait pas identifié. (Jean, directeur d'un espace protégé)

[We] met Guido [director of Alparc] to get his advice on the possibilities of applying to the European Commission for a project. Today, we have a pre-project for a future LIFE, with a list of people from different protected areas in the Alps. So, I [recommended to my assistant] to see Guido again and to find out whether there are other protected areas [with whom we can collaborate], since we will extend the project to the golden eagle. This way, we can see which other protected areas have already worked on the eagle, maybe we have not identified some of them yet. (Jean, director of a protected area)

So, it is clear that the network is a precious tool for its members, because it centralizes the information about "who did what". This information, collected in a database, is then shared among the members and can be used to search for potential partners. This role of information bearer is particularly appreciated by the members. Claudio, for example, emphasizes that Alliance in the Alps has "valid information instruments". Network managers are therefore important mediators between members. Jean describes this kind of support as logistical, but he adds that the network is important for its technical and administrative support, too.

One of the most important and appreciated services that the network provides is the translation of all the documents they publish into four languages (Italian, French, German and Slovenian) and interpretation between those same languages during meetings. Without this support, documents and meetings would simply be inaccessible for a great number of members who are not fluent in the four Alpine languages that the network recognizes. More about the issue of languages will be discussed in the next paragraphs. By now it is important to underline that this service is a *conditio sine qua non* to allow members to get and give information; in other words, to allow the network to work and probably to exist as well.

The last point that I would like to mention is the call for mutual assistance that Alliance in the Alps organized for the municipality of Bohinj (Slovenia). Shortly after the decision taken in 2007 to join the network, along with Kranjska Gora and Bovec, Bohinj suffered from serious flooding that followed heavy rain falls. Local authorities estimated damage of around 8 million euros, a great amount of money for a municipality of some 5000 inhabitants. In view of the seriousness of the events, Alliance in the Alps decided to call for solidarity. The information was sent to all the members via the online journal *Flashinfo Alliance in the Alps*. Members could read that "the network wishes to support its members, which are facing difficult situations. That is the reason why we invite all the members to participate in our call for donations". The fundraising appeal allowed the network to collect 2680 euros for Bohinj (see fig. 22). Although this is not a great amount of money, it is a sign that networks can mobilize their members in case of urgent need. The wide communication capacity of the networks to keep members in contact is thus one of the network's strengths.

Spendenaufruf für die Gemeinde Bohinj/Slowenien

Am 18.September wurde die Gemeinde Bohinj durch Überschwemmungen schwer beschädigt. Alle sind herzlich eingeladen, sich bei der Spendeaktion zu beteiligen.

© LTO Bohinj

Die Gemeinde Bohinj (Slowenien) wurde am 18. September 2007 aufgrund schwerer Überschwemmungen für zwei Tage von der Aussenwelt abgeschnitten. Das Hochwasser hat vor allem die Infrastruktur beschädigt. Die Eisenbahn, die letzten Sommer das hundertjährige Jubiläum feierte und sich für das UNESCO Kulturerbe beworben hat, wurde an manchen Stellen völlig zerstört. Das Erdgeschoss der lokalen Grundschule wurde durch Wasser- und Schlammmassen verwüstet. Die größte Fabrik der Gemeinde musste für einige Tage schliessen, was mit finanziellen Einbussen einher ging. Zerstörte und instabile Strassen sowie die unterbrochene Bahnverbindung gefährden den Tourismus, welcher die Haupteinnahmequelle der Gemeinde ist.

Nach bisherigen Schätzungen betragen die Hochwasserschäden mindestens 8.5 Millionen Euro. Bohinj wird in den kommenden Monaten und Jahren viel investieren müssen, um die Schäden zu beheben. Das „Gemeindenetzwerk Allianz in den Alpen" möchte seine Mitglieder in schwierigen Situationen unterstützen. Daher laden wir alle Mitglieder ein, sich an der Spendenaktion für Bohinj zu beteiligen, die an der Mitgliederversammlung in Schruns angefangen hat.

Für weitere Infos: info@alpenallianz.org (de/fr/it/sl)

Fig. 22: Flyer of the fundraising appeal of Alliance in the Alps for Bohinj.

As this section has already pointed out, the networks are able to give, in four languages, technical information concerning how to implement sustainable development projects. They further provide a rich database in which members can find examples of projects as well as possible partners. Additionally, one of the network's strengths, as recognized by the interviewees, is the possibility of belonging to a group of municipalities that fight together for a common goal, namely sustainable development at the Alpine scale. Thus, by promoting sustainable development, the network has a strong ecological component, which will be explained in the next section.

6.3.4 Ecological dimensions of the network

In the 4th issue of the Austrian magazine *Vorum, Forum für Gemeindeentwicklung in Vorarblerg* (Vorum, 2005), two mayors highlight the sustainable dimension of the Alliance in the Alps network. Bürscher, mayor of Grossraming (Austria), says that "through membership of Alliance in the Alps we had to learn how to increase our awareness of the features of nature and environment". In this regard, Bürscher concentrates on how to care about the cultural landscape and, using this argument, how to communicate with the population about sustainability. In the same issue, Hilti, the mayor of Schaan (Liechtenstein), sees the aim of the network as a possibility to collaborate in order to "develop an Alpine habitat in a sustainable way, following the objectives of the Alpine Convention, i.e. by searching for an equilibrium between the economic interests and the protection of nature".

As Hilti underlines, the principles of sustainable development anchored in the Alpine Convention should be the guiding thread for the actions and projects that the networks' members undertake. Indeed, the first principle of the Charter of Alliance in the Alps states: "We [members of Alliance in the Alps] want to be model municipalities for a sustainable development in the Alpine space, following the ideal of the Rio Agenda 21". Consequently, in the light of the importance that the networks attach to sustainable development, I expected to hear many mentions of it in the interviews.

Therefore, I was surprised to find only a few (explicit) references to sustainable development. These occurred only in the interviews Nicolas and I conducted with members of Alpine Pearls, while for the other networks, if it occurred at all, it was only mentioned indirectly, for example when members spoke to me in general terms about the projects they implemented in their municipality. Those projects are, of course, moving in the direction of sustainable development; however, it was rare for the word "sustainable development" to be mentioned openly. I assume this is because it is taken for granted by the majority of the members, and they do not consider it worth mentioning. As Mathieu says: "The fact that we are an Alpine Pearl means that we respect the environment." The way in which Mathieu pronounced those words seems to suggest that "sustainable development goes without saying". Nonetheless, big was my astonishment when I re-analysed the interviews and did not encounter any explicit reference to it. More in-depth research on how the members conceive sustainable development would probably help to find the reasons for this silence on the matter.

As I said, only members of Alpine Pearls refer to the ecological dimension of the network. Peter, the president of Alpine Pearls, is probably the most explicit in that regard:

[Es ist] nicht zu vergessen warum wir eigentlich kooperieren. Wir kooperieren nicht weil wir einfach nur gemeinsam Tourismus machen, sondern weil wir eben diese umweltfreundliche Mobilität unterstützten. (Peter, Präsident von Alpine Pearls)

[It should not be] forgotten why we cooperate. We do not cooperate just because we all are in tourism together, but because we support eco-friendly mobility. (Peter, president of Alpine Pearls)

Hence, Peter points out that the cooperation built thanks to the network is based on the aim of implementing sustainable development for Alpine Pearls in the field of transport. For him, the cooperation is based on that principle, and the network is there to "show the way"to achieve it, as Luc says:

Nous avons tous les atouts sur place, et comment arriver à valoriser ces atouts, tout en les préservant, là on est vraiment dans le développement durable. La mobilité dans les Perles des Alpes, c'est un des volets sur lesquels on travaille pour arriver à cela. (Luc, employé d'une municipalité)

Here we have all the assets. The question is how to give value to those assets while preserving them. This is actually the point of sustainable development. Mobility in the Alpine Pearls is one of the dimensions on which we work to reach that aim. (Luc, employee of a municipality)

The possibility given by the network to find tools in order to keep a balance between an attractive tourist offer and the preservation of the environment seems to be particularly important to the members of Alpine Pearls. Among the three networks studied here, Alpine Pearls is the most concerned with tourism. The network, thus, becomes a sort of label of sustainable development, which can be used in marketing terms in order to attract tourists. The network, as Antje, an employee in a tourist office, said, gives "a possibility to the members to position themselves differently on the market", i. e. to play the sustainable development card. This position should be clear to the potential guest, who should understand, as Normann (an employee in a tourist office) says, that by choosing an Alpine Pearl destination for his or her holidays, he or she is doing something for the environment and can be sure that in five or 10 years there will still be clear drinking water.

So, Normann suggests that tourists, by choosing their holiday destination, are aware of Alpine Pearls' commitment to sustainable development, since the Pearls are attentive to the natural capital they possess. It is precisely this capital that attracts tourists, and hence the importance of preserving it in order not to lose guests:

Und [wir machen] eigentlich wirklich alles im Sinne des sanften mobilen Tourismus und der Natur, die wirklich unser Kapital ist. Wenn wir diese zerstören, dann werden wir automatisch auch unsere Gäste verlieren. Da sehen wir uns auch ein bisschen in der Verantwortung hier etwas zu tun. (Viktoria, Mitarbeiterin im Fremdenverkehrsbüro)	And [we actually do] it following the vision of a soft mobile tourism and of nature, which is truly our capital. If we destroy it, we will automatically lose our clients. So we sort of feel this responsibility to do something. (Viktoria, employee in a tourist office)

The importance of maintaining a preserved environment is therefore of crucial importance for the tourist destinations that joined Alpine Pearls. Implementing the principles of sustainable development is seen as one means to reach this objective.

The survey among members of Alliance in the Alps also shows the importance they attach to sustainable development, as fig. 16 illustrates. However, during the interviews sustainable development and the preservation of the environment did not emerge as important subjects for discussion. The members of Alliance in the Alps only rarely mentioned it. This probably does not mean that its members do not care about this issue: It could simply be a sign that sustainable development is taken for granted and that being part of the network is already a guarantee that local authorities are committed to sustainable development. Yet, while this is certainly true for members of the network, it is debatable whether it is true as well for the single members inside a region. In fact, the examples from the two regional members in Ticino show that only the two regional secretaries are really persuaded on this issue, while the majority of mayors are busy running the daily life of their municipalities.

6.3.5 Economic dimensions of the network

The last, but not the smallest, factor to which the members of the networks refer is the economic one. As already discussed in part for the pragmatic dimension, networks are an interesting platform where members can find potential partners in order to answer calls for projects and, hopefully, receive money from different funding agencies (European Union, national states, private foundations, etc.) in order to implement them. The economic aspect of networking, thus, is not insignificant. The mayor of Saint-Martin (Switzerland) declares in *Vorum. Forum für Gemeindeentwicklung in Vorarlberg* that to be part of the Alliance in the Alps network "greatly" contributed to obtaining "substantial" financial support for his municipality (Vorum, 2005). This aspect is also underlined by the people I interviewed. Two of them, Julien, director of a protected area, and Peter, president of Alpine Pearls, are convinced that the odds of receiving funds are higher if the partners' configuration is international, hence the interest of getting in touch with protected areas and municipalities from other Alpine countries:

J'ai entendu que pour les crédits européens l'aspect transfrontalier va être de plus en plus prioritaire. Donc pour nous c'est une bonne carte à jouer avec le Parc National du Grand Paradis et pour mobiliser les crédits sur des actions communes. (Julien, directeur d'un parc)

I have heard that, for European funds, the cross-border aspect will have more and more priority. This means that in our case we hold a good card with the Grand Paradiso National Park in order to receive funds on common actions. (Julien, director of a protected area)

Julien is probably right, in the sense that protected areas or local authorities that can show that they have previously taken steps in transnational collaborations are likely to be chosen for further projects. Robert confirms Julien's perception by saying that:

Il y a plein de programmes, des programmes Interreg, dans lesquels, lorsqu'il s'agit de chercher des communes susceptibles de servir de cobayes, tu vas te tourner plutôt vers des communes qui sont déjà habituées à travailler de manière interdisciplinaire ou en réseau, dont on sait qu'elles seront capables d'aller de l'avant sur un projet-pilote. (Robert, animateur)

There are so many programmes, Interreg programmes, where, when one is looking for local authorities likely to serve as a "guinea pig", one first looks at those municipalities already used to working in an interdisciplinary way or in networks. So we know that they are capable of going ahead in a pilot project. (Robert, animator)

Indeed, the capacity of networks to secure funds that are reallocated to their members is quite an attractive feature of the network. To give a concrete example of the amount of money involved, Alliance in the Alps received three main grants for projects: with the Interreg programme Dynalp (2003–2006), 2.16 million euros were distributed to its members; in 2006, 1.7 million euros were allocated by the MAVA foundation for the Dynalp 2 project (2006–2009); finally, since 2009, the same foundation has sponsored a grant of 800000 euros for a programme dealing with climate change. This sum of money, although constantly diminishing through the years, is an important incentive for municipalities to be active in the network. However, while Leonardo, a former intern in a region member of Alliance in the Alps, welcomes this financial contribution, he raises a pertinent question about long-term perspectives:

Io credo anche che adesso in questi ultimi anni Alleanza nelle Alpi è funzionata, perché c'era questo grosso progetto Dynalp. Sono arrivati dei fondi, e quindi è chiaro che a tutti interessa essere membri. (Leonardo, ex stagista in una regione membro di Alleanza nelle Alpi)

I think that in the past few years Alliance in the Alps has worked well, because there was this big Dynalp project. Funds arrived, so clearly everybody has an interest in being a member. (Leonardo, former trainee in a region member of Alliance in the Alps)

Although Leonardo did not pursue his line of thought, it is obvious that he would have continued by asking himself what will happen to the network when Alliance in the Alps is no longer able to collect further funding. Fig. 18, which shows the years of membership of Alliance in the Alps' members, indicates additional elements to support Leonardo's theory. In fact, as has been already discussed, there is a clear correspondence between the acquisition of new funds by the network and the arrival of new members. So there are doubts as to whether the network would be able to continue growing if funds were no longer allocated.

Another way of obtaining economic advantages by being a member of a network is well described by Fabio, regional secretary. During the interview he explained his strategy for offering his region the possibility to profit from the network. Fabio tried, thanks to two networks of which the region is a member, Alliance in the Alps and another non-Alpine network that no longer operates, Destilink, to organize some network meetings in his region. Fabio calculates that, for example, for a three-day workshop he organized, he had to manage a budget of around 80000 Swiss francs.[120] Thanks to the exclusive use of products from local producers, 70 % of the money, in accordance with his calculations, remained in the region. In addition to a merely financial incentive, those meetings are interesting for other reasons, too. As Fabio says, they are normally organized during the tourist off-season, and they attract "sensitive persons" who are likely to return a second time not as delegates of their municipality but as tourists.

For more tourism-oriented networks such as Alpine Pearls, the network allows the creation of some scale economies. As Karmen, manager of Alpine Pearls, underscores, a network allows members to have 21 times (currently 27, since new members have joined the network) more resources, for example for the communication with the media. Viktoria, employee in a tourist office, appreciates this aspect of the network:

Es gibt auch noch einen Beitrag, den man in einem Marketingpool wirft und zu dem dann gemeinsam verschiedene Marketingaktionen gemacht werden. Für uns ist vor allem das Mediennetz sehr interessant, das sich dann auch in Deutschland, Österreich und Italien öffnet. (Viktoria, Mitarbeiterin im Fremdverkehrsbüro)

It is also interesting, because we are part of a marketing pool, where we can work on some marketing campaigns together. For us, it is particularly interesting to be part of a media network, because it opens up possibilities in Germany, Austria and Italy as well. (Viktoria, employee of a tourist office)

Obviously, via a common and stronger marketing tool, members of Alpine Pearls hope to gain guests in a kind of game in which one tries to find an equilibrium between competition and collaboration, as Antje clearly states:

120 At that time, when the Swiss franc was not as strong as it is at present, 80000 Swiss francs corresponded to around 50000 euros.

Sie [die andere Perlen] sind Mitbewerber, aber genau so ist es ein gemeinsames Netzwerk, in dem jeder davon profitiert, in dem jeder vom anderen profitiert. Weil es mein Gast lesen kann und nach Werfenweng oder nach Mallnitz geht. Aber genauso gut kann es der Gast in Mallnitz oder wo auch immer lesen [über uns und hier kommen]. In Berchtesgaden zum Beispiel. Berchtesgaden ist viel größer, also lesen es vielleicht viel mehr Gäste und kommen dann zu uns. Das wird man nicht messen können, aber das ist auch eine gegenseitige Befruchtung. (Antje, Mitarbeiterin im Fremdverkehrsbüro)

They [the other Pearls] are competitors, but we are also part of the same network, in which everyone can profit from the others. In fact, my guests can read and then go to Werfenweng or to Mallnitz, as the guests from Mallnitz can read [about us and come here]. If you take Berchtesgaden, for example, Berchtesgaden is much bigger, so that many people will read about the network and then they might come to us. It is impossible to calculate, but it is a mutual enrichment. (Antje, employee of a tourist office)

As already claimed by Antje with her example about Berchtesgaden, Normann (employee in a tourist office) also highlights that in such networks the "smallest" localities profit more than the bigger ones, since they will be presented in the same brochure written in five languages and distributed in every Alpine Pearl locality.

Thus, the members are aware of the diverse possibilities offered by the network to increase, although in different ways, their economic capital. Yet, it seems that this information does not go beyond the circle of the members, as is shown by the reasons mentioned by potential members for not joining the network.

6.4 Challenges of networking

Following the overview of the different features that networks can offer their members (political strength, rich socio-cultural encounters, technical support, spreading sustainable development and economic advantages), it is now time to discuss the challenges that pan-Alpine networks are facing, according to their members. The first challenge is linked to the actual lack of members' participation in the life of the networks. I will underline the barriers to participation in particular.

The second challenge concerns only Alliance in the Alps and is linked to its decision to favour regional memberships, which is not, as will be discussed, a risk-free choice. The third challenge that will be presented is the controversial issue about the number of members (i.e. should the network increase the number of members or not?). The last challenge, namely the question linked to structural dimensions, is the most topical, since it deals with recent concerns but also affects future developments of the pan-Alpine networks.

6.4.1 Participation at events organized by networks

Networks attach great importance to members participating in events. Nevertheless, the goal of having all members actively participate has yet to be reached. Networks put in place expensive means (notably for translation and interpretation in four languages) to make it possible for everybody who attends those events to understand what is being said and discussed. If those efforts are thoroughly appreciated by the members, they are not a sufficient incentive for local political actors to actively participate. This is but one example; other barriers to participation, as will be explained in a second part, also play an important role.

Coping with languages

From the very beginning, networks have paid a great deal of attention to the language factor and ensured that everyone can follow everything in their native language. The Alps are characterized by their cultural fragmentation, as evidenced by the different linguistic groups. Languages and language groups of the Alps include Slavic languages, the Rhaeto-Romanic language group comprising the minority languages Romansch, Ladin and Friulian, the German language group with Alemannic and Bavarian as well as the Walser dialect, which is still spoken in a few areas, and the Romance language group with French and Italian.

The official languages of the Alpine Convention, and of the pan-Alpine networks, are French, German, Italian and Slovenian. While it is undeniable that this diversity constitutes the cultural richness of the Alps, it is also true that because of this linguistic fragmentation we can question the capacity of the networks to manage high-level communications between the members. Although I expected the members to mention this factor, the survey conducted at the beginning of the research showed that the members do not feel that language represents a barrier to communication.[121] The time, distance and costs engendered by the Alpine topography seem to be much more important concerns.

On this issue, the majority of the interviewees told me they are more fascinated than frustrated with how communication between members works. Sabine phrases this sentiment very well:

Und durch diesen relativ hohen Anteil von Sprachkundigen [ist alles ziemlich einfach]. Die stellen sich natürlich ganz selbstverständlich immer wieder auch als Übersetzer zu Verfügung. Es wird oft nur so gesagt: "Komm bitte, übersetze!". Das funktioniert schon ganz gut. [...] Eigentlich finden es die meisten spannend. (Sabine, Betreuerin)

Thanks to the very high proportion of people who are talented with languages [everything is quite easy], they are always and naturally at the disposal of other people. Often people will just casually say: "Come, please, translate." This works quite well. [...] Actually, most people find it fascinating. (Sabine, animator)

Of course, sometimes communication is a bit difficult, but I would say that the great majority is satisfied with how communication works both during informal and formal meetings. Apparently, at the beginning, some attempts were made to introduce English as the official language. This was seen as the easiest solution, since only one language would be used. However, this did not happen. In fact, as Sabine remarks, "language awareness is relatively high" in the Alps, which relegates English to the status of an outsider's language. Antonio recalls his experience as follows:

L'esperienza che ho avuto io con Alleanza nelle Alpi è che magari si comincia il discorso in italiano, poi si passa al francese,

My experience with Alliance in the Alps is that maybe one starts saying something in Italian, and then they continue

121 In the questionnaire, municipalities were asked to enumerate the difficulties or limits they encountered with Alliance in the Alps. The question was open, and a blank space was left at the disposal of the municipalities.

e si finisce in tedesco. Però nelle relazioni comuni si riesce ad avere una comprensione. Serve un aiuto, forse nei linguaggi commerciali o piuttosto che negli affari, magari l'inglese è più standard, però nei rapporti normali credo che si possa riuscire a capirsi. [...] L'inglese è completamente estraneo all'arco alpino, [mentre] c'è un grande valore culturale nella parlata delle Alpi. (Antonio, sindaco)

in French, and they finish in German. But in the common dealings we understand each other. Help is perhaps welcome, in the commercial languages or for matters of business, but in the normal contacts I think we can understand each other. English is completely foreign to the Alps, while there is great cultural value in the Alpine languages. (Antonio, mayor)

Antonio confirms Sabine's remark on the attachment of the Alpine population to their linguistic richness. Members simply will not accept having to communicate in English, since English is completely unrelated to their world. Of course, this choice does come at a cost. Although Alliance in the Alps could not give me precise figures concerning the costs for translation and interpretation, I could get an idea of how much the network spends for this service.

In 2010, of the total expenditures of approximately 400000 euros, around 15000 euros went to translations. Furthermore, a sum of around 9000 euros has to be added for the translation of the main newsletters (*Kurzinfo* and *vomNetzwerkinfo*) and some 4500 euros per day for interpretation during meetings. Those figures give an idea of the heavy financial weight of translations and interpretations. Nevertheless, this financial effort is probably the only possibility the network has to give full access to its activity. This effort seems to be appreciated by the members, and the fact that nobody told me that they do not participate in meetings because of the language barrier is probably a good sign of the network's success in this area.

Barriers to participation

Nonetheless, although language is not mentioned by the interviewees as a deterrent, during the meetings that I attended I had the impression that it was rare to see "new faces". This impression is supported by conversations I had with mayors, animators, rangers and directors of protected areas. If I take the two regional members in Ticino (Switzerland) as an example, only one person, namely the secretary of one of the two members' region (of a total of 35 municipalities), actively participates in the meetings. The Grosses Walsertal counts six municipalities out of which only one mayor (plus the regional manager) regularly attends events organized by the network.

The situation seems better for Alparc, as shown by the figures its director, Plassmann, presented during the research seminar organized in Geneva.[122] The data he presented assert a low participation of directors and a high participation of employees of protected areas. Nevertheless, the interviews that Petra and I conducted with rangers during the Danilo Re event suggest that rangers would like to participate more than they are actually doing. In the following section, the possible hindrance to participation will be discussed.

In his research concerning the Walser Alps project, an Interreg project that set up a network of communities working to conserve the vestiges of the so-called Walser culture, Petite already underlines that the rhetoric promoted by networks concerning the crossing of borders hides the difficulties to overcome the distance separating the partners (Petite, 2009: 300). Obviously, distance (the Alpine arc measures 1 200 km) constitutes an inconvenient factor for communication. But, as Petite could observe for the Walser Alps project, not even the new communication tools are really solving the problems (Petite, 2009: 300). The results of the survey about Alliance in the Alps also correspond to the observations made by the geographer. To the question "Do you regularly use the website of Alliance in the Alps?", I received the following answers (see fig. 23):

Do you regularly use the website of Alliance in the Alps?	% (N = 88)
Never	17
At least once	16
Occasionally	51
Regularly	11
No answer	4

Fig. 23: Use of the Alliance in the Alps website.

122 "Mountain networks" research seminar (Geneva, 28.11.2007).

Clearly, consulting the network's website is not yet part of the members' habitual practices. Not even the network's newsletters and emails seem to be read on a regular basis, although their consultation demands less effort than attending meetings:

An die Bürgermeister kommt [die Newsletter] schon, aber meine Situation ist es jetzt so, ich lese dann ganz kurz, vielleicht interessiert mich das Thema und ich klick das an, dann lege es auf die Seite. Und ich gebe es auch von mir aus nicht weiter, weil ich mit den Tagesproblemen und so weiter beschäftigt bin. (Jakob, Bürgermeister)	Yes, we receive [the newsletter], but in my case, I read it quickly, maybe I am interested in the theme, and I click on the link, then I just put it aside. And I do not pass it to others either, because I am so busy with the everyday problems. (Jakob, mayor)

If "virtual participation" is already quite weak, "real participation", which demands time and financial investments, becomes even weaker. For some animators, however, actual participation is one of the points discussed during the first meeting with potential members. Sabine, for example, say that the members for whom she is responsible participate at least once a year in an international meeting organized by Alliance in the Alps. The other animators do not seem so precise about this issue, although everybody agrees on the importance of the personal meetings and exchanges. However, in reality it is difficult to motivate people to travel around the Alps in order to follow the networks' activities. During my fieldwork I realized that it was rare that new faces popped up during those events. This impression is confirmed by Maximilian, a relatively active Austrian mayor, who asserts that "more or less, one always meets the same people in those meetings".

This is not really surprising, as it became apparent in the interviews that the main obstacle to participation is the lack of time and available resources in terms of money and municipal personnel. Although almost every interviewee mentions this, the acuity of the problem seems to be partly dependent on the national outlook. Fabio underlines for example the difference between Ticino (Switzerland) and the "Mäder special case" (Austria):

Fare il sindaco di un piccolo comune vuol dire lavorare al fronte. Conosco benissimo il lavoro di segretario comunale, e bisogna far attenzione, perché è veramente gente che non ha un secondo libero del proprio tempo, cioè fanno la contabilità, danno retta alla gente, fanno il sociale, la pianificazione, devono fare di tutto. Questo deve essere messo in conto un po' di questo. Perché il Ticino è un po' più malmesso rispetto agli altri. [E' davvero difficile], a meno che non sei un comune come Mäder, dove [c'è una persona che ha il mandato di lavorare] per Alleanza dentro il comune. Qui sono tutti "poveri disperati", sono tutti volontari che devono prendere vacanza per andare al seminario su in Austria o nelle altre parti. (Fabio, segretario regionale)

To be mayor of a small municipality means to work at the forefront. I know the work of municipal secretaries very well, and you have to be careful, because they really do not have a free second; they do the accounting, they listen to people, take care of the social, of planning, they have to do everything. This has to be taken into account. Ticino is a bit poorly off compared with the others. [It is really difficult], unless you are a municipality like Mäder, where the person working for Alliance in the Alps has the specific mandate to work for it. Here they are all "helpless people"; they are volunteers, and so if they want to go to Austria for a seminar or to another place, they have to take holidays. (Fabio, regional secretary)

Fabio's remarks suggest a difference due to the national context. However, one must admit that the size of the Alpine arc is such that travel takes a long time for everybody (up to 12 hours in some cases) and is expensive. Thus, the chronic lack of time that local authorities are persistently struggling with is one of the most important constraints for attending events. And this is true for all three networks, without distinction.

Lack of financial means is cited less often as a reason for not participating. In only one case, which concerns a Slovene protected area, financial difficulties are mentioned as a deterrent to participation. Its director's words stress the difficult economic situation, which does not permit him to be more active in cooperation:

> We couldn't cooperate because of financial reasons. Although other parks kept inviting us for different meetings, we couldn't participate because of those reasons. [...] Every session, every excursion costs money, and we simply couldn't afford it. We had to take care of everything by ourselves, and we didn't have any funds for those kinds of things. When the financing is arranged within the framework of the regional park, then the service for the protection of the nature will be financed from the state's budget. Until now we have had to finance those things ourselves. We will probably get some money from the budgets of the municipalities for that kind of cooperation, which is certainly very useful and also desirable. Until now we haven't had those possibilities. (Goran, director of a protected area)

Goran actually started by saying "We feel guilty about this issue", which emphasizes the frustration he feels even more. Another kind of frustration, interesting to note even if probably not applicable to every case, concerns the internal organization of a protected area in Italy that Laura describes. She affirms that only a few people from her protected area participate in events, because the cooperation between parks involves only those who are informed inside the park. So, as she says, "[Directors] choose their friends and say: 'Come on, let's go to Africa', and they go. They come back, and nobody knows." In so doing, those who could not go only find out about the trip because they read about it in the reports. Laura admits being frustrated by this behavior, because she says that she can only dream of expanded cooperation in the Alps if there is already internal competition inside the park where she works.

Even if Laura was the only one openly criticizing how collaboration functions, her account shows that internal dynamics, especially in the case of large and hierarchical structures like protected areas, can be the cause of a selective participation of the staff in the events that networks organize. What is certain is that, in general, directors participate more in the Alparc's meetings than rangers, who have their "own" events, such as the Danilo Re Trophy.

Even if internal structure and organization can affect the choice of who is actually going to events, the major factor for non-participation remains a lack of time. This issue hinders not only the opportunity for face-to-face meetings but also "virtual" exchanges, as Ulrike explains concerning a particular example:

Ulrike (regional Sekretärin): Ein Schweizer hat mich mal angemailt und hat mich dann auf ein Projekt in der Schweiz verwiesen. Ich muss noch nachschauen. Ich habe kurz im Internet geschaut, aber jetzt wirklich nachtelefoniert habe ich nicht.	Ulrike (regional secretary): A Swiss wrote me an email, and he told me about a project in Switzerland. I still have to have a look at it. I had a brief look on the Internet, but I actually did not call him back.
Cristina: Warum hast du nicht?	Cristina: Why not?
Ulrike: Nicht dass es mir egal ist. Ich wollte das eigentlich schon länger besser anschauen, aber es ist immer etwas, was ich voran schiebe. Vielleicht habe ich im Sommer, mit den ruhigen Stunden, ein bisschen mehr Zeit.	Ulrike: It is not that it does not matter to me. I wanted to have a look, but it is always something that I think of doing later. Maybe I will have a bit more time during summer, when everything will be calmer.

As Ulrike suggests, it is not a lack of interest that makes it difficult to get in touch with people, but really time and a matter of priorities. This is also true for the information arriving from the network. As Jakob, a mayor, admits, even if he is concerned by the themes developed in the network's newsletters, he "*meistens leg[t] es auf die Seite*" ("mostly put[s] it aside") because of a lack of time. In fact, for a municipal administration, other urgent and important problems have priority on pan-Alpine concerns, as Manuele observes for the municipality where he works. There, administrators and politicians had to face a long and time-consuming process of *aggregazione comunale*, i.e. the merger between different municipalities into a single entity:

Durante questo processo di aggregazione, in un comune come il nostro, per i politici [Alleanza nelle Alpi] è sicuramente priorità 3 [che significa molto bassa], questo senza dubbio. Per cui adesso come adesso, o è qualcuno veramente sensibile a questo tema [e allora partecipa] se no scordati che partecipano a qualcosa del genere. (Manuele, impiegato comunale)	During this merging process, in a municipality like ours, for politicians [Alliance in the Alps] is priority 3 [which means really low], without any doubt. This means that now, if somebody is really sensitive to this theme [he or she participates], if not, you can forget about them participating in something like this. (Manuele, employee of a municipality)

Manuele highlights how personal motivation plays an important part: If someone believes in sustainable development and in the possibilities offered by the networks to get input for local politics, then he or she will travel farther and longer in order to meet other committed persons. If someone does not consider it a priority, he or she will not find the time to participate. This assertion is confirmed by the informal talk I had with a member of staff of the Ecrins Park, who asserts that "in the end, working in a transnational way is a choice and not only a matter of lack of staff".[123] And if the person is not interested in transnational meetings and does not have much time and resources, he or she can always, like Niklas, find other strategies to build cooperation, for example by restricting the sphere of activity.

123 Personal notes during the General Assembly of Alparc, Bled (Slovenia), 07.–12.10.2008.

In fact, when they decided to build a thematic walking path in his municipalities, Niklas tried to search for previous experiences in such a field. He admits that he did not search in the Alliance in the Alps' "best practices" Internet page but just had a look at the regional projects and made an excursion with some colleagues inside the cross-border region. Although Niklas was practising what the network offers, i.e. exchanging experiences and information, because of a lack of time he restricted his visits to a one-day excursions. By doing so, however, the *raison d'être* of a pan-Alpine network is still far from being implemented in the daily actions of local political actors.

6.4.2 Singular or regional membership

The possibility of joining Alliance in the Alps as regional members (i.e. a group of municipalities) and not as a single municipality is deemed by the network to be a promising solution for allowing mayors to be more present in the network's activities. In fact, as highlighted by Claire (manager of Alliance in the Alps) during the research workshop organized by our department in Geneva,[124] there are three main reasons why the network thought about allowing regional memberships:

- The membership fees for municipalities are lower if they join the network in a regional configuration;
- A certain number of issues related to sustainable development cannot easily be solved at the local community level; the regional level is often more adequate, for example for issues like transport;
- In a regional configuration, more than one mayor carries the responsibility of the membership, which means that they can organize themselves in order to share the time and resources devoted to the network.

124 "Mountain networks" research seminar (Geneva, 28.11.2007).

As the third point suggests, it would theoretically be possible for more people to participate in the events with regional memberships. In reality, things are more problematic, and the head of the network seems to be aware of that. In fact, as pointed out by Claire, one of the major problems that appeared with regional memberships is the lack of involvement of the lower level, i.e. the single municipalities comprising the region. This point is worth discussing in greater depth, as I could observe some important evidence related to this issue during my field work.

In the very first phase of the research, when I was collecting the questionnaires, I called the municipalities that had not replied. Somewhat surprisingly, a consistent number of them (especially in Switzerland and Italy) answered that they would have been really happy to answer, but that they simply could not since they thought, wrongly, that they were not members of the network.

All those municipalities have one common point: They all joined the network via a regional membership. This clearly illustrates a lack of information provided by the region to the municipalities. The different answers I received from Austria, where everybody seems to be better informed about and more active in the network, and Switzerland pushed me to conduct in-depth interviews in those two national settings. I chose the Vorarlberg *Land* (Austria) and Ticino Canton (Switzerland), since differences in the behaviour of single municipalities composing the regions seemed to be highest. Inside those regions, some municipalities decided to join Alliance in the Alps using the regional membership strategy: one valley in Vorarlberg (Grosses Walsertal) and two valleys in Ticino (Regione Malcantone and Regione Valli di Lugano).

From the beginning, it was interesting to note the different attitudes of the mayors of those two regional settings. In fact, only in the Ticino Canton did I find mayors who were unaware of being part of the network and had no idea what the network was about. During the interviews I tried to discern the causes of this difference in order to understand the challenges that this type of membership poses. Actually, even without speaking with the mayors, I noticed a different mindset, simply by entering the town halls. Whereas in the Grosses Walsertal I was "welcomed" by an Alliance in the Alps plaque hanging on their entrance doors, in Ticino I could only see this sign at the door of the regional office, but not at the municipalities' town halls. This observation, which to me at first seemed to be a simple curiosity, was actually a sign of a more substantial discrepancy that results in a different degree of identification with and knowledge about the network.

My observation about the different attitude between Austrian and Swiss regional members is, however, partly invalidated by Sabine's experience, an Austrian animator, who seems to be quite negative about the capacity of Austrian municipalities to work in a regional and coordinated way. In fact, she highlights that Austrians, compared with other areas, and ironically she cites Switzerland as an example, are much more attached to their municipality than to the region where they live. This attitude makes it difficult, in her opinion, to find effective regional decision-making processes:

> Bei uns ist es einfach diese Anbindung, dieses Gefühl "ich wohne im Ort", diese Identität ist wesentlich verwurzelter. Wir wohnen noch nicht in der Region, wir wohnen noch im Ort. [Dadurch] sind natürlich Entscheidungsprozesse, die einen Raum betreffen, der über die Gemeindegrenzen hinaus geht, viel schwieriger. (Sabine, Betreuerin)

> Here, in our place, this connection, this feeling of "I live in this locality", this identity, is much more embedded. We still do not live in the region; we live in our locality. [For this reason] decision-making processes that concern an area beyond the municipality limits are very difficult. (Sabine, animator)

Yet, it seems that, even without those strong regional structures, Austrian municipalities can find ways to organize themselves in order to, at least, share information. As two mayors of the Grosses Walsertal, Maximilian and Henrik, assert, every time somebody participates in an event organized by the network, they report to the others about what they have seen and discussed. Since the six municipalities comprising the region have monthly meetings, this information is shared through this channel. By contrast, the attitude of the mayors in Ticino seems to be different, and this already from the beginning, as Leonardo points out:

I comuni han sempre detto: "Se voi Regione volete entrare in Alleanza nelle Alpi, fatelo, potete anche rappresentarci, ma noi non abbiamo i mezzi per seguire questo progetto e poi non rientra nella priorità del comune, può rientrare nella strategia della regione". (Leonardo, ex-stagista in una regione)

The municipalities always said: "If you want to join Alliance in the Alps as a region, do it; you can represent us, but we do not have the means to follow this project, and it does not fall within the priorities of the municipality, it can fall within a regional strategy." (Leonardo, former intern in a region)

Leonardo's words indicate the mayors' distance from and non-identification with the network. They show a certain indifference about the issue, although they do not discourage the region from engaging in the project. They simply delegate the responsibility of the contacts between them and the network to the regional secretary and the regional animator of Alliance in the Alps, who happens to be, in the case of one of the regions in Ticino, the same person. Later in the interview, however, Leonardo admits how difficult it is to build a strong link between the local authorities and the network, given the mediated structure of the relationship. In fact, as he says, the region "is not really a representative of the municipalities". This leads to no proper feedback reaching the local authorities. This is probably one of the main reasons why some municipalities were not even aware that they are members of a network called "Alliance in the Alps".

The role of the regional secretary in this configuration is decisive. Fabio realizes that there are some advantages given to this mediated structure. In fact, as he says:

Il vantaggio è che se tu ne [Alleanza nelle Alpi] fai una parte della tua attività, di una tua visione, missione se vuoi, riesci poi a fare un filtro e a dare al comune veramente quello che gli serve. (Fabio, segretario regionale)

The advantage is that, if you manage to make [Alliance in the Alps] part of your activity, of a vision, of a mission if you want, you can act as a filter and give local authorities what they really need. (Fabio, regional secretary)

On the other hand, Fabio recognizes the difficulties of working this way. By acting as a filter, he receives much more information and knowledge than what he conveys to the mayors. He is therefore, and he admits as much, still searching for a way to balance what he gets and what he gives:

Il mio problema fondamentale è che mi rendo conto [...] che partecipando agli Interreg, alla rete, a Dynalp alla fine cresco io e devo capire come buttar fuori la mia crescita in conoscenza e in esperienza agli altri. Quello è un po' il problema. (Fabio, segretario regionale)	My fundamental problem is that I realize [...] that when I participate in Interreg, in the network's activities or in Dynalp, I am "growing", and I still have to understand how to pass this knowledge and experience on to others. This is basically the problem. (Fabio, regional secretary)

For him, it is particularly difficult to solve this dilemma, given the accumulation of his multiple professional and political activities. However, not all the responsibility of the non-involvement of local authorities can be placed on the shoulders of the regional secretary. In fact, even in the Grosses Walsertal, not every mayor regularly goes to the network's meetings. It is mostly the regional secretary, according to Niklas, who makes the trips. But, still, the mayors of the Grosses Walsertal are more updated simply because they try to inform themselves and are not waiting for the regional secretary to do it. In that sense, as Niklas admits, the mayors in the Grosses Walsertal are up to date on matters concerning the network:

Die Newsletter, genau! Dort wird es einfach stichprobenmässig mitgeteilt und eingebracht, und natürlich sind wir – da wir nicht direkt Mitglied sind – trotzdem durch eine zweite Schiene top informiert: [Region] – Ulrike – Allianz in den Alpen. Aber ich habe nicht das Gefühl, dass wir da irgendwie zu kurz kommen. Das glaube ich nicht. (Niklas, Bürgermeister)	The newsletter, exactly! There, the network shares and gives information on a sample of experiences. Of course, because of that, we are not "well-informed". In fact, we are not members in a direct sense, but we are on a second level: [region] – Ulrike – Alliance in the Alps. But I do not get the feeling that we are missing out. I do not believe that. (Niklas, mayor)

So, the regular meetings organized in the Grosses Walsertal where the mayors and the regional secretary can share their experiences with others is probably a more efficient way of communicating than by email, as Fabio seems to favour.

In fact, like Ulrike, Fabio also dispatches the information to the municipalities of the region. So, the mayors' lack of interest and curiosity is also an important reason why they are not well informed. In whatever configuration, as Sabine highlights, a great deal depends on the personality and goodwill of the mayors:

Es ist eine Wunschvorstellung, dass wir nur Regionen [in der Allianz] möchten, Realität ist, dass auch die Regionen abhängig davon sind, von diesen Köpfen aus der Gemeinde. (Sabine, Betreuerin)

It is an illusion to say that we want regions [in the Alliance]. The reality is that even the regions depend on these heads in the municipalities. (Sabine, animator)

This means, if one follows Sabine, that regional membership is not a solution for making municipalities participate more. If a mayor is interested in becoming a member of Alliance in the Alps, he will do so and probably be active in the network's activities. If the region does it for him, there will be no incentive to become involved in the project. So, one of the three reasons brought up by the head of the network for promoting regional membership finds no confirmation in reality. Therefore, even if the network has preferred the "regional solution" for the advantages it presents, this structure is not without inconvenience. Concerning the communication between the network and its members, for example, it adds a level to the information chain that can partly interrupt the transmission of knowledge.

Regarding the territorial adequacy for the implementation of sustainable development (another reason why regional membership is promoted by the network), the choice between a regional and a communal membership raises questions as well. The apparent contradiction in Leonardo's statement shows the difficulty of the choice. When asked the question, "Do you think that it makes more sense to join the network as a region or as a municipality?", Leonardo answered as follows:

Secondo me, dal principio della sostenibilità direi i comuni. Secondo me è meglio, perché il comune in fondo è legato al territorio. Se il comune aderisce è perché ci crede. Se però vogliamo portare avanti progetti regionali e delle vallate allora sono le regioni, quindi è difficile scegliere fra le due cose. Se tu vuoi coinvolgere la popolazione è meglio il comune evidentemente, però secondo me i comuni sono deboli e hanno poco tempo. E' fatto così tutto in maniera amatoriale. Un sindaco guadagna pochissimo, non

In my opinion, and if I think about sustainability, I would say that municipality makes more sense. I think it is better, because municipalities are more linked to the territory. If the municipality joins the network, it is because they believe in it, and it is much better. But if one wants to carry on regional projects or projects in a valley, then a regional membership is better. So, it is difficult to choose between the two things. If you want to involve the population, of course the municipality is better. But I think that local

ha il tempo di portare avanti anche Alleanza nelle Alpi. E' un peccato, però è veramente difficile. Secondo me le regioni sono meglio. (Leonardo, ex-stagista in una regione)

authorities are too weak and don't have the time. Everything is done in a "recreational" way. A mayor earns very little; they don't have the time to take care of Alliance in the Alps. It is a shame, but it is really difficult. In my opinion, regions are better. (Leonardo, former intern in a region)

So, as Leonardo seems to suggest, there is no simple solution. Everything depends on what the network wants. Regional membership is certainly an attractive choice, notably for the implementation of some projects, which demand large-scale coordination (such as transport or water management, for example). But, the risk is higher of losing contact with both the single members of the region and with the population. A mayor is much closer to the local reality than to a region. Additionally, the risk is higher that the information stays at the regional level without trickling down to the local level. As for the next challenge that networks are facing, i.e. the number of members they wish to have, that question should be carefully considered so that it fits the network's strategy.

6.4.3 Number of members: quality or quantity?

Another controversial and discussed point, besides the type of membership, is the ideal number of members. The first interview for this research took place in Mäder, in June 2007. Bernard Debarbieux, Rainer, Elke (secretary of the network Alliance in the Alps) and I sat around a table discussing various issues related to Alliance in the Alps. One of them had to do with the size of the network. Rainer Siegele suggested that ideally the network should cover 10% of the total Alpine local authorities inside the Alpine Convention's area, 10% of the Alpine territory and 10% of the population. This would translate to around 600 members, which means double the actual number. Rainer believes such a number would allow more political weight and the members being able to better implement the principles of the Alpine Convention.

Others share this assumption that a larger number of members leads to increased political influence, as Maximilian opines:

Ich glaube, dass die Verantwortlichen hoffen, dass wenn man eine bestimmte Größe und Stärke hat und man mehr Mitglieder hat, dass man dann ernst genommen wird. (Maximilian, Bürgermeister)	I think that the network leaders hope that when we've reached a certain size and power, i. e. when the number of members increases, we will then be considered serious partners. (Maximilian, mayor)

Jakob, a mayor, is also convinced that the political impact *(Schlagkraft)* will be stronger if the number of members increases, and if they stand up all together to inform the political authorities of common problems. If all members act and react jointly, the impact would be even stronger, Jakob says, because the voices are international and not coming just from one country. The president of Alpine Pearls also points out that only by having a "critical size" the network will be able to collaborate with private firms such as railway companies.

Members of Alpine Pearls agree on the importance of quantitative growth, although some of them point out the danger in terms of "exclusivity", a reaction that seems to be peculiar to this network. The people concerned with this issue seem worried that if too many municipalities joined the network, fewer tourists would choose their locality for their holiday destination. This reaction is not surprising, since the aim of Alpine Pearls is not only the implementation of a soft mobility politic but also a campaign to attract environmentally friendly tourists. The fact that the network's Internet site is also translated into Japanese shows this tourism marketing orientation.

In that sense, the network is viewed as a label with the motto "sustainable holidays in tune with nature" that clients should recognize. Therefore, increasing the number of members is a "double-edged sword" *(zweischneidiges Schwert)*, as Viktoria admits. Viktoria's worries are shared by Antje, who explains:

Die Frage ist auch, was es bringen würde wenn jeder dazu kommt, wahrscheinlich nichts mehr. Entweder ist es eine Spezialisierung und Positionierung oder es ist das nicht. (Antje, Mitarbeiterin eines Fremdenverkehrsbüros)	The question is what it will bring. If every municipality joins, probably nothing more. Either it is a specialization and a positioning, or it is not. (Antje, employee of a tourist office)

So, a larger number of members is good, but only under certain conditions. This is also Normann's opinion:

Wir sind der Ansicht, dass es sehr gut wäre, wenn einige Perlen noch dazukommen würden. Die Kooperation lebt nur deshalb, weil verschiedene Orte in verschiedenen Alpenländern zusammenkommen. Aber es hat keinen Sinn, dass wir zum Beispiel drei, vier Südtiroler Perlen wieder mit dazu nehmen. Wichtig ist, dass in Frankreich welche sind, in Österreich, in Deutschland, in der Schweiz, in Italien und so weiter. (Normann, Mitarbeiter in einem Fremdverkehrsbüro)	Our opinion is that it would be really good if some other Pearls joined. The cooperation functions only because different places in different Alps are joining. But it makes no sense that, for example, three or four Pearls from South Tyrol join. What is important is that there are some from France, from Austria, from Germany, from Switzerland, from Italy and so on. (Normann, employee in a tourist office)

Normann suggests that competition and collaboration is welcomed, but only internationally. Within a country, he does not judge it indispensable to have new members. In addition to this question of labelling, which I could see only in the interviews with members of Alpine Pearls, another aspect, namely quality, is shared by the two networks. The president of Alpine Pearls explains as follows his argument, after he stated that he wishes to get more members for its network:

Trotzdem ist es noch wichtiger die Qualität zu halten, um glaubwürdig zu sein. Wenn wir dem Kunden, dem Gast, versprechen, dass wir eine gute Qualität zum Beispiel im Bereich der Mobilität haben, und dann stimmt das alles nicht, dann haben wir ein Problem. (Peter, Präsident von Alpine Pearls)	Quality is even more important for our credibility. If we promise our client, our guest, good quality, for example in the area of mobility, and this is not the case in reality, then we have a problem. (Peter, president of Alpine Pearls)

However, the concern for quality seems to be, for members of Alpine Pearls, part of a tourist strategy: "We cannot disappoint our clients with something that we promised, and then they do not find it once they arrive", Peter seems to suggest.

For Alliance in the Alps, the matter of quality seems to have other points. Especially the animators, who are working close to the local level, are realizing that the network, due to its size, has to change strategy and concentrate more on the quality of its work with municipalities than on the search for new members. During the animators' meeting, which took place in January 2008, everybody, without exception, agreed on the necessity of thinking more about "quality" (the quality of the animation, but also quality in terms of each member's sustainable development strategy) than about "quantity". As Leonardo, a former intern of a region, asserts, more members would no longer be manageable *(gestibile).* This is especially true considering the actual structure of the network, seen by some animators as too "amateurish". Thus, the network's actual size, which is already significant, demands some structural changes of the networks. Additional members will only intensify the structural problems explained in the next section.

6.4.4 Structural challenges

The question of the type of membership (singular or regional), as well as the issue of the possibility of increasing the number of members, appears at a moment when Alliance in the Alps seems to have reached a turning point. Inside the network, important questions are asked about the general structure that, for some people, should be reconsidered, given the number of members (now more than 300). The current animation's structure no longer allows a satisfactory service to the municipalities. In January 2008 I was invited by the animators of the network to present some of the first results of my research at their internal meeting.[125]

The discussions that preceded and followed my presentations revealed a general discontent concerning the existing structure of the network and their role inside it. There was a general consensus about a need to professionalize and structure the figure of the animator, which has for a long time been characterized by a lot of volunteering work. As one animator underlined, "Those times [described by another animator as 'pioneering'] should be over".

125 The meeting took place in Schaan (Liechtenstein) on the 31.01.2008.

Some animators proposed the creation of a new position, a person working 50 % for the network and taking on the role of "strategic consultant". This would allow a closer link between the managers of the network and the animators, and, according to the proponents of this solution, to have a person in charge of international contacts.

The importance of rethinking the role of the animators is also due to the fact that in some countries (for example, Switzerland and Germany) the funds for their work are not assured. Therefore, in their opinion, the strategy should be to find a way to stabilize their positions, which means finding new funds. One of the possible solutions considered by the animators was to ask the MAVA foundation, which already funded the Dynalp 2 project, to change its funding strategy. In fact, the animators seemed to agree that it is more urgent to create a stable strategic position for helping their work, or, at least, for activities that all municipalities could benefit from, than to fund further projects for some chosen municipalities, as happened with the Dynalp programme. However, their wish did not come true. In fact, in 2010, the MAVA foundation decided to grant 300000 euros to 20 local projects dealing with climate change in a programme called "dynAlp-climate".

However, this target funding, which does not reflect the needs expressed by the animators, is exactly the kind of project that the MAVA foundation, represented by its consultant responsible for the "Swiss and Alpine arc" programme, is looking for. In January 2009, Bernard Debarbieux and I were invited to the meeting of the "Reflection team Dynalp 2", a yearly meeting that took place during the Dynalp 2 project and brought together about 10 people, representatives of Alliance in the Alps, CIPRA and the MAVA foundation, as well as some experts.

On that occasion, the representative of the foundation made known the internal restructuring process that would lead to a different attitude of the foundation towards the projects it is willing to support financially. The foundation would henceforth decide the themes in which it wants to invest, while until then, those had been proposed from the "bottom up". As a second step, he underscored the necessity of increasing the network's visibility both within the foundation and amongst the broader population. For example, he welcomed the network's initiative of producing a DVD presenting short portraits of the projects that municipalities implemented thanks to the MAVA funds. This makes the network more visible and therefore more likely to receive further funding.

This evolution contains hidden (and not unproblematic) consequences: In fact, MAVA's new attitude reduces the network's leeway in terms of decision-making. Firstly, they will not have complete freedom in deciding the topics they want to work on. Secondly, in order to increase visibility, the network will be obliged to invest in concrete projects that can be flaunted, for example by producing films or brochures. The need expressed by the animators for the creation of a position working on strategic plans would therefore not be included in this new orientation. In 2009, the MAVA foundation gave a credit of 300000 euros to the network for 20 "visible" projects, which shows that the animators' expectations to reinforce the structure of the network's animation could not be fulfilled, at least not via this channel of funding.

Structural uncertainties are another issue Alliance in the Alps has to face. Here, the major problem is that only elected mayors can sit on the network's steering committee. During my field work I witnessed two concerns regarding this issue, which I discussed with an animator.

Firstly, the condition of being an elected mayor in order to be part of the committee obliges members to keep the function inside the town council, which is not everyone's wish, especially in a long-term perspective. For others, the question does not even depend on the resolve of the mayor but on the political and legal possibilities. The law in Friuli (Italy) does not permit a person to be elected mayor for more than three consecutive terms. Antonio, mayor of Budoia from 1995 to 2009, could therefore not be a candidate in the 2009 elections and consequently could no longer be admitted to the network's steering committee. Concerned about the perspective of losing one of the most active persons in the network, a solution had to be found: Antonio was integrated into the animation team. However, this first instance with the ending of a political mandate raised questions about the network's structure.

Secondly, this question also touches the lower levels of the network's structure, i.e. its members. The commitment of local authorities to sustainable development is a crucial motivation to join the network. When local authorities change, due to elections, one of the consequences can be that the municipality drops out from the network, because the new mayor does not share the same priorities, as was the case in Piedmont:

Wir hatten auch im Piemont einen Bürgermeister, der Feuer und Flamme war, der überall dabei war. Als er nicht mehr gewählt wurde, ist die Gemeinde aus dem Gemeindenetzwerk ausgetreten, weil niemand verstanden hat, was man da drinnen soll. (Andreas, Direktor von CIPRA International)	In Piedmont we had a mayor who was really fired up, he was everywhere. When he was not re-elected, the municipality quit the network, because nobody could understand why the municipality should participate. (Andreas, director of CIPRA International)

6.5 Conclusion

The Alpine Convention was created with the aim of solving common problems (such as traffic, population ageing, environment, etc.) that the Alpine region was facing. All the problems identified in the Alpine Convention have a common characteristic: They do not stop at national borders. Inside the (legislative as well territorial) confines of the Alpine Convention, tools were developed to make it effective. Networks were created in order to have structures in place that could anchor the general principles of the Alpine Convention at the local level.

However, when can the Alpine region be considered a referent for collective action, as Debarbieux suggested? The geographer proposes two answers: Firstly, this occurs when a region perceived as a mountain entity is used to define and problematize governance issues (taking into account the region's mountainous character) that give rise to institutional reconfigurations at that scale (Debarbieux, 2009). Secondly, this takes place "when, at the scale of pre-existing entities, and possibly within the population directly concerned, there are resources and room for manoeuvre that enable local actors to promote an institutional configuration, and, if need be, one based on identity, which is complementary to existing configurations" (Debarbieux, 2009).

Concerning the first remark, as I discussed, networks of local political actors are one of the main actors, together with the administrative structure of the Alpine Convention and NGOs committed to pan-Alpine issues that are working at an institutional reconfiguration of the Alpine region. By linking local authorities in order to implement the principle of the Alpine Convention at the local level, networks are institutions that contribute to what Debarbieux describes as the first condition for mountain regions to become a referent of collective actions. In fact, networks are one of the major institutions contributing to a problematization of pan-Alpine issues. They have been conceived as appropriate tools to deal with common problems that local Alpine authorities are facing. All together they create a web of links inside the Alpine Convention's map and therefore contribute to delimiting a new space of action that is independent from the national states.

Inside this new space of action, networks, as the interviewees highlight, are able to find appropriate tools to share information and practical recommendations and are an astute strategy for municipalities to receive financial support for a project's implementation. The projects carried out at the local level contribute to building the Alpine region: not an "abstract region", but a region corresponding to the mountain range's limits and devoted to sustainable development. So, by carrying out projects inside a new institutional framework, the Alps can really be considered a referent area for collective action.

One can affirm that all the links between municipalities that are considerably spread out through the Alpine region form a "real" network capable of making people communicate over a remarkably big geographical area. Hundreds of municipalities are linked in a spread out modality transcending national, linguistic and geomorphologic barriers. It is the map of the Alpine Convention that provides the "regionness" of the project. The map thus functions as the areal container of the networked initiatives.

The second issue brought up by Debarbieux, however, has to be considered differently. Out of the roughly 6000 municipalities in the

Alps, about 300 are members of Alliance in the Alps and 27 of Alpine Pearls. This means that a vast majority of the Alpine municipalities are not part of this reconfiguration system. Additionally, if one considers that only a few members of the network are really active in the networks' activities, one has to admit that the second condition highlighted by Debarbieux for the mountain ranges to become a referent for collective action is most likely not yet fulfilled. In fact, it seems that the room for *manoeuvre* that would enable Alpine actors to sustain the new institutional configuration is quite narrow.

This point was underlined in the discussion held with the interviewees concerning the current political weight of networks of local political actors. Members are not convinced that pan-Alpine networks have the power to steer national and supranational decisions concerning mountain policies. This last consideration leads us to the question of the social and political significance of the Alpine region, which will be discussed in the next chapter.

7. Socio-political significance of the pan-Alpine region

The previous chapter focused on the way pan-Alpine networks are performing in the Alpine region. I explained how those networks function by presenting their different facets and respective challenges. As discussed, networks of local political actors are among the most interesting and innovative established structures giving concrete significance to the pan-Alpine project. In this chapter, I specify some elements to judge the socio-political significance of the pan-Alpine project. This corresponds to the fourth shape of Paasi's regional institutionalization theory, i.e. the "establishment of the region". It is worth recalling that the other three steps are: the assumption of a territorial shape, the development of a symbolic shape and the establishment of institutions (Paasi, 1986).

Networks, by definition, combine two basic elements: at least two points in space and a "line" connecting them. A phone call between the mayor of Budoia and the mayor of Mäder can be considered a way of performing a network. In theory, networks can, theoretically, work without the intervention of any traditional administrative entity: Budoia's mayor does not have to ask for authorization from Rome in order to get in touch with his colleague in Austria. However, the question can be raised whether, how and in which measure networks are really free from any influence of the state apparatus and can decide independently on the future developments of the pan-Alpine region. The analysis on this topic can give a measure of the current significance and importance of the pan-Alpine region, compared with that of the traditional territorial grid.

Another important way in which to measure the socio-political significance of the pan-Alpine project is to assess its impact outside the small circle of the very active actors. Questions can be raised concerning, on the one hand, the implication of the networks' base in the pan-Alpine issues and, on the other hand, the Alpine inhabitants' awareness of these issues. This is what Paasi considers the fourth step of the regional institutionalization process, i.e. when the region achieves a proper spatial structure and is part of the social consciousness of society (Paasi, 1991: 246).

7.1.1 The role of the traditional administrative grid on networks

Despite the three networks' diverse characteristics, the transnational links between local political actors unites them all. The political actors involved in these projects cope with different geographical levels: local, regional, transnational and European supranational. In this context, the national level seems hidden or left to one side. It was rarely mentioned in the interviews with network members, whose discourse simultaneously concentrates on the local and pan-Alpine levels. Notwithstanding the political actors' willingness to work together at the pan-Alpine transnational level, and despite the national level not playing an important role in defining the statutes and missions of these networks, it remains an issue that sparks lively discussion.

As Debarbieux writes in the preface to the special issue of the *Journal of Alpine Research* dedicated to mountain regions as referents for collective action, "the constitution of trans-border regions invariably requires that the state and sub-state entities [be] involved in their emergence" (Debarbieux, 2009: 15). This, as the geographer points out, is because it is exactly at these levels that "resources and levers for action are found" (Debarbieux, 2009: 15). Debarbieux affirms that, thus, the organizational configurations of mountain regions have to be considered as still linked with those entities that continue to function within frameworks inherited from the past, i.e. national, subnational, local (Debarbieux, 2009: 16). So, as Leitner *et al.* acknowledge, networks are intrinsically related to hierarchy, since they cannot be "created out of thin air as an alternative to hierarchical modes of governance" (Leitner *et al.*, 2008: 288). Therefore, new modes of governance are created, Leitner *et al.* continue, as part of their struggle with other scales in order to "overcome and challenge constraints posed by hierarchical modes of governance" (Leitner *et al.*, 2008: 288) but without cancelling them.

Keating stresses that while states are generally not contested, they are more permeable, so that they nowadays give space for more cross-border functional co-operation on different matters such as the economy, environment and culture (Keating, 2003: 11). Moreover, he highlights the symbolic importance that states still exert. Taking the example of the three pan-Alpine networks, I want to show how the traditional grid is still present, and not merely symbolically, in discourses and practices of local political actors, despite the transnational way of cooperating. This gives support to the concept of regional assemblage (see chapter 2), that defends the idea that central governments are to be considered "actors as part of 'regional' assemblages, not bodies that sit over or apart from regions" (Allen & Cochrane, 2007: 1172).

Symbolically, this chapter can be illustrated by a photograph I took in 2009 during the Alpine Crossing, a "soft mobile trip in the Alps" that the Alpine Pearls network organized for the media (see as well Del Biaggio, 2009b on this event). This picture shows the birthday cake offered by the network to the participants in the journey. Emblematically, besides the names of the then-members of the network (in red) and of the largest cities (in black), the national flags also decorated the cake. Additionally, the borders drawn with chocolate are that of the national states and not those of the Alpine Convention (fig. 24).

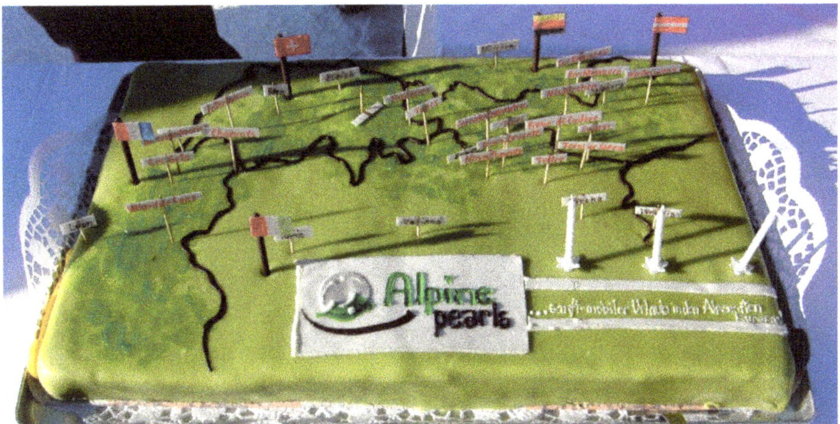

Fig. 24: Alpine Pearls' birthday cake served on the top of the Bischling (Austria) on the occasion of the Alpine Crossing 2009, 30.01.2009 © Cristina Del Biaggio.

This attachment to the state symbol is a sign that networks are still linked to national imagery, even in such particulars as the decoration of a birthday cake. National influence is present even in more important issues, and it may be noted that, despite the willingness to focus on more horizontal relations, the administrative, political and ideological structures of the networks and their members are not able (and probably not even willing) to completely disconnect from the national level, even if it does not play an essential role in carrying out the projects of these networks, as I will explain in the following sections.

Three types of interference of national structures on the network will be discussed in the following sections. They can be qualified as *structural* (because of the presence of national structures in the networks' arrangements), *political* (because of the influence that national politics have on networks) and *cultural* (because of the influence of national mentalities on the ways of networking).

7.1.2 Structural influence: the presence of national organizational structures

Although, for the Alpine Convention, the state structure is imposed, given that it is an international treaty, the networks could have chosen a method of operation without state interference. However, almost certainly for reasons of efficiency, national considerations strongly influence the organizational structure of the networks: the distribution of seats in the network's decision-making bodies is carried out according to this method (see chapter 4).

Moreover, policies promoting national interests sometimes take precedence over transnational considerations. Voting decisions during assemblies are now and then based on partisan interests and in defence of national privileges. This occurred, for example, during Alparc's General Assembly in Bled (Slovenia) in October 2008, when participants were first discussing and then voting on the opportunity to give voting rights to the general secretary of the Alpine Convention on issues concerning Alparc. Different people presented valid arguments for and against this proposal.

Yet, those arguments systematically followed (and this is probably not a coincidence) a "national line": For example, Austrians would all have to find good arguments against that proposition; Italians would have to argue the opposite. By contrast, if someone voiced an argument that was theoretically and practically convincing for both sides, the movement of people in the room in search of allies and the words whispered in the ears of their own countrymen were obviously filled with strong partisan and national interests. This attitude irritated some participants, who pointed out, while in discussions with me outside the assembly room, that the interests of the Alps go beyond petty politics.[126]

These national considerations are also followed in the distribution of strategic seats inside the network, as Branko, director of a protected area, explains to Petra Arnuš:

> Branko (director of a protected area): For many years, the president was always French. It was deemed "unhealthy" if there were only French presidents. They said it would be nice if someone else took this position. Mr. Vogel took over after that. He was elected during an official meeting, although we arranged this ahead of time, but he was formally confirmed during a meeting.
>
> Petra: Was this choice dictated by a sort of balance?
>
> Branko: Yes, balance. And first, he didn't want to be president; he said the French would resent it, because here you have to be careful, the German and the French are very sensitive! There must always be a balance. And then Mr. Vogel took over, and they said: "Let's have two vice presidents to balance it out a bit!" The French didn't want to name a vice president, because for so many years they had a president. So Mr. Ottino came, the director of Gran Paradiso [Italy], and this was quite successful.

For Alparc, at least, national considerations in the distribution of important seats seem to play a significant strategic and political role: Mr. Vogel is not only Mr. Vogel, he is also a German representative of protected areas, and this counted as a central issue in his election. The general atmosphere for Alliance in the Alps seems to be more relaxed concerning this issue. The network's steering committee is also

126 Sentiments expressed by two park representatives with whom I talked after a "sensitive" vote at the last General Assembly of Alparc (personal notes, 10.10.08)

"nationally balanced", but elections for the presidency and vice presidency are more consensual and guided by other considerations than just national belonging.

In 2009, when Antonio had to renounce his seat as vice president because he could not run for re-election as mayor of his municipality, Giraud took over the empty seat. Although I was not present during the negotiation and the vote, the choice of a French representative was certainly not dictated by partisan interests alone. The network always suffered from a lack of French municipalities among its members. This trend has reversed in recent years, thanks to a considerable increase in the number of French members. The decision to give political responsibility inside the network to the mayor of L'Argentière-la-Bessée, which became a member in 2006, has been a signal to support this development and to underline the political weight of the network towards the French national and regional administration.

Alpine Pearls is too young to establish a policy on the alternation of the presidency, and Peter Brandauer has served uninterruptedly in this capacity since the start. Yet, the board is also structured according to national considerations (one nation, one seat), although there is one exception: South Tyrol also holds one seat.

Besides the distribution of the seats at the head of the networks, another interesting point has to be mentioned. Within Alliance in the Alps, Switzerland and Austria have set up national sections (see chapter 4). This structure has enabled these countries to benefit from funding provided by their state governments in addition to the subscriptions of members.[127] Moreover, the motivation of the network's national structure, as Debarbieux and Rudaz highlight, aims to strengthen the position of the network as a key player in the institutional meetings dealing with the Alps. In doing so, they try to influence the Alpine states so that they reinforce their cooperation and take concrete steps to implement the objective of sustainable development in the Alps (Debarbieux & Rudaz, 2008: 504).

127 According to Rainer, the creation of national agencies provides increased financing for Alliance in the Alps. He thinks that in the future this national level may enable an increase in political power (personal notes, interview with C. Del Biaggio and B. Debarbieux, Mäder, 11.06.2007).

Yet, during the General Assembly of the Swiss Alliance in the Alps, which took place in Flühli (Switzerland) on the 15th of June 2006, a mayor highlighted, not without a worried tone, that the establishment of national sections of Alliance in the Alps is moving the network away from its original purpose. He remembers when, at the very beginning, somebody had the idea to "illuminate" the Alps with hundreds of flashlights so that the Alpine arc could be seen on a satellite image. This is, in his view, how the network should work, and it illustrates the original intent of the network, i.e. to show that the Alps are transnational. He compares this project to the Swiss idea of building, in 1998, the network's Swiss national section, which he says contradicts the pan-Alpine project.[128]

The networks' national structuring, indeed, shows the difficulty they have in "thinking pan-Alpine". Although local political actors are aware of defending Alpine interests and promoting transnational projects, the national anchorage is still very present in their minds. Moreover, the Alpine states are not only structuring the networks' organization but, in some cases, also influencing their activities, as I explain in the next section.

7.1.3 Political influence: the weight of the national state on networks' activities

As was broadly discussed in chapter 4, the decision to establish national sections within Alliance in the Alps allowed it to receive extra financial means from their national governments. However, and this is the other side of the coin, due to this structuring certain choices are subjected to pressure from the services providing the funding. In Switzerland, the Federal Office for Regional Development (ARE) has promised increased subsidies on condition that the number of network members increase.[129]

128 Field notes from the General Assembly of the Swiss Alliance in the Alps, which took place in Flühli (Switzerland) on the 15.06.2006.

129 During the General Assembly of the Swiss Alliance in the Alps, it was mentioned that the Federal Office for Regional Development is ready to invest more money on condition that the Swiss Alliance in the Alps find new members (personal notes, 15.06.2007).

As discussed in the previous chapter, however, not everyone (the animators, in particular) agrees that the membership needs to increase. Yet, the magnitude of the means that the ARE provides to Alliance in the Alps seems to be of crucial importance to the network's members and board. So, as Goodwin points out, the role of national states in sponsoring and supporting networks in regional assemblages does not have to be denied (Goodwin, in Allen & Cochrane, 2007: 1171).

The modality of this support, however, can raise questions about its consequences. For example, during the General Assembly of the Swiss Alliance in the Alps, lively discussions took place on how to deal with the conditions that the ARE imposed. This account shows how networks can be influenced by choices dictated by national officers, pressure which is not always appreciated, although network members accept it in the name of increased financial means.

Additionally, during that same assembly, the then-president of the Swiss section, as well as other members of the network, expressed his worry concerning the imminent departure of the referent for Alliance in the Alps working for the ARE. Would the people who took over that role still support the network? In view of the importance of the ARE's financial support, they were concerned about future developments within the ARE's staff. This confirms the degree of dependence and the problem of long-term relationships of the network's Swiss section towards the national administration's office.

The Swiss administration's "close look" at the network's activity is not the prerogative of all the Alpine states. As stated by Price, an expert in questions linked to mountain policies, "while Italy has the largest proportion of the Alps, its capital is distant and the mountains are not of great significance in national policy-setting" (Price, 2000: 193). Since Rome does not have particular interests concerning mountain issues in general, they do not influence the activity of the networks either. In other countries, such as Slovenia, the influence of the Ministry of the Environment on the life and activity of protected areas seems to be quite important, as Branko highlights in an interview conducted by Petra Arnuš:

Petra: You didn't want to have influence from Ljubljana, from the ministry?

Branko (former director of a protected area): No. These were years, when we were angry with Ljubljana, because there was no money. They made fun of us, lots of paperwork and only bureaucracy, terrible!

Petra: In general, does the ministry now have influence over protected areas?

Branko: Well, it's calmer now, it's alright now.

Petra: Does it depend on the minister in power?

Branko: Of course. The actual minister, Mr. Podobnik, and his personnel are much more open. The former minister said clearly that protected areas do not mean anything to him! [...] We understood each other well, but there was no money. He said that there is no nature for him, that nature is not a priority.

So, in fact, this shows how the influence of ministries is both dreaded and regretted when this signifies less financial means at the disposal of the protected areas. In fact, less funds also means fewer possibilities to participate in the network's life, as Goran asserts (see chapter 6).

Luc and Mathieu, employees of two Alpine Pearls member municipalities, raised other issues concerning the differences between France and Austria:

Luc (employé d'une commune): La principale barrière, c'est le cadre réglementaire. Au départ, nos autres partenaires avaient un petit peu de mal à comprendre comment on fonctionne [en France]. Notre marge de manœuvre est très limitée par rapport au ministère autrichien qui va directement sur le terrain et qui comprend bien les enjeux.	Luc (employee of a municipality): The main barrier is the regulatory framework. At the beginning, our partners had some difficulties understanding how things work [in France]. Our latitude is really limited if you compare it with that of the Austrian ministry, which goes directly to the field and understands the issues.
Mathieu (employé d'une commune): C'est-à-dire qu'en face de nous on avait soit effectivement des représentants de ministères autrichiens ou d'un *Land*. C'est dire que les [membres autrichiens du réseau], automatiquement, s'ils ont une question, ils frappent au bureau à côté et la règlementation peut évoluer. Nous, on est une simple commune, le jour où on a une question, on lève la main, on envoie un courrier et parfois on nous répond.	Mathieu (employee of a municipality): This means that in front of us we had either representatives of the Austrian ministry or of a *Land*. This means that if the [Austrian members of the network] have a question, they knock on the office next door, and the legislation can evolve. We are just a municipality. The day we have a question, we raise our hand, we send a letter, and sometimes we receive an answer.

Luc: Et même si on a un conseiller régional qui est un peu sensible à la question, il doit se battre en assemblée du conseil général et faire remonter [les revendications].	Luc: And even if you have a regional councillor who is sensitive to the question, he has to fight during the assembly of the general council so that the higher levels hear [the demands].

In another instance, the centralized French national framework was also mentioned for precluding the freedom of action of the members of the Alliance in the Alps. During the General Assembly of the Swiss Alliance in the Alps, a Swiss mayor cited the cost of membership as part of the difficulty to convince new municipalities to join the network. On that occasion, he made a comparison to the French case. In fact, as he highlights, the binding French administration framework makes it impossible for municipalities to take decisions such as becoming a member of Alliance in the Alps.

The Swiss mayor, probably exaggerating a bit, illustrated this difficult context by saying that French mayors have to call Paris for every 20 euros they want to spend. Although the Swiss mayor may have referred to something he really witnessed, one cannot be sure whether this kind of remark in fact results from the representations and stereotypes that everybody has on the ways national administrations function. Probably, the reality is a combination of the two explanations. Whatever the case may be, even if there is a strong willingness to join the network, the hierarchical administrative structure does not allow it.[130] This shows the strong influence that the French central state exerts on the municipality level. In comparison with Austria, where the influence of the state on the municipality is welcomed by the members, the situation in France is deemed more negative.

During the "mountain networks" seminar that was held by our department in Geneva,[131] the French question was debated at length. Two employees of CIPRA, whom I will call Antoine and Janine, provide further details about the difficulties they encounter while dealing with French municipalities. Antoine underlines the incompat-

---

130 Personal notes from the assembly.

131 Geneva, 28.11.2007.

ibility between the French administrative culture and what he calls a "network culture". In fact, in his opinion, the French political system forces them to think and to act at the very local level, with difficulties to engage in co-operations, even if those are thought to connect neighbouring local communities.

Janine blames the centralized system for these complications. According to her, solutions traditionally come from Paris, and a mayor who wants to implement something in his or her community must wait for the answer from Paris. By contrast, municipalities and regions that are part of a federal administrative system do not have to turn to the capital but can communicate directly with their neighbours in order to solve common problems.

Antoine says he wonders why some "reckless" French local representatives commit themselves to sustainable development issues and decide to join Alliance in the Alps. He stresses that this solution is quite risky for them, and although they know it is a difficult path, they nonetheless choose to follow it. This is even more courageous if one considers that local citizens do not really see the benefit of working with other local communities, since they have the impression that it is a waste of time and money. In fact, it is difficult for them to think and understand that their own problems can be solved by following the examples of other countries. Thus, these two interventions show the difficult and dissuasive mechanisms that the French political system causes.

However, a counterexample has to be mentioned: The intervention of a French minister was decisive in the founding of the Alparc network. Julien, director of a protected area, remembers how Michel Barnier, France's then-minister of the environment, wanted his country to play an innovative role and promote new types of actions for the environment. For this reason, he organized, in partnership with Les Ecrins National Park, the first International Conference of the Alpine Protected Areas in 1994. At the time, France was ready to provide financial means to support the idea and the creation of a network of protected areas. Julien recalls that Les Ecrins National Park was not the initiator of the idea; in fact, it was first proposed by the ministry and then taken up by the park:

Je pense que d'abord on a été sollicités [par le Ministère], et qu'on y a répondu sans interrogations sur la suite à donner. [...] La France était prête à se positionner et à financer pour lancer "les choses» sans attendre qu'un financement international se mette en place, ou que les structures internationales se mettent en place. (Julien, directeur d'un espace protégé)

I think that we were first approached [by the ministry], and that we responded to that proposition without hesitation. [...] France was ready to position itself and to finance "things" without waiting for international funding, or the international structures, to be put into place. (Julien, director of a protected area)

Other interviewees, especially those who see the advantage of fruitful collaborations in order to stimulate a debate on the issues of sustainable development, consider the closeness of the regional and national administration to the network something positive. By contrast, as stated by Luc and Mathieu, the lack of interaction between the network's members and the upper levels of the administration causes stasis that afflicts all French municipalities.

The president of Alpine Pearls welcomes the tight links that the network has established with the Austrian Federal Ministry of Agriculture, Forestry, Environment and Water Management, also known as the *Lebensministerium*. The close relationship that Peter has with the ministry has, in fact, allowed some network projects to become reality, which has a positive influence on the network:

Vor allem hat natürlich das Lebensministerium ein besonders enges Verhältnis zu *Alpine Pearls*, da sie Mitgründer waren. Die Abteilungsleitung des Ministeriums hat schon seit vielen Jahren das Thema der umweltfreundlichen sanften Mobilität zu einem wichtigen Thema gemacht. Man sieht das auch jetzt an der Unterstützung, die sie unseren Aktivitäten zukommen lassen, wie zum Beispiel für die von uns veranstaltete Medienreise. Das österreichische Ministerium ist ein wichtiger Ansprechpartner, und unterstützt uns weiterhin, obwohl wir hier von den 21 Mitgliedern nur vier aus Österreich haben. Da gibt es eine ganz besondere Beziehung, die man auch pflegen muss. (Peter, Präsident von Alpine Pearls)

First of all, the *Lebensministerium* has a particularly close relationship to Alpine Pearls, because it is a co-founder of the network. The minister has for many years now been a supporter of soft mobility. We can see how they still support our activities, for example our media tour. The Austrian ministry is an important partner, and it is still supporting us, even though only four of the 21 members are now Austrian. There is a special relationship that we must maintain. (Peter, president of Alpine Pearls)

These different experiences show the various ways in which national administrations are said to influence, positively or negatively, the pan-Alpine networks' activities. Not everybody considers the influence exerted by national administrations to be negative. On the contrary, in some cases, it is a *stimulus* for further development of the network. However, as the Swiss demands on Alliance in the Alps demonstrate, national administrations can require conditions that are not desired by the network and therefore undermine the independence of the network with regard to its strategic choices.

7.1.4 Cultural influence: the existence of national mentalities

Sabine, animator for Alliance in the Alps, also highlights the different national political systems affecting cooperation. However, she later explains how this is primarily a matter of different national organizations and approaches in project management. As she points out, there are "cultural" differences that have to be taken into account and should be respected:

Sabine (Betreuerin): [Man sollte sich fragen:] "Woher kommt man?" und auch die politischen Systeme der anderen Länder betrachten. Es gibt in Frankreich ein anderes politisches System, andere Möglichkeiten für Gemeinden als in Österreich.	Sabine (animator): [One should ask:] "Where do we come from?", also in terms of the underlying political system. In France there are other political systems, different opportunities for municipalities than in Austria.
Cristina: Was sind denn die Schwierigkeiten, die du siehst?	Cristina: And what are the difficulties that you see?
Sabine: Für uns ist es die Hauptaufgabe zu erkennen, dass es Unterschiede gibt. Wir sind ein internationales Netzwerk und wir dürfen nicht alle das Gleiche machen. Es gibt diese Unterschiede, auch der Kultur oder der Art, wie man mit Projekten umgeht. Deutschland ist viel klarer strukturiert. Österreich bewegt sich ein bisschen spielerisch. In Frankreich herrschen ganz starke Hierarchien.	Sabine: Our main task is to recognize that there are differences. We are an international network, and not everybody should do things in the same way. Those differences exist, also in culture and in the way one manages projects. Germany's structure is clearer. Austria moves a bit playfully. In France, there are really strong hierarchies.

The debate on the interaction between different national cultures was previously brought up by Petite in his Ph.D. dissertation, which deals with an Interreg Project called Walser Alps. Petite affirms that if these kinds of projects have the aspiration of transcending national borders, national cultures still stand out. As the geographer highlights, local political actors often refer to different "mentalities" in considering and managing projects (Petite, 2009: 301).

The following quotation shows that those differences are not only perceived when trying to jointly implement some concrete project but also result from a more general attitude towards how different actors think and conceive transnational co-operation. While Karmen underlines this different approach *(Herangehenweise)* of managing concrete projects, Peter continues by emphasizing more general cultural differences with which one has to cope by showing a tolerant attitude and referring to the common objective of the network, i.e. the promotion of soft mobility. Peter seems to highlight that the shared ideal of soft mobility can help in diminishing the importance of cultural barriers:

Karmen (Manager von Alpine Pearls): Die Herangehensweise, wie man Projekte strukturiert ist unterschiedlich [...]. Das hat man einfach zu lernen. [...]	Karmen (manager of Alpine Pearls): The attitude about how one manages projects is different. [...] One simply has to learn. [...]
Peter (Präsident von Alpine Pearls): Es ist einfach eine Tatsache, dass der Alpenraum aus verschiedenen Kulturen besteht und da muss man einfach auch lernen damit umzugehen und auch diese Toleranz zu haben. [...] Natürlich, in unserem speziellen Fall bei Alpine Pearls ist die extreme Unterschiedlichkeit der Mitglieder eine ganz große Herausforderung. [...] Da ist es wichtig, einfach die gemeinsame Klammer zu stärken und nicht zu vergessen, warum wir kooperieren. Nicht zu vergessen, dass wir nicht nur kooperieren weil wir einfach gemeinsam Tourismus machen, sondern weil wir diese umweltfreundliche Mobilität unterstützen.	Peter (president of Alpine Pearls): It is a fact that the Alpine space involves different cultures, and one must simply learn to cope with it, to have tolerance. [...] Of course, in our particular case at Alpine Pearls, the extreme diversity of the members is a great challenge. [...] It is important to strengthen shared interests and not to forget why we co-operate. We do not co-operate just because we all do tourism, but because we support this environmentally friendly mobility.

Mathieu, an employee of an Alpine Pearls member municipality, also juggles with two different levels: the Alpine and the national. He stresses that there are different "tourist cultures", but that there is a common "Alpine side" as well. This "common side", while not clearly mentioned, is probably what Peter avers to be the belief and the commitment for promoting sustainable development in general and soft mobility in particular. Mathieu, however, does not address the contents of these "different tourist cultures", and his statement appears to be neutral, a simple observation on distinct ways of dealing with tourism. A German mayor, citing France as an example, was more critical of the different national attitudes towards sustainable development:

Nicolas: Ich war letzte Woche in Morzine und die beklagten sich über die fehlende Sicht des [Alpine Pearls] Konzeptes, es sei nicht klar definiert.	Nicolas: I was in Morzine last week and complained about the missing visibility of the [Alpine Pearls'] concept, which is not clearly defined.
Hans (Bürgermeister): Ja, ich glaube das ist ein typisches französisches, also eher ein französisches Problem, dass man mit diesem Begriff nicht so viel anfangen kann, wage ich zu sagen. Hier im deutschen Teil haben diese Dinge (Umwelt, sanfte Mobilität), ein wesentlich größeres Gewicht, als das in Frankreich der Fall ist.	Hans (mayor): Yes, I think that this is a typically French, well, quite a French problem. They do not know how to deal with this concept, I daresay. Here in Germany, everything that has to do with those things (environment, soft mobility) has a much bigger weight than it has in France.

These national differences are felt not only in the tourism sector but in other areas as well and in the other networks. Markus, an employee in a protected area, explains the differences he experiences while working on the protocols of the Alpine Convention. Markus refers here to issues linked to water:

Im gesamten Alpenraum sind die Mentalitäten für die Umsetzung der [Protokolle der Alpenkonvention sehr unterschiedlich]. Wenn man von den Wasserrahmenrichtlinien ausgeht, dann sind die Franzosen wieder ganz anders als wir in Bayern oder in Deutschland, in Slowenien sind diese noch anders. Also	In the entire Alpine space, mentalities with regard to the implementation of the [protocols of the Alpine Convention are really different]. If one takes the Water Framework directives, for example, then the French are really different from us in Bavaria, or in Germany, in Slovenia or elsewhere. Concerning these areas, there

es sind in diesen Bereichen schon wieder sehr, sehr viele Unterschiede. (Markus, Mitarbeiter eines Schutzgebietes)	are many, many differences. (Markus, employee of a protected area)

As for the Swiss mayor who claims that French representatives greatly depend on decisions taken in Paris, one cannot be sure whether Markus' impressions correspond to "real" cultural differences or represent a more stereotypical attitude. The dilemma is open to debate and cannot, of course, be solved. What is interesting here is not solving this dilemma but observing that local political actors consider these cultural differences to be true.

It is interesting to note that Markus started to talk about those differences by first stating that they depend on the "basic prerequisites" of the different parks, probably to underline that "not everybody starts from the same point". It recalls what Mirko, the employee of a Slovene municipality, in quite a polemical and critical tone, says about the Alliance in the Alps network's demand for "harmonized development". He sees the fact of considering a common goal in terms of sustainable development as a problematic way to impose choices without taking into account what Markus calls the "basic prerequisites":

> It's a fact, when they talk about "harmonized development". For us who have a lower development, "harmonized development" actually means that we are still developing. [...] The French, the Swiss and the Austrians built up the whole Alps already [...] with artificial settlements. Then they said: "Now it's enough. Now you, Slovenes, won't do anything either." Now we are blocked. They will not build anymore, because they don't need it, but we can't build, even though we would need it. (Mirko, employee of a municipality)

Mirko feels, as other Slovenians probably do, that national specificities in the field of sustainable development are not taken into account by the network. He criticizes the attitude of the French, the Swiss and the Austrians towards Slovenians, because the representatives of these three countries do not allow Slovenia to develop as it wishes. Although Mirko says the three countries have the same attitude and belong to the same group of "developed countries", others such as Luc highlight differences between them. Luc, for example, stresses some differences between France and Austria and describe them as problematic. He affirms that there is a gap, also dictated by national differences, between the ways sustainable development and tourism should be promoted:

Les autrichiens poussent de plus en plus pour que [Alpine Pearls] devienne une marque, l'idée de *package* vraiment: proposer des transports, de l'hébergement et que tout cela soit intégré. C'est vrai que nous, on est à la traîne là-dessus, parce qu'il n'y a pas eu une appropriation au niveau local. Les offices du tourisme n'ont pas bien cerné le concept. Ils avaient même tendance à le rejeter, parce que cela faisait peur. Donc voilà, petit à petit, il y a un contraste qui se creuse entre les stations des Perles des Alpes. On n'est plus vraiment sur les mêmes orientations. (Luc, employé d'une commune)	Austrians are pushing more and more for [Alpine Pearls] to become a label. It is really the idea of a package: to propose an integrative system of transportation and accommodation. It is true that we lag in that field, because there has not been an appropriation of this at the local level. Tourist offices did not really understand this concept. They even tend to reject it, because it scares them. So, that's it, slowly contrast is widening between the Alpine Pearls' stations. We are not on the same trajectory anymore. (Luc, employee of a municipality)

From time to time, these different expectations place members at a distance from each other to such an extent that some do not see any valid reason to keep believing in the project. This happened to a French member, who decided to quit the network. The arguments that the employee of this municipality brought up are similar to the ones expressed by Luc and concern a lack of understanding of the general concept of the network.

All the facets linked to the issue of the perceived "national mentalities" I introduced in this section were also presented by two French political scientists in an article in which they present a North-South Alpine divide. Gerbaux and Zuanon explain that, on the one hand, German-speakers defend the idea of a conservative mountain based on the respect of cultures. On the other hand, Latins believe in balanced and controlled growth. The North-South divide is not a simple stylistic device, since some experts have wondered, not without reason, whether the applicability of a global model of development for the entire Alps makes sense (Gerbaux & Zuanon, 1995: 108–109). This question leads to the following section, in which the issue of a global model for development is debated. I will concentrate on the discourses of interviewees on an institutionalized Alpine region. This requires an investigation into whether those anchored national divides, mentioned by different local political actors, are, for others, surmountable in order to build an Alpine region in the future that would have its own legal and political structures.

7.2 An institutionalized Alpine region

The issue can be raised whether the actors involved in the pan-Alpine networks take the discussion on Alpine borders even farther, i.e. whether they think it is pertinent to imagine a sort of Alpine perimeter capable of defining public policies independently from national influences, for example in the area of political decisions concerning sustainable development issues. In order to spark the debate, I openly asked the interviewees whether they could imagine an "Alpine state" being built in the future. According to the information gathered during interviews, this seems to be a remote, and, most of the time, not even a desirable, likelihood. As Sabine admits, the administrative borders are still too strong to think that they could be completely erased. However, she wishes they could at least be more permeable:

Ich denke, dass unsere Verwaltungsgrenzen, die wir im europäischen Raum haben, sehr stabil sind. [...] Größere Strukturen und Gebilde zu bauen, wird immer schwieriger. [Wir] akzeptieren, dass wir in einem Staat leben, aber gleichzeitig [wäre es gut, wenn] es mehr Austausch [gäbe]. Das wäre schon durch eine Angleichung der Strukturen erreicht, aber das muss nicht unbedingt gleich miteinander gekoppelt sein. Aber diese Grenzen im Kopf, die müssen aufgehoben werden. (Sabine, Betreuerin)	I think that our administrative borders, such as they stand in the European territory, are very stable. [...] It will be ever more difficult to build bigger structural entities. [We] accept that we live in a national state, but at the same time [it would be good if] there [were] more exchanges. This would already mean an adjustment of the structures, but it does not mean that we have to be coupled with one another. But those borders have to be abolished in people's heads. (Sabine, animator)

The networks, and this is the opinion of a great number of the interviewees, help diminish the weight of national borders. For example, an Austrian mayor declares that thanks to the networks the importance of the regional boundaries are growing in importance, while, in parallel, those of the national states are losing power. However, like Sabine, Erik does not believe that those transnational initiatives are a first step towards an Alpine space with growing political weight. Additionally, he underlines that this is not the aim of local political actors:

Je ne sais pas si cela serait très bénéfique. [...] C'est important d'avoir des Etats souverains. [...] Je pense que cela ne soit pas forcément un objectif de créer une superstructure alpine qui couvre le massif. (Erik, employé d'Alparc)	I do not know whether this would be beneficial. [...] It is important to have sovereign states. I think that the creation of an Alpine superstructure covering the massif is not necessarily an objective. (Erik, employee of Alparc)

So, Erik adds an additional component to the reflection, namely the desirability of an Alpine political structure. Based on the argument of the existing structural links between the mountain and the plain, Peter gives some reasons of why this would not be advisable:

Ich glaube, dass auch die einzelnen Bevölkerungsteile sehr stark miteinander verbunden sind, d.h. die Menschen in den Bergregionen mit denen in den flacheren Gebieten. Ich glaube, dass man das einfach gar nicht so trennen kann. Aus meiner Sicht ist das auch nicht sinnvoll. (Peter, Präsident von Alpine Pearls)	I think that the individual parts of the population are really strongly connected to the plains. I think that we cannot separate those two entities. I do not even consider it reasonable. (Peter, president of Alpine Pearls)

Although arguments against the constitution of an Alpine political entity can differ from one person to the other, we must keep in mind that the great majority of the people do not deem the creation of an Alpine state to be a promising idea. There are of course exceptions, which I present in the following paragraphs. Four persons among my interviewees clearly supported this idea, although the first, Antonio, initially said that it did not make sense to establish an Alpine border, since this would mean creating another state. However, he then continues:

Antonio (sindaco): [Ma] che le Alpi siano senza frontiere sarebbe una bella cosa, no?

Cristina: Sarebbe possibile?

Antonio: Non lo so. Potrebbe essere interessante, perché verrebbe ritenuto che l'arco alpino, al di là dei valori culturali che ha, gioca anche un grande ruolo ambientale. Quindi che sia un'unica regione, questo potrebbe essere molto interessante, però, fra 100 anni?

Antonio (mayor): [But] the Alps without borders, this would be a nice thing, no?

Cristina: Would it be possible?

Antonio: I don't know. It could be interesting, because this would mean that the Alpine arc would be recognized, beyond the cultural values it has, for the important environmental role it plays. So, if it were a unique region, this could be really interesting, but, in 100 years?

So, Antonio's words show an interest in building an Alpine region without borders, but without thinking about the constitution of a sovereign state. The other three supporters of this idea are keener on conceiving a more radical project:

Sandro (collaboratore della Convenzione Alpine): L'ideale sarebbe di avere una macro-regione in Europa. Non ci arriveremo mai, però.

Cristina: E' una visione che è portata avanti dalla Convenzione delle Alpi o da te personalmente?

Sandro: Ufficialmente non è pensabile. [...] Dal punto di vista naturalistico lo è già, è una regione biogeografia, però non è una regione geopolitica, anzi, gli Stati l'hanno divisa. Le Alpi un tempo erano più unite di oggi, perché poi i confini degli Stati hanno tolto potere politico alle Alpi. [...] Il mio sogno è quello di vedere le Alpi come un grande Stato alpino, ovviamente non ci arriveremo mai.

Cristina: E' solo il tuo sogno, o pensi che sia il sogno di diverse persone attorno a te?

Sandro: Non abbastanza. Ti posso nominare un po' di persone, però sono pochi, anche perché è un'utopia, chiaramente, è un'utopia. Ovviamente oggi domina il modello statale.

Sandro (employee of the Alpine Convention): The ideal would be to have a European macro-region. But we will never reach that goal.

Cristina: This is an idea that is supported by the Alpine Convention or only by you, personally?

Sandro: Officially you cannot support this idea. From the point of view of nature, it is already a biogeographical region, but it is not a geopolitical region. On the contrary, states divide it. The Alps were more united once upon a time than today, because the national borders took away political power from the Alps. [...] My dream is to one day see the Alps as a great Alpine state, but it will never happen.

Cristina: Is it only your dream, or is it shared by others around you?

Sandro: Not enough. I can give you some names, but we are few, because it is a utopia, clearly a utopia. Nowadays, the state model dominates.

If Sandro thinks in terms of a politically structured Alpine region, it is just in terms of "dream" or "utopia", as he says. He is convinced that this is only a figment of the imagination of a few idealistic individuals. The subject comes out in casual discussions among colleagues who are also friends. In two other instances, I received information about this issue. It is interesting to note that both persons who talked openly to me about that are not involved in the pan-Alpine project. In fact, they are not even members of any transnational network or employees of pan-Alpine associations.

The first discussion I had about this topic was with Gianluca, the mayor of an Italian municipality, who is promoting sustainable development with a number of local projects but did not consider becoming a member of Alliance in the Alps. The discussion with him took place in June 2008, during the "Alpine Week". Gianluca started by observing that the pan-Alpine networks are still too weak and are just the business of a few sensitive persons. They do not move, in his eyes, as a unique entity. And, when I asked him what he meant by "unique entity", he explained that he saw an Alpine macro-region that would have proper administrative structures and specific modalities of government. He used the expression "Alpine Europe" to illustrate what he imagines. This concept would, in his eyes, define a geopolitical region and not only a geographical one.

This idea seems to fit perfectly with the opinion of a member of the regional Council of Trentino-Alto Adige / South Tyrol, who contacted me via email after the publication of my article in the *Journal of Alpine Research* (Del Biaggio, 2009). The deputy wrote that he always believed that the Alps should be set up as an autonomous region within Europe in order to be able to give concrete answers to the problems mountain regions are facing, and that he hopes that one day there will be an Alpine Euroregion, with specific elected bodies, legal competences and funds for the development of mountain political affairs.

These last two examples challenge the view that it is inside the network that one can find the most radical people who support and promote the notion of the creation of an Alpine geopolitical region. The debates within the networks are not centred on whether or not it is advisable to constitute an Alpine government and parliament.

Instead, the discussions concentrate on the question of whether networks should be considered a technical or a political tool. Yet, they are never radical when they consider the networks a "political tool"; they never imagine the possibility of constituting a politico-administrative region *per se*, as I will demonstrate in the next section.

7.3 Should networks be a political or a technical tool?

In order to comprehend the socio-political significance of the pan-Alpine project, it is important to understand whether there is a will to give political weight to the pan-Alpine networks. As already discussed, networks of local political actors were first established to facilitate the exchange of information amongst members (municipalities, protected areas, etc.). Yet, with the growing number of members, some actors see a further aim for those networks: the possibility of contributing to the creation of a pan-Alpine political region. The tension between these two configurations will be presented and discussed in this section.

As stated in the previous chapter, the president of Alliance in the Alps supports the proposal that the association will increase its political clout if it succeeds in representing 10 % of the Alpine population, 10 % of the territory and 10 % of the communes. But, as we have seen, opinions differ on this point since, for some, this quest for more members should not be an objective in and of itself. In fact, for those individuals, it is more important to have very active and committed members. The discussion at the General Assembly of the Swiss section of Alliance in the Alps in 2007,[132] which I present here, clearly reflected this tension.

---

132 This assembly was held in Flühli (Switzerland) in June 2007. The ideas presented here are from my personal notes and from the minutes (*Allianz in den Alpen GV-Protokoll*) of the meeting taken by an animator and kindly made available by the network.

The assembly started with the presentation of the results of the internal working group created on the insistence of the Swiss group for mountain regions (SAB). After the presentation, a mayor underlined the importance for Alliance in the Alps "to politicize and to make its mark" (*politisieren* and *profilieren*). According to this mayor, this is even more important in Switzerland than in the other Alpine countries, because Switzerland has not ratified the Alpine Convention. For him, Alliance in the Alps should help the Alpine Convention gain importance in Switzerland, independently from the ratification of the protocols.

After this intervention, the then-president reacted by proposing another possible strategy, namely that Alliance in the Alps focus on obtaining extra funds for its operations. He asked the audience to give their opinion on what should be more important: exerting political force for the Alpine Convention or finding additional and more stable financial support. This question led to an animated debate that took up almost the entire meeting. Opinions diverged among the attendees, split them into two distinct groups and focused on three different topics that should be the network's focus of attention. The person moderating the debate suggested these three points, the first element of which can be defined as more political, while the other two are generally understood to be more technical:

- Defence and support of sustainable development in general and the Alpine Convention in particular;
- Support in the networking activity between municipalities;
- Technical support (by the animator) to implement projects.

The moderator asked the attendees to express their opinion on these topics with the aim of finding a shared vision. Despite long discussions, this consensus was not reached. On the contrary, I had the impression that distinctive and incompatible visions were splitting the group into two factions: those supporting a more technical role for the network on the one side and those defending a more political line on the other side. If for some the network helps in discussions and in the preparation of projects (and thus to obtain funds for municipalities and for the network itself), others said Alliance in the Alps should flex its political muscle and make a clearer stance in order to ensure that the Alpine Convention becomes a truly powerful instrument and, as a mayor pointed out, "to ensure that the Alps are able to survive".[133]

The defenders of a stronger political role stress the importance of more open support for the Alpine Convention. Some have even identified the implementation of the Alpine Convention as the "fundamental mission" (*Grundauftrag*) of the network. They justify this stance by saying that without the will to implement the Alpine Convention, there are no possibilities of achieving the aim of sustainability. In fact, as some continued, municipalities that are willing to walk the path through sustainability are still taunted (*ausgelacht*). They feel that it is important to let a political awareness grow amongst the population so that the Alpine Convention can be better understood and supported; especially in Switzerland, where the political context is not favourable towards a larger acceptance of the Alpine Convention. Some members therefore think that the mission of the network is to act in order to defend this international treaty.

The other half of the group was more sceptical about the ability of Alliance in the Alps to help the convention's political acceptance. For them, the efforts should be put into strengthening the essential figure of the animators, who are the ones able to facilitate the exchange of experiences. Their role is seen as critical to the smooth functioning of the network and should therefore receive more means and attention. For the supporters of the technical vision, this is more important than the defence of the Alpine Convention as such. Thanks to the animator, other important goals can be reached, such as the support to and motivation of municipalities to implement projects of sustainable development and the cooperation and networking between Alpine local authorities.

But, in any case, there was a general fear that the financial support that Switzerland gives to the network could end if interest for the convention diminished. Thus, the participants, even those leaning more on the technical dimension, admit that the ratification and general acceptance of the Alpine Convention influence the future

133 Personal notes of the General Assembly of Swiss Alliance in the Alps, Flühli (Switzerland), 15.06.2007.

development of the network. In fact, this second group is worried that less financial means for the Alpine Convention's activities would lead to fewer resources for the technical support the network gives its members. Others remember that it is through networking and lobbying that public attention can be raised about the convention. This shows that the two aspects, political and technical, are probably two sides of the same coin.

It is interesting to note that these two factions do not only differ in their opinion of what should be the role of Alliance in the Alps, but that they also diverge in terms of their geographical references: While the defenders of a more political line refer to the Alps, the supporters of the network's technical role are more concerned with a smaller level, i.e. the municipality. The incapability of mayors, as I could see during this meeting, to include both roles (the political and the technical) and levels (the municipality and the Alps) in their discourses and practices shows the difficulty to attain the very basic aim of the networks, i.e. the implementation, at the local level, of a pan-Alpine concept of sustainable development.

Nevertheless, the debate surrounding this question is particularly interesting and important, since it concerns defining the role of the networks, the primary objective of which is the sharing of experiences. However, certain actors seem to want to go beyond simple exchanges on technical points and to develop a political voice for the Alps that reaches beyond its frontiers to the level of the state and even the European Union. In order to achieve this aim, however, the role of another actor should be mentioned: the broader Alpine population. In fact, real political influence can be attained only with the population's involvement in the pan-Alpine project. This issue is debated in the next section.

7.4 Socio-political significance of the pan-Alpine project: the participation of the base in the networks and in the population

In an article that was published in 1995 and focused on the Alpine Convention, Gerbaux and Zuanon wrote that there was still one major problem to solve concerning the treaty, namely the participation of the population in the elaboration and decisions concerning it (Gerbaux & Zuanon, 1995: 112). The situation probably has not changed much since the 1990s and, as the two authors conclude, "the question of the 'democratic deficit' still remains" (Gerbaux & Zuanon, 1995: 112).

In 2011, Simon still writes that although local communities and Alpine populations are quoted in the Alpine Convention as being actors of Alpine sustainable development, the reality is that only a few amongst them even know about the existence of this text (Simon, 2011: 17–18). For Church and Ramcilovic, ignorance about the treaty is even more generalized, as, "the knowledge of the existence of the Alpine Convention, even in the Alpine area and even among experts in environmental issues or European integration, seems extremely limited and uneven" (Church & Ramcilovic, 2008: 14). This observation is shared by the people I interviewed:

Le constat que je fais aujourd'hui sur la thématique alpine, c'est qu'il y a des choses, mais qui sont globalement peu connues du grand publique, et même, je me mets dedans, encore moins les protocoles [de la Convention Alpine]. Si tu me demandes, quels sont les Protocoles de la Convention Alpine, je suis incapable de te les donner. Donc ils sont globalement peu connus et peu portés par les Etats et les gens ne se les ont pas appropriés. (Jean, directeur d'un espace protégé)	The observation I can make today concerning the Alps is that there are things, but they are generally little known by the larger public, and, in this I include myself as well, they know the protocols [of the Alpine Convention] even less. If you ask me what the protocols of the Alpine Convention are, I am unable to tell you. They are on the whole little known and little supported by the states, and the people did not adopt them. (Jean, director of a protected area)

After reading this remark, it becomes questionable whether the layperson would know anything about the Alpine Convention if even the directors of protected areas are unfamiliar with it. I think it says a lot about the awareness of the population concerning pan-Alpine issues. In this section, however, the focus will not be on the Alpine Convention but on the networks' activity. In the previous chapter, I discussed the issue of barriers to members' participation in events organized by the networks, and I already pointed out that few members can be considered to be really active. Undoubtedly, networks have benefited from the organizational skills of some particularly active and enthusiastic members. The issue that I would like to raise here is whether (and how) these people can spread their enthusiasm to less active members (following section) and then also to the broader population (subsequent section).

7.4.1 Networks' capability of getting less active members involved

The handful of enthusiastic personalities that animate the networks were decisive in setting them up. They have always been very involved in network projects and are still active in their management structures today. Their contribution is an asset for the efficient operation of the networks, but it also conceals a major risk, that of the personalization of these associations. The president of Alliance in the Alps is aware of this and therefore wishes to have a larger group of management personnel, so that acquired knowledge and expertise do not disappear when a member resigns. For the moment, according to him, those at the head of the network would be difficult to replace.[134]

I raised, quite provocatively, this issue during the interview I conducted with the director of CIPRA International. I told Andreas that I had the impression that when I went to the networks' meetings there were always the same "10 friends" showing up and that I felt

134 Interview conducted by Cristina Del Biaggio and Bernard Debarbieux with Rainer Siegele (Personal notes, Mäder, 11.06.2007).

that this could be compared to a kind of "airtight club". I asked him to react to this. Here is his answer:

Grundsätzlich ist das einfach menschlich. Netzwerke basieren nur auf Vertrauen. Wenn du zu jemandem kein Vertrauen hast, dann kannst du nicht kreativ sein, keine gemeinsamen Ziele verfolgen. Dann funktioniert die Zusammenarbeit nicht. Mit Leuten, mit denen du auf gleicher Wellenlänge bist, mit denen machst du Projekte aus Freude und mit Begeisterung. Aber muss man auch aufpassen, dass man Leute nicht ausschließt, weil sie nicht zum exklusiven Club gehören. Aber ich glaube, dass das (Netzwerk) nicht so exklusiv ist. Das ist kein geschlossener Kreis. Da kommt man schon rein, das ist nicht hermetisch abgeschlossen. Aber grundsätzlich ist diese Beobachtung absolut richtig. Ich glaube, dass es auch so sein muss. Man muss immer aufpassen, dass man keine ausschliessende Handlungsweise etabliert. Aber grundsätzlich finde ich das gut. (Andreas, Direktor von CIPRA International)

Basically, it is simply human. Networks occur only via trust. If you do not trust somebody, then you cannot be creative, you cannot have a common agenda. In this case, collaboration does not work. With people with whom you are on the same wavelength, you carry out projects joyfully and enthusiastically. But then you really have to be careful that you do not exclude people, because they are not part of this exclusive club. However, I think that it is not so exclusive. It is not a closed circle. It is possible to come inside; it is not hermetic. But, generally, your observation is absolutely correct. I think that it should be like that. But one always has to be careful not to exclude people, although I think that in principle this is good. (Andreas, director of CIPRA International)

Through Andreas's words one can feel the tension between, on the one hand, the necessity of creating a small group of people working on the same "wavelength" and, on the other hand, the danger this attitude causes.

Besides this small group of people, whom I will qualify as leaders, some other dynamic community officials are regularly present at the meetings and are trying to implement ideas in their own municipalities. These individuals broaden the group of active members somewhat. As Robert observes, they are the ones that benefit the most from the network activity:

Il y a des degrés de dynamisme très variable. Les spécificités du microcosme communal politique font que certaines

There are different degrees of dynamism. Because of the specificities of the local political microcosm, some munici-

communes en tirent mieux profit que d'autres. Cela dépend encore une fois des personnes. [Dans une commune], on a le maire et le secrétaire communal, les deux personnes qui ont le pouvoir, sont toujours là et ils en tirent le meilleur avantage possible. En revanche dans d'autres communes c'est ["seulement"] un délégué. Il n'y a qu'un qui y va et si ce n'est pas une personnalité qui a beaucoup d'influence dans le conseil communal, les choses se passent de manière moins dynamique. [...] Tout ça pour dire que c'est très variable d'une commune à l'autre. (Robert, animateur)

palities benefit more than others. Once more, this depends on the individuals. [In a municipality], the mayor and the town secretary, the two people with power, are always there, and they are able to make the most of it. By contrast, in other municipalities it is ["only"] a representative of the town. There is only one person attending, and if he or she is not an influential personality in the town council, things happen in a less dynamic way. [...] All this is to say that it really varies from one municipality to another. (Robert, animator)

Besides this handful of leaders, to whom one should add a few dozen active members, all the others are in fact "silent members". In some cases, nobody in the local administration is informed about the membership, as I witnessed when I called the municipalities to have them answer the questionnaire. This occurred in particular in the municipalities' members of Alliance in the Alps via a regional membership. In those contexts, the respondent simply told me: "I am sorry, I cannot answer, because we are not members" (see chapter 6). Others, especially in Italy, asked me to hold the line while they connected me to the municipal library or the local police station, obviously not the right places to ask questions about Alliance in the Alps. Debarbieux and Rudaz had similar experiences when they attempted to get in touch with the municipalities: "in most cases the person at town hall who was reached did not know that the municipality was in the Alliance in the Alps network" (Debarbieux & Rudaz, 2008: 513).

The difficulties I encountered obviously have to do with the management of the network's internal communication. As discussed in chapter 6, there are motivations of time and costs in the non-participation of members to the life of the network. This leads to critical consequences. What Manuele, one of the people in a local administration who truly knew very little (actually no more than that such a "thing" called Alliance in the Alps existed) told me about the network is interesting with regard to this issue:

Per noi, per fortuna che c'è Fabio [segretario regionale] che ci dà una mano a gestire il tutto. Per esempio, noi abbiamo un comune di 1300 abitanti, esce da un'aggregazione che abbiamo fatto due anni fa, adesso come adesso la mole di lavoro è tanta. Io farei un po' fatica a seguire tutto, lui conosce la tematica in dettaglio, conosce le persone con cui dialogare, e di conseguenza è un po' privilegiato. Trovo che per fortuna che c'è Fabio. (Manuele, segretario comunale)	For us, thankfully, there is Fabio [regional secretary], who helps us manage everything. We have a population of 1 300 inhabitants, we just finished a two-year process of territorial restructuring (*aggregazione*), and now we have a tonne of work. I would have problems following everything. He knows all the details, he knows the people to contact, and therefore he has a privileged position. I have to say that we are lucky Fabio is there. (Manuele, town secretary)

If it is true that, as Manuele says, "thankfully, there is Fabio, who can follow the Alliance in the Alps dossier", it is also true that in doing so, everything is entrusted to one person: Fabio. This example shows that in some regions even the most active person cannot manage to bring enthusiasm and knowledge to other people working for the municipalities and/or for the region. Although it is practical and reassuring that one person is always active in the network's activities, the risk is high that this person becomes the only one familiar with the network and the benefits that one gets from it. Thus, they display an inability of conveying information and enthusiasm from the network to its base.

Having analysed the reasons for the (non-)involvement of the less active members of the network, in the next section I widen the circle a bit more to get an insight into the involvement of the population in the pan-Alpine project.

7.4.2 *Capability of the networks to raise awareness amongst the population*

Having a small active group at the head of these networks also raises the question of whether it is possible for them to broaden awareness in their base, meaning amongst their members (as discussed previously) but equally in the general population. Even if my research does not focus on Alpine residents and no interviews were conducted with them, the question of the Alpine population's awareness of the pan-

Alpine project can be determined by how local leaders explain the ways in which they involve (or not) residents in the networks' activities.

From regions on paper to regions of social practice

It is important to underline the role of the population in the rise of the pan-Alpine region in order to understand the social and political impact of the project. In fact, without the participation of the broader population in trans-border spaces, it is hard to shift from "regions on paper" to "regions of social practice" (Paasi, 2002: 200), as Paasi calls them.

The possibility of speaking about an "Alpine region of social practice" requires an examination of the population's (real and desired) level of implication. Although it is important to involve the population in the networks, according to 75 % of questionnaire responses received from the members of Alliance in the Alps, the interviews conducted with the mayors and network officials show that the meaning of this issue is perceived in very different ways, even though the major pan-Alpine networks officially stress the importance of the population's involvement in their activities. In fact, Alparc underlines the "environmental education and the awareness-raising targeting the general public"[135] as two key components of its role. The charter of Alliance in the Alps includes a principle called "participation of the population", which clearly underlines that the network wishes for the population to play an active role in the implementation of the pan-Alpine goals at the local level.[136]

After a discussion on the different "national behaviours" of the members, I asked Sabine (animator of Alliance in the Alps), to tell me what makes them similar; she answered clearly: "The same wish for development with the residents." Yet, this seems to be quite at odds

135 <www.alparc.org/our-actions/mountain-education/topics/communication-and-awareness/item/240-environmental-education-and-awareness-raising>, consulted on the 04.05.2015.

136 The Chart can be found, in French, here: <http://alpenallianz.org/it/fr/sur-alliance-dans-les-alpes/charte>, consulted on the 13.07.2015.

with what she declares later in the interview. For the moment, this observation actually gives additional weight to the official statements of the networks and to the declaration of Norer, who asserts in a 2002 article that there is a "general consensus" (without really specifying to whom this "general" is applied) in support of the idea that the population should be more involved in pan-Alpine issues (Norer, 2002: 27).

Borders exist in the minds of the population

It is worth looking into the capability of local leaders to pass pan-Alpine awareness and engagement on to the population. The interview with Andreas, director of CIPRA International, provides some insights into how people perceive and experience trans-boundary regions:

Andreas (Direktor von CIPRA International): Die Grenzen existieren in den Köpfen.

Cristina: Inwiefern?

Andreas: Man spürt es einfach auf kleinräumiger Ebene. Ich lebe ich in einer Grenzregion. [...] Bei der Raumplanung ist es da besonders sichtbar. Alle sagen immer: "Das muss man natürlich grenzüberschreitend planen". Aber niemand tut es. [...] Nehmen wir z.B. das Thema Nachhaltiges Bauen. Es gibt [Gruppen] im Vorarlberg [Austria], die kennen sich alle. Und es gibt solche in Graubünden [Schweiz], die kennen sich jahrelang. Aber schon bei 50 km Distanz kennen sie einander nicht. [...] Und das sind einfach Grenzen. Ich bin in einem Dorf 30 km von hier aufgewachsen. Da waren Liechtenstein und Vorarlberg geographisch so nahe. Aber das war für mich unendlich weit weg. Feldkirch [Österreich] ist das auch bei uns jetzt noch. Von Maienfeld [Schweiz] geht man gerne am Samstag auf dem Markt nach Chur [Schweiz], das sind 20 km.

Andreas (director of CIPRA International): Borders exist in the minds.

Cristina: What do you mean by that?

Andreas: Simply, one can feel "small spaces". I live in a border area. [...] For issues concerning spatial planning, it is particularly visible. Everybody says: "Of course, we have to plan space accordingly." But nobody does it. [...] For example, in the field of sustainable building, there are [groups] in Vorarlberg [Austria], and they all know each other. There are others in Grisons [Switzerland], and they've known each other for years. But even if these groups are only 50 km away, they do not know each other. [...] And these are in fact borders. I grew up in a village located 30 km from here. Geographically, Liechtenstein and Vorarlberg are so close. But it was so incredibly far away for me. Feldkirch [Austria] still seems that way. From Maienfeld [Switzerland], we like to go to the market in Chur [Switzerland]; that's 20 km. Feldkirch is also

Feldkirch ist auch 20 km entfernt, naja, 35 km. Aber nach Feldkirch zu gehen sieht soooo weit aus, aber nach Chur zu fahren ist normal. Nach Zürich, St. Gallen, die alle viel näher liegen als Feldkirch oder Bregenz [Österreich], [fährt man]. Das bleibt [in den Köpfen], aber nicht in der Politik, da ist es weniger so, weil nun, eben mit den Projekten, Grenzen überschritten werden. Man denkt "grenzüberschreitend", da kann man nicht mehr zurück.	20 km away, well, 35 km. But Feldkirch is sooooo far away, while Chur is normal. To Zurich, St. Gallen, everything is closer than Feldkirch or Bregenz [Austria]. This remains the case, although in politics it is less so, because now we cross borders, even with the projects. One thinks "cross-border"; we cannot go back anymore.

What Andreas seems to suggest is that while politically cross-border projects are now a reality in trans-border regions and political leaders can think in terms of trans-border regions, this is still not the case for the broader population, which prefers driving longer rather than crossing a national border to go shopping, for example. Andreas speaks in terms of "mental maps" that are, within the population, still strongly linked to the traditional administrative grid.

Networks do not reach the masses

What is observable during network meetings is that the most dynamic members attach particular importance to the participation (often festive, made up of good food and a reasonable amount of wine and beer) of the population. Mäder (Austria) and Budoia (Italy) are excellent examples of this. However, generally, the population today knows very little about the networks, as the majority of the mayors and employees of tourist offices admit. As Jakob (Austrian mayor) summarizes: "Alliance in the Alps does not reach the masses."

Antje, employee of a tourist office, admits that the public does not know any more than just "*a bit* about the network", even though the municipality has organized workshops and presentations to inform the inhabitants. It seems, however, that when the citizens know anything about the network, it is just the bare minimum, which is probably no more than simply being aware of its existence. The interview with Mathieu provides some interesting insights:

Mathieu (employé d'une commune): [Les habitants] ont eu l'information dans les bulletins municipaux, même dans le cadre des brochures de nos offices du tourisme, dans le cadre des horaires de transports en commun. Systématiquement, il y a le logo. Là-dessus il y a eu un travail. Il y a eu une signature officielle de l'adhésion à l'association qui a été faite, relayée dans la presse. Cela, ils le savent.

Mathieu (employee of a municipality): [The inhabitants] got the information via the municipality bulletins, the brochures of our tourist offices and the public transport timetables. There is always the logo. There is an effort made in that regard. The official signature of the membership was advertised in the media. This, they know.

Nicolas: Est-ce que cela a-t-il suscité un certain intérêt de leur part?

Nicolas: Has this aroused public interest?

Mathieu: Pas spécialement. La population est plus intéressée à son petit confort quotidien et à savoir si le gymnase où il y a le judo va pouvoir rouvrir bientôt, plutôt que de se dire: "Il y a une démarche qu'on fait pour l'environnement". Je suis un peu sévère en disant cela, mais c'est un peu cela.

Mathieu: Not really. The population is more interested in its daily comfort and in knowing whether the gymnasium will be open soon rather than thinking: "Something is being done for the environment." I am a bit harsh in saying that, but that's about where things stand.

Two other members of Alpine Pearls see the tangible projects that the municipalities implement as the only ways that raise awareness of the network amongst the population, because they have an impact on their daily lives. Only as long as the projects affect their everyday activity, the population shows some interest in them. And even when the citizens benefit "indirectly" from one of the network's actions, for example from the better public transport that it helped to establish, it does not increase the general awareness of the network, as Hans admits. The municipality informs the population through those kinds of projects, while general communication about the network remains a "side issue" *(Randthema)*:

Nicolas: Informieren Sie auch die Einheimischen über die Projekte des Netzwerkes?

Nicolas: Do you inform the residents about the network's projects?

Hans (Bürgermeister): Über die Projekte, selbstverständlich, wobei jetzt die Informationen über *Perlen der Alpen* nur ein Randthema ist. Für uns sind die

Hans (mayor): Of course, we inform them about the projects; the information about Alpine Pearls is only a side issue. For us, the essential question is

Fragen der Buserreichbarkeit, der Anschlüsse zwischen Bus und Bahn, die Möglichkeiten, die sich im Bereich der Spaßmobilität, Mountainbike oder dem Wandern bieten, entscheidend. Da sind wir natürlich in enger Abstimmung, und verkehrsbedingte Bereiche müssen wir immer mit den Bürgern abstimmen, das ist natürlich klar.

bus accessibility, the connections between buses and trains, and the offers in terms of mobility or sports, such as mountain biking or hiking. For these kinds of questions, we have to be close to the population, since all issues linked to mobility have to be decided with the population. This is clear.

From Hans's declaration, it seems apparent that although the population is openly involved in the decisions concerning specific projects, it is not implicated in the general framework in which those decisions are taken. This is probably one of the reasons why, as Hans continues, Alpine Pearls is not comprehensible (*greifbar*) to the inhabitants. Luc speaks in terms of a gap (*rupture*) between the politicians and the population:

Le gros enjeu c'est de faire comprendre à la population... C'était [Perles des Alpes] une démarche qui a été faite sans mobilisation de la population locale, sans véritable travail de sensibilisation. Il n'y a pas eu de démocratie participative avec la population autour de ce label. On a un petit peu un souci avec cela, on a une petite rupture entre les politiques et la population, par rapport à Perles des Alpes. La majorité n'a pas bien compris l'intérêt de ce réseau, parce qu'ils ne le retrouvent pas dans leur quotidien. Par contre, les actions concrètes comme les navettes, qui ont pu être mises en place, ont eu une véritable appropriation de la population de ces services. (Luc, employé d'une commune)

The big issue is to make the population understand... It [Alpine Pearls] was a process undertaken without the mobilization of the local population, without raising consciousness. There was no democratic process. We have a small worry concerning that; there is a gap between the politicians and the population concerning Alpine Pearls. The majority does not understand the relevance of this network; they don't see it in their daily life. However, the population has really taken ownership of concrete actions, such as the shuttle buses, that have been introduced, or other such services. (Luc, employee of a municipality)

If we perceive a deficiency of information from the municipalities towards the population, the observation of Maximilian (mayor) must also be taken into account. Every relationship, even that of the municipalities with their inhabitants, is formed by at least two entities, and hence the responsibility cannot be placed on the shoulders of just one of them. So, if it is undeniably true that the municipalities do not put much effort into informing the public, Maximilian's view that the population does not show any interest is probably true as well. Whatever the case may be, it is too easy to put all the blame for the gap between politicians and population on town officials, as Maximilian avers:

Ehrlich gesagt, die Bevölkerung bekommt nichts mit, oder nur ganz wenig, vielleicht wenn jemanden zu unserer Exkursion kommt. [...] Das könnte vielleicht schon ein Minus sein, dass eigentlich nur der Bürgermeister informiert ist. [Aber] mich fragt auch niemand wegen der Tafel an der Tür [mit dem Logo von *Allianz in den Alpen*], mich hat noch niemand gefragt, was das ist. (Maximilian, Bürgermeister)	Honestly, the population does not understand, or understands very little, maybe only if somebody comes along on an excursion. [...] This could be a minus, that only the mayor is informed. [But] nobody asked me about the plaque [with the logo of Alliance in the Alps] that hangs at the door; until now, nobody had asked me what this is. (Maximilian, mayor)

It is difficult to explain the population's lack of interest in the networks. Antje is the only interviewee who says it is important to remember that the concept of soft mobility in particular, and sustainable development in general, is not consensual, and that this could be a reason why a part of the population feels that the network's activities do not concern them. She declares that there are people who "do not want to be soft mobile" and do not see any economic interest for the municipality in pushing tourism development in this direction. This is probably a valid hypothesis that should be taken into account, although it cannot completely explain the attitude of general indifference as perceived by the persons I interviewed.

This attitude seems to be acknowledged in other trans-border contexts, as Häkli witnesses in the Pyrenees (Häkli, 1998) or, more recently, Strüver in the Dutch-German cross-border region (Strüver, 2004). Strüver speaks of "soft barriers to cross-border cooperation in mind" (Strüver, 2004: 27) to explain why people do not participate in the cross-border regionalization process.

About the necessity to involve the population

There is a general consensus amongst the interviewees that the population does not know much (or knows nothing at all) about the activities of the pan-Alpine networks. However, and this is more interesting for the purpose of this section, there is less consensus about the necessity of involving the larger public. Some, like Andreas, think that, generally speaking, if one is dealing with sustainable development, the population should be integrated into the decision-making process:

Wir [CIPRA] haben immer dazu gesagt, man redet so theoretisch von Nachhaltiger Entwicklung, aber muss man auch versuchen, an die Leute heran zu kommen. (Andreas, Direktor von CIPRA International)	We [CIPRA] always thought that if one speaks about sustainable development, theoretically one has to try to involve the people. (Andreas, director of CIPRA International)

Another key actor of the pan-Alpine project, who represented CIPRA in a meeting between "Alpine experts" devoted to the Dynalp 2 project, declared: "In 10 years, when the Alps have to face even more problems than now, we will need a stronger civil society, also with the help of networks such as Alliance in the Alps."[137] For him, civil society will be a central actor in the further development of the region, and he identifies Alliance in the Alps in particular and the networks in general as major political actors capable of mobilizing the population. During my five years of research, this was one of the first times that I heard an official of the pan-Alpine institutions recognizing the networks as potentially strong political actors able to stimulate civil society to partake in the debate about pan-Alpine sustainability.

In order to reach this aim, the question of the involvement of the population has to be raised and better discussed inside the networks. Yet, until now, the participation of the inhabitants does not seem to be acquired, as the interviewees admit. Henrik, for example, regrets how difficult it is to galvanize people:

137 Personal notes taken during the "Reflection team Dynalp 2" meeting (Schaan, Liechtenstein, 06.12.2007).

Henrik (Bürgermeister): Man kann die Leute [nicht dazu bringen], aktiv [zu werden], das bleibt alles auf Bürgermeisterposten sitzen.	Henrik (mayor): One cannot activate people. Everything stays in the hands of the mayor.
Cristina: Glauben Sie, dass die Bevölkerung [Ihrer Gemeinde] weiß, dass *Alpenallianz* existiert?	Cristina: Do you think that the inhabitants of [your municipality] know that Alliance in the Alps exists?
Henrik: Kaum, das kann ich sagen. Hinsichtlich der Information der Bevölkerung wäre nicht nur seitens der Allianz, sondern auch von unserer Seite aus noch viel möglich. Das stimmt.	Henrik: Hardly, I can say. Actually, it would be possible, from our side, and not only from Alliance, to inform the people. This is true.
Cristina: Glauben Sie, es wäre wichtig?	Cristina: Do you think it would be important?
Henrik: Ja. Ich würde das gutheißen. Ich würde nicht sagen: "Es ist wichtig, man muss das unbedingt machen", aber ich würde es gutheißen. Das wäre sicher nicht schlecht, dass das nicht nur auf Gemeindeebene bleibt, sondern auch in die Bevölkerung kommuniziert wird.	Henrik: Yes. I think it is good. I would not say: "It is important, we absolutely have to do it", but it would be good. It would not be bad, if it did not stay only at the municipality level but reached the population as well.

So, for Henrik, it would at least be "good" to inform the population. For others, this is not the network's task, which first of all should reach all the decision-makers. Two interviewees, a mayor and an animator, share this opinion. Niklas, mayor, explains his position as follows:

Cristina: Wäre es auch wichtig, den Leuten zu erzählen, dass Allianz in den Alpen existiert?	Cristina: Would it be important for people to know about the existence of Alliance in the Alps?
Niklas (Bürgermeister): Ganz spontan, ich glaube es nicht, dass es wichtig ist, dass die Leute das wissen. Die Allianz in den Alpen ist einfach ein Austausch zwischen den Gemeinden und ein Netzwerk, von dem ich Informationen beziehe, das ist eigentlich der Bevölkerung egal. Sie erwarten einfach, dass es weitergeht, dass es gut gemacht wird. Woher die Entscheidungen stammen, das, glaube ich, ist denen egal. So ganz spontan, glaube ich nicht, dass es ganz wichtig ist.	Niklas (mayor): I do not think that it is important that people know. Alliance in the Alps is simply an exchange between municipalities. For the population, where I get my information from is not important. They expect that it goes farther, that everything is done well. Where I got the information from is actually all the same to them. So, I do not think that this is important.

Niklas and other actors do not see the involvement of residents as a main concern. Normann, for example, clearly notices that the "criteria checklist" *(Kriterienkatalog)* of the network includes the involvement of the population, but that this is not put into practice in the networks' activities. To justify that, Normann says that this is not a priority for the municipality where he works, since, for him, the first priority is to involve the base of the Alpine Pearls' network, i.e. the hotels:

Ich denke, dass es bei uns es erste Priorität ist, die Beherbergungsbetriebe noch stärker über diese Kooperation zu informieren. [...] Das ist, glaube ich, die erste Priorität. Die zweite ist dann im Tal und dann in dem Gemeindegebiet und jede einzelne Person damit zu konfrontieren. Aber das ist eine zweite Sache. (Normann, Mitarbeiter in einem Fremdenverkehrsbüro)	I think that for us the first priority is to widely inform hotels about this cooperation. [...] This is the first priority. The second is to inform the municipalities and then every single person. But this is a secondary task. (Normann, employee of a tourist office)

Sabine, animator of Alliance in the Alps, shares the same idea. Thus she underlines that the network has been established as an instrument to support political decision-makers and not to increase awareness amongst the population:

Das Netzwerk ist gegründet worden, damit politische Entscheidungsträger voneinander lernen, damit sie sich auszutauschen. Damit diese Entscheidungen nicht ganz abgehoben von der Bevölkerung getroffen werden, kann es jede Gemeinde in die Bevölkerung zu tragen. [...] Wenn sie die Zeitung lesen würden, da steht fast in jeder Ausgabe ein kurzer Artikel über das Netzwerk. Aber weil es ihren Lebensbereich nicht wirklich aktuell berührt, wird es überlesen. Das hat mit der Berührung zu tun. Würden wir einen Schüleraustausch machen oder einen Spielwettbewerb, dann wüssten sie davon, weil es	The network has been established so that political decision-makers can learn from each other, and so that these decisions are not completely detached from the population: Every municipality can present them to its population. If they read the local newspaper, they will find in almost every issue a small article about the network, but because it does not directly affect their lives, they skip it. This has to do with feelings. If we organize a student exchange programme or a sports competition, then they would know, because this affects them. [...] [But the network] has been created for municipalities, actually only

sie betrifft. [...] [Aber das Netzwerk] ist für Gemeinden gegründet, es können auch nur Gemeinden Mitglied werden, nicht die Bevölkerung. [...] Das wäre schon ein Ziel für mich, dass wir auch Instrumente entwickeln, dass es den Gemeinden erleichtert, die Bevölkerung zu einzubeziehen, aber das ist nicht die Aufgabe des internationalen Netzwerkes. (Sabine, Betreuerin)

municipalities can become members, not the population. [...] It could be an aim for me to develop instruments so that the municipalities can more easily reach the population, but this is not our task, not the task of the international network. (Sabine, animator)

It was actually Sabine who said that she can observe a shared wish by all members to involve the inhabitants in their activities. Obviously, she observes a general wish of the members to involve the population, but she does not see it as a strategy for the network itself. For her, this should not be the task of the international network but a tool to be developed, if need be, at the municipality level. During the "mountain networks seminar"[138], Peter, another animator of Alliance in the Alps, raised the following question: "At what level do we want to establish this network?" Peter answered the question himself: at the level of the community leaders and not at the level of the population. The two animators, Sabine and Peter, thus share the same opinion: It is at the municipality level that the network has to concentrate its efforts.

This calls to mind what I wrote about in the introduction of this chapter, i.e. the importance of the involvement of the population in order to create a pan-Alpine region that is not only "on paper" but a "region of social practice", as Paasi suggests (Paasi, 2002: 200). Sabine and Peter, who each occupies a strategic position inside the network, seem to support the idea that the Alps are still far from being "a region of social practice" and that there is no universal agreement on how to accomplish this.

138 "Mountain networks" research seminar (Geneva, 28.11.2007).

On the importance of emotions

Sabine's remarks raise another important issue. She speaks about the role of *Berührung* (a word that in English can be translated as "feeling" or "emotion") when she speaks of the links that the network establishes with the population. Sabine claims that only with specific activities is it possible to reach "the hearts" of the inhabitants. Since those kinds of activities are not organized by the network, Sabine concludes that it is impossible for the network to reach the population emotionally.

The question of "emotions" was also raised during a dialogue between Michael (president of Alparc) and Guido (director of the Task force protected areas) at the seminar on mountain networks, which took place in Geneva in 2007.[139] Michael introduced the topic by stating that if one wants to run a project, it is important to create emotions around it and to find symbolic value bringing together the group. Guido answered this declaration by adding that the issue of emotions and symbols is really important for protected areas.

Guido explained his thoughts by saying that, traditionally, protected areas and the network have worked on emblematic species, such as chamois or eagles, for example. When the network decided to change its strategy in order to be more efficient from an ecological point of view, they started working on indicator species, such as insects, which are not highly symbolic and do not "touch" society. Recently Alparc started working on the ecological network, which does not seem to be a great catalyst of attention either. The population's attention, as per Guido's reasoning, is still focused only on symbolic species. In his opinion, this contributes to creating a gap between the networks and the population. Thus, in accordance with Sabine, Guido finds a common basis of this evolution: emotions. But if in Guido's tone one could hear some regret about not being able to reach the population anymore, Sabine does not seem anxious to push the network to find a solution for that.

While there is a universal consensus about the general population's lack of involvement in issues concerning the pan-Alpine net-

139 "Mountain networks" research seminar (Geneva, 28.11.2007).

works, there is also disagreement on whether or not the project should remain at the level of the decision-makers or be extended to the general population. A conclusion that can be made from the interviews my colleagues and I conducted is that the heads of the networks (also via their publications in official documents) welcome broader participation by the population. In the case of Alliance in the Alps, however, the animators are keener on devoting efforts to increase participation at the municipal level. Local authorities and employees of tourist offices seem to be divided on this issue and cannot find common ground. This reveals an obvious gap between the leaders on the one hand and the base of the network on the other.

If, as Häkli writes, in the long term, the success of major political regional innovations depends on their appeal among the "ordinary people" (Häkli, 2008: 477), then, at least in the Alps, the path to reach this aim still appears to be long. This even as the cover of a DVD that CIPRA produced in 2010 (see fig. 25), as the final result of the Dynalp 2 project, clearly shows the region composed of "ordinary faces", as if the pan-Alpine region was supported and represented by the "man on the street".

Fig. 25: Cover of the DVD "Storytelling", 2010 © CIPRA International.

So, the results show that there is a clear gap between two levels: the leaders on the one side and the basis of the network on the other. In fact, the entire networking process looks like it is firmly in the hands of the leaders (Del Biaggio, 2009) without any involvement from the population. This is corroborated by other experts of the Alps, who admit that "the involvement of the local population at first seems to be a secondary objective [for partnerships as networks]" (Debarbieux & Rudaz, 2008: 513). This confirms what Häkli discovered in the Pyrenees, which is that the majority of the population living on both sides of the mountain range is still linked to their local political community and do not see the development of the trans-boundary region as a whole (Häkli, 2004: 65–66; see as well Strüver, 2004).

This situation is problematic for the goal of reaching general acceptance and support for the Alps and creating "major political regional innovations" (Häkli, 2008: 477), but it does not stop leaders involved in the making of the region from initiating new processes that will stimulate the future of the Alpine region.

7.5 Conclusion

Having analysed in previous chapters how networks of local political actors work, I used this last chapter of the book to examine what significance this work has for the Alpine region. As discovered, the traditional administrative divisions still play an important role – in a concrete but also in a symbolic way – on the networks' activities. Thus, networks are not taking decisions independently from the national states to which they refer. The majority of the members, who in any event cannot imagine that an Alpine "superstructure" will ever be built, do not see this role of the national state as something threatening.

The most important conclusion that must be drawn is that only a small fraction of the networks' base and the population participates in the networks' activities and, consequently, they contribute little to the idea of the Alpine region. As such, it is not possible to consider the Alpine region as a region of social practice.

8. Conclusion

Throughout this research, I focused on the operation and significance of a recent phenomenon occurring in the Alps. Since the 1990s, a number of networks of political actors have been established. These connect different social and political actors (mayors, regional secretaries, managers of protected areas, rangers, ski resorts, scientists, ecological activists etc.) inside the Alpine region as defined by the Alpine Convention. Some of them were created as a strategy to provide consistency to an international treaty signed in 1991 by the ministries of the environment of Germany, Austria, France, Italy, Liechtenstein, Monaco, Switzerland and the European Union. The Alpine Convention, which is also the name of the treaty, aims to promote sustainable development in the Alpine area and to protect the interests of the people living inside it. Given that the focus is on sustainable development, the research embraces environmental, social, economic and cultural dimensions.

Criticism towards the convention, which some consider to be merely a piece of paper without any real implementation in the territory, pushed political actors and organizations, first and foremost the *Commission Internationale pour la Protection des Alpes* (CIPRA), to find solutions to reach the local decision-makers and put into practice what the Alpine Convention promised: sustainable development in the entire Alpine arc. That is why, often thanks to the impetus of CIPRA, networks of local actors were established, which gave rise to a number of initiatives that Debarbieux identifies as a sign of a veritable "pan-Alpine activism" (Debarbieux, 2008a: 40). The plan was to create new ways of exchanging experiences and know-how in the field of sustainable development inside the Alpine region. The Alpine Convention defines the region as an area of approximately 200000 km² comprising part of the territory of eight countries, roughly 6200 municipalities and some 13 million people.

In coding the interviews and analysing my field notes, I concentrated on three issues that correspond to the three main aims of the research presented in this book: uncovering the modalities of how the Alpine region is *objectivized*, *identified* and *institutionalized* through a process of social spatialization.

Firstly, I analysed how local political actors and the networks in which they are involved objectivize and identify with the Alpine region. As a result of the decision to divide the concept of identity into two sub-concepts, objectivation and identification, I was able to explore the significance of a many-faceted concept, which Bray characterizes as "indefinable" (Bray, 2002: 14); a concept that signifies too much in its strong sense, too little in its weak sense or nothing at all because of its intrinsic ambiguity (Brubaker & Cooper, 2000: 1). That is why the decision to divide the concept into two distinct but complementary meanings helped to ensure the comprehension of a notion that has become a meaningless "catchword" (Häkli 2001: 115) in the past few decades.

Secondly, I turned my attention to the operation and challenges of the new model of cooperation chosen by Alpine political actors: networking. I analysed in depth this new mode of cooperation inside the Alps, something that no other scientists had done before.

Thirdly, because the pan-Alpine activism led to the creation of *ad hoc* institutions, I evaluated the socio-political significance (or institutionalization) of the pan-Alpine region. I discuss the main results with regard to these three issues in the following paragraphs.

8.1 The Alps: a unique entity based on its natural and cultural diversity

My first hypothesis was that Alpine political actors create pan-Alpine institutions because they have a shared belief of what the Alps are (and should be). In the first part of chapter 4, I explored how local political actors observe, think and refer to the Alps. Without surprise, there is a consensus among the networks and their members that the Alps, due to their topography, constitute a unique biogeographical region within Europe. This distinctive natural structure supports a unique Alpine nature and culture. Through its mountains, rivers, animals and plants, but also its inhabitants, the Alps are seen as a distinct region inside Europe.

Geomorphology is the material "substance" delimiting the Alpine region. Although taken for granted by the majority of the interviewees (the Alps are the "rock" upon which they are built), the confines of the Alps are not uncontested. Based on the same argument that some use to justify the existence of a region called the "Alps", others claim that, for those exact same geological reasons, the boundaries of the Alps as drawn by the Alpine Convention are inaccurate.

Hence, the distinctiveness and uniqueness of the Alps is authenticated by the rock that constitutes them, even by those who contest the Alpine delimitation. However, it is also explained by the presence of a common culture, a likeminded mindset or "mentality". Interestingly, it is also validated by the natural and cultural diversity marking them: different natural settings (the limestone plateaus of the pre-Alps, mounds of schist, granite needles, moraine and scree, farmed terraces, marshes and wetlands, dark pine forests and vibrant larch forests, flower-strewn meadows and grassy pastures, lofty cliffs, breathtaking caves and valleys, tumbling waterfalls, roiling mountain streams and gentle rivers, etc.); as well a diverse linguistic heritage (from Slavonic to Germanic, with Alemannic German, Bavarian and the Walser dialect, to the Rhaeto-Romance languages – Romansch, Ladin, Friulian – and Gallo-Romance and Italo-Western dialects), and different architectural heritages. Moreover, it is characterized by different national legal systems, natural conditions, population distribution, agriculture and forestry, different state and development of the economy, of the volume of traffic and of the nature and intensity of tourism.

Consequently, the following question was raised in this book: Why do local political actors and networks use diversity to claim unity? How it is that the claim for diversity does not abolish the goal of a united Alpine region that networks and local political actors want to build? Considering the Alps as a root metaphor (Häkli, 2005), it helped to find an answer to this seemingly paradoxical situation: The

Alps are built on such "solid" rock (taking it as a metaphor, as well as the mineralogical phenomenon), that they transcend differences and unite what is objectivized as different. Thus, the Alps cannot "crumble" if somebody proclaims their differences, because they are "obviously there". Diversity can thus be used to justify the Alpine peculiarity and specificity.

In order to study more in depth the question of the Alps' objectivation, I adapted the grid of "five recurrent images of the Alps" that Petite proposed in his Ph.D. thesis (Petite, 2009) to my case study. I analysed the image of the Alps linked to the concept of sustainable development, to nostalgia, to the economic and tourism capital of the Alps and to ecology. I will present the main results here.

Firstly, even though local actors continually refer to sustainable development, they do so not only in terms of the protection of nature but also in terms of an equilibrium that must be found between the social, economic and ecological issues; the idea of the Alps associated with spectacular natural and cultural landscapes still dominates during an analysis of the iconography that the networks produce. In that sense, the notion of the Alps that the networks convey does not diverge from the one present in the collective imagination.

Secondly, probably because networks of local actors were established in order to find innovative solutions to the problems of the Alpine arc, they are not lingering in a "glorious past" that no longer exists but are drawing attention to future-oriented ways that would implement sustainable development. Past activities and beliefs are not held in the highest regard, but sometimes they are used in a way that combines them with contemporary solutions.

Thirdly, only the members of one of the three networks analysed in this research seem to be aware and to put forward the economic and tourism capital that the Alps represent for them. Not surprisingly, it is the network whose aim is the promotion of sustainable tourism: Alpine Pearls. For the pearls constituting the network, it is crucial to preserve the natural conditions of the Alps so that they can, be attractive to tourists, not only today also in the future.

Fourthly, the principle of sustainable development, the basis upon which the Alpine Convention is constructed, can be considered as a concept that the networks share. It is used as a reference for the actions undertaken by their members. The belief that sustainable development is the key concept that will allow the Alps to overcome the threats that they are facing is uncontested at present. This was not the case in the past, when CIPRA first tried to collaborate with municipalities. Now, sustainable development is undoubtedly a shared concept that shapes action.

The ways described above, in which local political actors and networks objectivize the Alps, also have an impact on the actions they undertake. In fact, actions implemented under the aegis of the Alpine Convention are carried out with reference to these images, which can be considered shared representations circulating amongst and internalized by pan-Alpine political actors.

This point confirms in part one of the initial hypotheses: By sharing experiences and know-how, local political actors also share a common representation of the Alps. However, the three networks and their members do not focus on the same features while objectivizing them. What they have in common is that they all mobilize pan-Alpine projects following the same aim: to guarantee a sustainable future for the Alps.

8.2 A strategic identification with the Alps

My second hypothesis was that Alpine actors not only create pan-Alpine institutions because they have a shared belief of what the Alps are, but also because they feel part of it. In the second part of chapter 4, I argued that shared representations of the Alps are built by confronting one's own representations with those of others living and working in what local political actors define as being the same region: the Alps. In that sense, pan-Alpine identity is not built in opposition to that of "others" who live outside the Alps, but as a sort of collective contract based on the awareness that if united it is possible to build a strong regional entity capable of being a heart inside the European Union's administration, and that it is easier to solve common problems if standing united.

This final point was central to the interviews I carried out. In fact, a number of interviewees from the three networks highlighted that they chiefly identified with the Alps because they all share the same mountain environment and the same problems this causes to mountain communities. Accordingly, the members of the pan-Alpine networks can be considered a sort of "community of problems", which underlines the fact that networks are able to create a community of people sharing the same problems (such as avalanches, aging of the population, traffic, implications of climate change etc.). So, the Alps consist not only of "material substance" but also of problems caused by this "stuff".

Recognizing this, some local political actors decided to act in order to solve these problems, which, in their opinion, cross national borders. They say they believe that becoming members of networks of local actors is the best approach to achieve their goals.

However, the motivations behind why local political actors decide to take part in transnational initiatives are very diverse.

The most important point that needs to be underlined here is that even the most active network members did not become involved in pan-Alpine initiatives because they strongly identify with the Alpine region. In most cases, their involvement is dictated by pragmatic reasons: getting funds, solving local problems or promoting sustainable development. By choosing the Alps as a point of reference for their projects, local political actors reveal what can be qualified as a "strategic identity" (identification with the Alps is a tactic to achieve a goal) more than an "affective identity". Moreover, the research showed that the Alpine scale is not the sole scale of involvement; local political actors are often implicated in other territorial scales.

There are diverse ways of identifying with the Alps, and analysis demonstrated that the one essential characteristic for local entities to become involved in pan-Alpine projects is the presence of local social and political actors capable of transforming individual identities into collective ones. It is in fact crucial that people working inside the municipality understand the benefits of being members of a network in order to steer the municipality's strategy towards the Alps. These people's presence is necessary not only to become member but also to make the inhabitants feel part of the project.

Examples from another network, Alparc, which connects Alpine protected areas, also show that it is not always necessary to identify with the Alps in order to become involved in pan-Alpine networks. One can be more attached to the idea of network (regardless of whom they are linking) than to the notion of engaging in a pan-Alpine project. However, as interviews with rangers working in protected areas show, strongly identifying with the Alps, as is mostly the case, does not automatically mean that they get the possibility to participate in pan-Alpine projects. Structural problems make it difficult if not impossible for the majority of them to take advantage of the activities offered by the networks.

We are able to conclude that local political actors who identify with pan-Alpine projects (e.g. by partaking in pan-Alpine networks) do not necessarily identify with the Alps as such. Additionally, identifying with the Alps does not automatically lead to involvement in pan-Alpine projects, and involvement in pan-Alpine projects does not necessarily mean an affective identification with the Alps. Therefore, identification and involvement are not necessarily evolving in parallel.

The initial hypothesis, namely that identification with the Alps is the engine that drives social and political action, is not completely supported by the facts; even when they do not strongly identify with the Alps, local political actors can decide to act at the pan-Alpine scale. What is confirmed, however, is the importance of the presence of particularly active persons (whom I identified as leaders) in the decision-making process undertaken by municipalities to participate to a networking activity.

8.3 The Alps as a hyphen: functioning and challenges of networking

Given the theme of the research I undertook, I was particularly attentive to one kind of involvement inside the Alpine region: networking. As I discussed above, political actors do not necessarily decide to join pan-Alpine networks because they identify with the Alpine perimeter. If not identification with the Alps, what reasons do members cite for joining the networks? This part of the research answered my main question: How do networks of local political actors actually work? This constituted the first part of chapter 5, while in the second part I outlined the challenges with which they have been confronted. Finally, I specifically raised the issue of the socio-political significance of pan-Alpine networks of political actors.

8.3.1 Functioning of networks

The survey I conducted amongst the members of the Alliance in the Alps network shows that municipalities decide to join the network mainly for what I defined as "practical reasons", as demonstrated in the first part of chapter 5: to find solutions for local problems, to exchange know-how on sustainable development and to find financial means and partners for implementing projects. As the interviews conducted with the municipalities that did not join the network show, however, these arguments are valid only if there is a person in the municipality capable of convincing the rest of the people working in the municipality of the validity of the approach and the pertinence of the Alpine perimeter.

In general, the interviews indicate that networks fulfil three functions. Firstly, they are a political answer to local social problems (loss of identity, loss of links with the territory). Secondly, they are a political response in a changing political context. Interviewees emphasize that with the establishment of the European Union there is a need to cooperate in order to be "heard" at higher territorial levels.

Thirdly, they are a social answer in a changing political context. The European Union, as some aver, "opened the minds". These last two points highlight that the European Union prepared the ground socially, politically and economically (through the funds it allocated for trans-boundary projects) for an increase in networking activity.

Concretely, I highlighted five dimensions of networking: political, socio-cultural, pragmatic, ecological and economic.

Concerning the first (political) dimension, members have used networks in the past as a real political lobbying instrument or as a more general source of inspiration to convince the local, national and/or supranational administrations to steer their policies in a certain direction. Yet, not everybody is convinced that the current political influence that pan-Alpine networks are able to exert towards the other political institutions carries much weight. Interviewees are divided on this issue.

Networking also entails a socio-political dimension. Through networking, people from different countries and with different socio-cultural backgrounds meet, more or less regularly, in different locations in the Alps to discuss their problems, their strategies and their aspirations. Networks place emphasis on personal meetings, in conjunction with good virtual communication with their members. Conferences, excursions and workshops are organized to give members the possibility to meet face to face. Along with good wine and food, the pleasant atmosphere during these meetings guarantees fruitful exchanges and durable contacts, as some members proclaim.

There is also a pragmatic way to think of networks. In that sense, networks are seen as practical tools that enable municipalities to learn from each other and to take advantage of experiences others have had before. What some members underline is that one does not meet only "banal" municipalities but also ones that are future-oriented and serve as a source of inspiration.

Members can thus be inspired by what others have already done or are doing, especially in the field of sustainable development. Thus, networks add an ecological dimension as well, that of sustainable development composed of three branches: social, economic and ecological sustainability. Given that the Alpine Convention is the main referent for the networks, members pay particular attention to environmental issues. Yet, when speaking with members of the networks, only rarely did they explicitly mention sustainable development; in such cases, it was almost always only the members of Alpine Pearls who did so. I expect that this is because sustainable development is "obviously" a driving concept in their daily activity and therefore not worth mentioning. Yet, not everyone shares the idea that ecology is as important as the economy when thinking and implementing projects in a municipality. Some Slovenian interviewees said that, in view of the socio-political and economic situation in their country, ecology should be put set aside in favour of the country's economic development.

Even if the great majority of the members share the idea that sustainable development is the concept that drives their actions, this does not mean that networks do not also comprise an economic dimension, the last dimension to be uncovered through my interviews. The most important means that members have of taking financial advantage of their membership are the credits that they receive from European and private funds to implement local and trans-boundary projects.

8.3.2 Challenges of networking

Bearing in mind the increased importance (shown by the growing number of members) of the networks, it was worth researching the challenges that they now face and those they will face in the future. In the second part of chapter 5, I exposed the four I could detect.

Firstly, it is generally agreed that only a few members actively participate in events organized by the network. The 190000 km^2 that the Alps comprise are not always easily reachable. Networks have not yet found an optimal solution to increase members' participation to the events they organize. Participation demands time and resources, not only financial, but also (and probably mainly) in terms of personnel. The different languages spoken in the Alpine arc, however, do not present an obstacle to participation.

Secondly, Alliance in the Alps introduced and encouraged regional memberships. In recent years, not only individual municipalities could join the network, but groups of municipalities could also do so. Although they had thought this would solve some of the problems they encountered (for example, non-participation by members), in reality this gave rise to new problems, such as the non-involvement of the municipalities composing the region. Astonishingly, some of them are not even aware that they belong to a network.

Thirdly, should networks increase membership or increase the quality of the services offered to their members? This is a dilemma that has not yet been solved. This question is of interest within Alliance in the Alps and also at Alpine Pearls, albeit for different reasons. Some members of Alpine Pearls are afraid of losing the exclusiveness of the label if more municipalities join the network. Nonetheless, its president is pushing for more members, because, in his view, this will enable the network to strengthen its political weight and consequently to increase the potential for partners likely to fund projects. The president of Alliance in the Alps shares this vision: More members mean more political power. Yet, in this network as well, not everybody agrees with the president; some would like to improve the quality of services offered to the members instead. A divergence of vision on this issue separates the head of the network from its base.

Fourthly, the expressed desire of Alliance in the Alps' animators to bank on quality instead of quantity suggests that structural changes need to be made. In fact, in order to help municipalities better implement projects of sustainable development and facilitate the exchange of ideas between municipalities, it is necessary to professionalize certain positions and give stable structures to people who could take on the role of strategic consultants delivering such a service. However, such wishes do not always match those of the funding agencies, which want to support "visible" projects. Hiring a coaching staff for members of the network cannot be put in brochures, cannot be photographed, cannot be filmed and therefore does not interest funding agencies.

8.4 Socio-political significance of the Alpine region

In my final hypothesis I argued that, while there is clear involvement by a number of social and political actors for the Alpine cause, and *ad hoc* institutions have been created in order to support this action, the Alps cannot yet be considered a real institutionalized region, as Paasi defines it (Paasi, 1986). The reason for this is that the Alpine region has not acquired an established status in the spatial structure and social consciousness of the society (Paasi, 1991), which corresponds to the fourth shape of institutionalized regions that he identified (Paasi, 1991).

Analysis of the discourses of the members and the managers of pan-Alpine networks of local actors contributed to understanding whether and how the Alpine region reached (or not) the fourth stage of regional institutionalization.

Networks connect different points in a given space. Pan-Alpine networks of local actors, as their name suggests, are circumscribed inside the perimeter of the Alps as delineated by the Alpine Convention. They could theoretically exist without the influence of any traditional administrative entity. However, I detected different ways in which traditional administrative institutions (such as the national state and the regional administrations or the European Union) affect the decisions made by networks. The national state, as far as I could see, is present in different ways in the networks' activities. It influences the networks structurally, politically and culturally.

In fact, the networks' decision-making bodies are constructed following the national establishment: executive boards, for example, are constituted according to national criteria. The political influence of the national state differs geographically and in its nature but is still present in the networks' daily management. French members of Alpine Pearls regret that the French administration does not have any influence over its members and praise the intervention of Austria, while Swiss members of Alliance in the Alps are simultaneously worried about the possible non-involvement of the state administration concerning the network and sceptical of the pressure that it exerts on them. In some cases, depending on the results of elections, the administration can spur or inhibit the participation of members inside the network. Finally, members highlight different ways, which some define as "mentalities", of managing projects and conceiving networks, depending on national settings.

The Alpine region exists on maps and has its own borders and institutions, but it is still bound by traditional spatial arrangements that influence the networks' structure and functioning. Nonetheless, does the Alpine region have an "established status in the social consciousness of society" (Paasi, 1991)?

Before answering this question, it is useful to remember some facts. Alparc connects more than 900 protected areas of more than 100 hectares each, covering approximately 25% of the territory defined by the Alpine Convention. These protected areas are automatically members of the network. They are part of Alparc by virtue of the fact that they are protected areas within the Alpine Convention's boundaries. Alliance in the Alps currently counts around 300 municipalities; Alpine Pearls has 27 tourist resorts. To become a member of Alliance in the Alps or of Alpine Pearls, municipalities need to follow a procedure of adherence and pay for membership. The number of members, although constantly on the rise, has to be evaluated in relation to the size of the Alpine region as defined by the Alpine Convention. The convention consists of 6200 municipalities. This means that, all together, the two networks (Alpine Pearls and Alliance in the Alps) comprise around 5% of the municipalities. I made this calculation not to disparage the work of the networks but to put it in context. Five percent of municipalities is not a very high percentage, especially when taking into account that amongst this 5% very few are really active, as I observed during my research. Although these very active people are essential to keep the network active, close observers of the networks, including these active persons themselves, are aware that knowledge and expertise would disappear if network "leaders" resigned, because they are difficult to immediately replace.

Another point that has to be mentioned is that too many people from the base rely on the participation and involvement of the active individuals. This means that only a small group of people are really up-to-date regarding what is happening at the Alpine scale. Neither the Internet nor the efforts by the networks to provide information to their members seem to reach the base.

If networks encounter difficulties to reach the entire membership adequately, they encounter even higher obstacles when they have to reach the larger population. Although the inhabitants of the municipalities and regions benefit from the projects that their mayors implement (for example, increased public transport frequency, the opening of a new cultural centre, the construction of walkways etc.), they are not aware that the necessary funds for implementing them result from membership in a network. Many researchers and social and political local actors observe the non-involvement of the population. And even if there is consensus on this observation, there is disagreement on whether or not the population should be involved. For some, in fact, the *raison d'être* of the networks is to reach decision-makers and not inhabitants. As I stated in the introduction, one of the aims of this research was to measure the social and political consequences of three processes (objectivation, identification and institutionalization). After an analysis of the discourses and practices of the political actors involved in networks, one has to conclude that the pan-Alpine region is not (yet?) a "region of [daily and broad] social practice" (Paasi, 2002).

On a more political scale, the collected interviews provided no evidence that local political actors supported the idea that the Alpine region should acquire proper political structures, such as a parliament and a government. In that sense, there is no indication that, even without the concept, members of the pan-Alpine networks or employees of the different pan-Alpine institutions would support the establishment of a bioregion, as intended by Berg and Dasmann (Berg & Dasmann, 1977).

Again, according to information gathered during interviews, very few people see the possibility of constituting an institutionalized region able to make decisions independently from the national states. These persons, as they themselves admit, form a small minority, and they agree that the constitution of an Alpine state is at present no more than a dream, and at best a utopian idea. The majority of the people involved in this project, including those I have qualified as a "minority", would support the less revolutionary idea that the establishment of an Alpine state is not necessary and that the traditional administration grid does not have to be contested. Nevertheless, members of the network seem to desire more porous national borders in order to facilitate collaborations between cross-border communities.

Although the Alpine region did not become a "region of social practice" and hence did not reach the "institutionalized" stage, as Paasi intends, it nevertheless can be considered a "regional assemblage" (Allen & Cochrane, 2007), a territorial entity partly escaping political borders. In fact, national borders do not really exist in the daily practices of the networks, nor do they exist in the social, political and scientific work undertaken by organizations and associations promoting the pan-Alpine region. The border they relate to is the one defined by the Alpine Convention, which corresponds to what they consider its "natural" borders: The Alpine region stops right where the mountain does.

At the same time, decisions made by the Alpine political actors are influenced by considerations stemming from the national administrations and from the European Union, which do not necessarily lie within the Alpine perimeter. Without the financial resources ensured by the European programmes, for example, a number of projects would not have been implemented in the Alps, and the networks would have no reason to exist. In that sense, "regional assemblage" is the right concept for the Alpine region, because it escapes political boundaries (national borders, as well as the delimitation of the Alpine region as such) and captures the idea of a region as a meaningful territorial entity.

This meaningful territorial entity, the Alpine region, is also shaped by the networks of local actors that refer to it in their daily practice. These networks, which Götz defines as the "ambassadors of the Alpine Convention" (Götz, 2011), were created to solve common problems inside the Alpine perimeter. By connecting scattered points of the Alps (municipalities and protected areas), they contribute to building the skeleton and giving substance to the idea of the Alpine region. Accordingly they demonstrate the validity of Burkeley's intuition: Scales and networks can be mutually constitutive rather than mutually exclusive (Burkeley, 2005).

By analysing the functioning of these networks and the challenges they face, I could demonstrate, on the one hand, that it is useful to think of the new regional world in terms of a co-construction of networks and territories and, on the other hand, that this new regional world is built by actors that have a similar way of objectivizing the region but a different attitude concerning their identification with it. The case of the Alpine region helps to answer in the negative the question that Häkli raises, namely whether regional networking creates "spatially disembedded political formations" (Häkli, 1998: 98). I can argue that regional networks instead contribute to making the already dense and intricate territorial assemblages even more complex.

References

Aberley, Doug, ed. 1993. *Boundaries of Home: Mapping for Local Empowerment.* New Society Publishers.

—. 2006. "Interpreting Bioregionalism. A Story from Many Voices." In *Bioregionalism*, edited by Michael Vincent McGinnis, 13–42. Routledge.

Agnew, John. 1999. "Regions on the Mind Does Not Equal Regions of the Mind." *Progress in Human Geography* 23 (1): 91–96.

—. 2001. "Regions in Revolt." *Progress in Human Geography* 25, 1: 103–10.

Ahner, Dirk. 2010. "The Concept of European Macro-Regions." In *Metropolises and "their" Alps*, edited by Michal Didier, Peter Eggensberger, Silvia Jost, Thomas Scheurer, Dirk Ahner, Christian Ude, Nicolas Evrard, Paul Messerli, and Monika Seynsche, 112–13. Munich: Bayerische Akademie der Wissenschaften.

Alexander, Don. 1999. "Bioregionalism: The Need for a Firmer Theoretical Foundation." *Trumpeter* 13 (1): –.

Allen, John, and Allan Cochrane. 2007. "Beyond the Territorial Fix: Regional Assemblages, Politics and Power." *Regional Studies* 41 (9): 1161–75.

Allen, John, Doreen Massey, and Allan Cochrane. 1998. *Rethinking the Region.* London: Routledge.

Alpine Pearls. 2006. *Bericht Alpenperlen-Ticket*, tech. rep.

Amin, Ash. 2002. "Spatialities and Globalisation." *Environment and Planning A* 34: 385–99.

Anderson, Ben, and Colin McFarlane. 2011. "Assemblage and Geography." *Area* 43 (2): 124–27.

Arnus, Petra. 2007. "Pan-Alpine Networks: The Case of the Alpine Network of Protected Areas (ALPARC)."

Atkinson, David, Peter Jackson, Neil Washbourne, and David. Sibley, eds. 2005. *Cultural Geography: A Critical Dictionary of Key Concepts.* London: I.B. Tauris.

Avanza, Martina. 2007. "Les 'purs et durs de Padanie'. Ethnographie du militantisme nationaliste à la Ligue du Nord (Italie), 1999–2002". Paris: Ecole des Hautes Etudes en Sciences Sociales.

Avanza, Martina, and Gilles Laferté. 2005. "Dépasser la 'construction des identités'? Identification, image sociale, appartenance." *Genèses* 61: 134–52.

Balsiger, Jörg. 2006. *Regional Identity and Resource Policy Effectiveness: Sustainable Mountain Development in the Swiss. Alps and California's Sierra Nevada.* <http://userpage.fu-berlin.de/ffu/akumwelt/bc2006/papers/Balsiger-Regional Identity.pdf>.

—. 2007. "Transnational Social Movements." In *Encyclopedia of Governance*, edited by Mark Bevir, 2:986–87. Thousand Oaks: Sage publications.

—. 2008. *Regional Sustainable Development in the European Alps [Regionalism Reconsidered: Earth System Governance and the Alpine Convention]*. Working Paper MWP 2008/23. San Domenico di Fiesole: European University Institute.

Bätzing, Werner. 1994. "Die Alpenkonvention – Ein Internationales Vertragswerk Für Eine Nachhaltige Alpenentwicklung Auf Dem Mühevollen Weg Der Politischen Realisierung." In *Gefährdung Und Schutz Der Alpen*, edited by Herbert Franz, 185–204. Wien: Österreichische Akademie der Wissenschaften.

Bausch, Thomas, Thomas Dax, Umberto Janin Rivolin, François Parvex, Sergeja Praper, and Martin Vanier. 2005. *Alpine Space Prospective Study. Sustainable Territorial Development in the Alpine Space: Towards Long Term Transantional Cooperation*. Alpine Space Programme 2005.

Becker, Howard S. "Generalizing from Case Studies." In *Qualitative Inquiry in Education*, edited by Elliot W. Eisner and Alan Peshkin, 233–42. New York and London: Teachers College Press, Columbia University.

Berg, Peter, and Raymond Dasmann. 1977. "Reinhabiting California." *The Ecologist* 7 (10): 399–401.

Bialasiewicz, Luiza. 2002. "Upper Silesia: Rebirth of a Regional Identity in Poland." In *Region, State and Identity in Central and Eastern Europe*, edited by Judy Batt and Katarina Wolczuk, 111–32. London: Routledge.

Bialasiewicz, Luiza, Paolo Giaccaria, Jones Alun, and Claudio Minca. 2013. "Rescaling 'EU'rope: EU Macroregional Fantasies in the Mediterranean", *European Urban and Regional Studies*, 20: 59–76.

Blacksell, Mark. 2006. *Political Geography*. London: Routledge.

Bourdieu, Pierre. 1980. "L'identité et La Représentation. Eléments Pour Une Réflexion Critique Sur L'idée de Région." *Actes de La Recherche En Sciences Sociales* 35: 63–72.

Bozonnet, Jean-Paul. 2002. "Un Siècle D'imaginaires Dans Les Alpes. Mutation Du Récit D'ascension et Fin de L'initiation Institutionnelle." In *L'Espace Alpin et La Modernité, Bilans et Perspectives Au Tournant Du Siècle*, edited by Daniel Grange, 339–54. Grenoble: PUG.

Bradshaw, Michael J. 1990. "New Regional Geography, Foreign-Area Studies and Perestroika." *Area* 22 (4): 315–22.

Bray, Zoe. 2002. "Creating a Unified Urban Space astride a State Frontier: Opportunities and Challenges. The Case of Bidasoa-Txingudi, Basque Country." *Die Erde* 133: 55–67.

Brenner, Neil. 2001. "The Limits to Scale? Methodological Reflections on Scalar Structuration." *Progress in Human Geography* 25 (4): 591–614.

Brubaker, Rogers, and Frederick Cooper. 2000. "Beyond 'Identity.'" *Theory and Society* 29: 1–47.

Bryman, Alan. 2004. *Social research methods*, Oxford University Press, New York.

Bulkeley, Harriet. 2005. "Reconfiguring Environmental Governance: Towards a Politics of Scales and Networks." *Political Geography* 24 (8): 875–902.

Campbell, Donald. 1975. "Degrees of freedom and the case study." *Comparative Political Studies* 8: 178–185.

Castells, Manuel. 1999. *Le Pouvoir de L'identité*. Paris: Fayard.

—. 2004. *The Power of Identity. The Information Age: Economy, Society and Culture*. Vol. II. Cambridge, MA; Oxford, UK: Wiley-Blackwell.

Cerulo, Karen A. 1997. "Identity Construction: New Issues, New Directions." *Annual Review of Sociology* 23: 385–409.

Church, Jon Marco, and Jon Marco Church. 2010. "Environmental Regionalism: The Challenge of the Alpine and Carpathian Conventions". New Orleans.

Cuypers, Stefan E., and Céline Randier. 2009. "L'application juridique de la convention sur la protection des Alpes: La situation en Allemagne, Autriche, France, Italie et Slovénie." *Revue Européenne de Droit de L'environnement* 13 (1): 3–32.

De Groote, Nicolas. 2009. "Les Alpes, Identité Montagnarde Politisée. Exemple du Partenariat Alpine Pearls, Réseau Transnational Servant de Cadre Díaction Politique pour une Identité Collective Partagée." University of Geneva.

Deas, Iain, and Axel Lord. 2006. "From a New Regionalism to an Unusual Regionalism? The Emergence of Non-Standard Regional Spaces and Lessons for the Territorial Reorganization of the State." *Urban Studies* 43 (10): 1847–77.

Debarbieux, Bernard. 1999. "Le territoire: histoires en deux langues. A Bilingual (his-)story of Territory." In *Discours scientifiques et contextes culturels: Géographies françaises et britanniques à l'épreuve postmoderne*, edited by Christine Chivallon, Pascal Ragouet, and Michael Samers, 33–46. Bordeaux: Maison des sciences de l'homme d'Aquitaine.

—. 2004a. "De l'objet spatial à l'effet géographique. Présentation Générale." In *L'effet géographique: Construction sociale, appréhension cognitive et configuration matérielle des objets géographiques*, edited by Bernard Debarbieux and Marie-Christine Fourny, 11–33. Grenoble: Maison des sciences de l'homme-Alpes.

—. 2004b. "The Symbolic Order of Objects and the Frame of Geographical Action." *GeoJournal* 60: 397–405.

—. 2006. "Prendre Position: Réflexions sur les ressources et les limites de la notion d'identité en géographie." *L'Espace Géographique* 4: 340–54.

—. 2008a. "Construits identitaires et imaginaires de la territorialité: variations autour de la figure du 'montagnard.'" *Annales de Géographie* 660–661: 90–115.

—. 2008b. "Mountains: Between Pure Reason and Embodied Experience. Philippe Buache and Alexander von Humboldt." In *High Places: Cultural Geographies of Mountains, Ice and Science*, edited by Denis Cosgrove and Veronica della Dora, 99–121. I.B. Tauris.

—. 2008c. "Cultures and Politics in the Present-Day Alps." *Journal of Alpine Research | Revue de Géographie Alpine*, no. 96–4 (December): 45–52. doi:10.4000/rga.590.

—. 2009. "Preface." Edited by Anne Sgard. *Journal of Alpine Research* 97 (2): 11–18.

Debarbieux, Bernard, and Gilles Rudaz. 2010. *Les faiseurs de montagne*. Paris: CNRS Editions.

Del Biaggio, Cristina. 2007. "Pan-Alpine Networks: The Case of Alliance in the Alps".
—. 2009a. "Point d'arrivée de l'Alpine Crossing." *Revue de Géographie Alpine* –. <http://rga.revues.org/index949.html>.
—. 2009b. "The Institutionalization of the Alpine Region: An Analysis Based on a Study of Two Pan-Alpine Networks (Alliance in the Alps and Alparc)." *Journal of Alpine Research* 97 (2): 99–115.
—. 2015. "Territory beyond the Anglophone tradition." In *The Wiley Blackwell Companion to political geography*, edited by John Agnew, Virginie Mamadouh, Anna Secor and Joanne Sharp, 2nd edition, Wiley-Blackwell: 35–47.
Delaney, David, and Helga Leitner. 1997. "The Political Construction of Scale." *Political Geography* 16 (2): 93–97.
Devriendt, Arthur. 2011. "Travaux d'agencement en géographie (Deleuze et les autres…)", blog *Technogéographie*, URL: <http://technogeo.hypotheses.org/370>.
Di Méo, Guy. 1998. *Géographie sociale et territoires*. Paris: Nathan.
—. 2002. "L'identité: Une médiation essentielle du rapport espace/société." *Géocarrefour* 77 (2): 175–84.
—. 2004. "Composantes spatiales, formes et processus géographiques des identités." *Annales de Géographie* 638–639: 339–62.
Diffenderfer, Mark, and Dean Birch. 1997. "Bioregionalism: A Comparative Study of the Adirondacks and the Sierra Nevada." *Society & Natural Resources* 10: 3–16.
Dornel, Laurent. 2002. "La frontière (le voisin) et l'étranger. Les enjeux identitaires d'un conflit frontalier." *Revue d'histoire du XIXe siècle* 24: –.
Dubar, Claude. 2000. *La crise des identités. L'interprétation d'une mutation*. Paris: Presses Universitaires de France.
Eisenhardt, Kathleen M. 1989. "Building Theories from Case Study Research." *Academy of Management Review* 14 (4): 532–50.
Escobar, Arturo. 2001. "Culture Sits in Places: Reflections on Globalism and Subaltern Strategies of Localization." *Political Geography* 20: 139–74.
Eysenck, Hans Jürgen. 1976. *Case Studies in Behaviour Therapy*. Routledge and Kegan Paul.
Fall, Juliet. 2006. "Lost Geographers: Power Games and the Circulation of Ideas within Francophone Political Geographies, Progress in Human Geography." *Progress in Human Geography* 30 (6): 195–216.
—. 2009. "Beyond Handshakes: Rethinking Cooperation in Transboundary Protected Areas as a Process of Individual and Collective Identity Construction." *Journal of Alpine Research* 92 (2): 73–84.
—. 2003. "Planning Protected Areas across Boundaries: New Paradigms and Old Ghosts." *Journal of Sustainable Forestry* 17 (1/2): 81–102.
—. 2005. *Drawing Lines, Constructing Spaces*. Hants: Ashgate.
—. 2007. "Designing Framework Conventions to Promote and Support Transboundary Protected Areas: Theory and Practice from the Carpathian Convention." In *Biodiversity Conservation and Protected Areas. The Italian and*

Ukrainian Legislation, edited by Gianfranco Tamburelli, 101–17. Milano: Giuffré.

—. 2011. "Natural Resources and Transnational Governance." In *The Ashgate Research Companion to Border Studies*, edited by Doris Wastl-Walter, 627–41. Ashgate.

Fitzthum, Gerhard. 2011. "Entretien avec Klaus Töpfer, 'N'attendons pas qu'une nouvelle dynamique nous tombe du ciel'." *Alpenscène* 95: 14–16.

Flyvbjerg, Bent. 2001. *Making Social Science Matter. Why Social Inquiry Fails and How It Can Succeed Again*. Cambridge: Cambridge University Press.

Fourny, Marie-Christine. 1999. "Affirmation identitaire et politiques territoriales des Villes Alpines." *Revue de Géographie Alpine* 87 (1): 171–80.

Frenkel, Stephan. 1994. "Old Theories in New Places? Environmental Determinism and Bioregionalism." *Professional Geographer* 46 (3): 289–95.

Gallois, Lucien. 1908. *Régions naturelles et moms de pays: étude sur la région parisienne*. Paris: Ed. du CTHS.

Gerbaux, Françoise. 2001. "La montagne entre nature, histoire et sociétés." *Revue de Géographie Alpine* 89 (2): 21–27.

Gerbaux, Françoise, and Jean-Paul Zuanon. 1995. "La voie étroite de la convention Alpine." *Revue de Géographie Alpine* 83 (2): 101–12.

Gervais-Lambony, Philippe. 2004. "De l'usage de la notion d'identité en géographie. Réflexions à partir d'exemples Sud-Africains." *Annales de Géographie* 113 (638–639): 469–88.

Gibbert, Michael, Winfried Ruigrok, and Barbara Wicki. 2008. "What Passes as a Rigourous Case Study?" *Strategic Management Journal* 29: 1465–74.

Gilbert, Anne. 1988. "The New Regional Geography in English and French-Speaking Countries." *Progress in Human Geography* 12: 208–28.

Gillet, François, and Hervé Gumuchian. 1996. "I – Quels objectifs pour le Forum-Alpin '96? / Ziele des AlpenForums '96 / Quali sono gli obiettivi del Forum-Alpino '96?" *Revue de Géographie Alpine* 84 (4): 9–12.

Giraut, Frédéric. 2008. "Conceptualiser le territoire." *Historiens & Géographes* 403 (Dossier 9: "Construire les territoires"): 57–67.

Gonon, Emmanuel, and Frédéric Lasserre. 2003. "Une critique de la notion de frontières artificielles à travers le cas de l'Asie Centrale." *Cahiers de Géographie Du Québec* 47 (132): 433–61.

Götz, Andreas. 2001. "Das Netzwerk Allianz in den Alpen. Aufbruch in den Alpengemeinden." *Jahrbuch des Vereins zum Schutz der Bergwelt* 66: 65–70.

Häkli, Jouni. 1998. "Cross-Border Regionalisation in the 'New Europe' – Theoretical Reflection with Two Illustrative Examples." *Geopolitics and International Boundaries* 3 (3): 83–103.

—. 2001. "The Politics of Belonging: Complexities of Identity in the Catalan Borderlands." *Geografiska Annaler* 83 B (3): 111–19.

—. 2004. "Governing the Mountains. Cross-Border Regionalization in Catalonia." In *Cross-Border Governance in the European Union*, edited by Oliver Kramsch and Barbara Hooper, 56–69. London and New York: Routledge.

—. 2005. "Who Is the Finn? Globalization and Identity in Finland." *Finnishness in Finland and North America: Constituents, Changes, and Challenges* 9 (2): 12–26.
—. 2008. "Re-Bordering Spaces." In *Handbook of Political Geography*, edited by Kevin Cox, Murray Low, and Jenny Robinson, 471–82. London: Sage.
Hamm, Bernd, Barbara Rasche. 2002. *Bioregionalismus. Ein Überblick*. Universität Trier.
Hamel, Jacques, Stéphane Dufour, and Dominic Fortin. 1993. *Case Study Methods*. Sage Publications.
Hannerz, Ulf. 1996. *Transnational Connections*. London: Routledge.
Harrison, John. 2012. "Configuring the New 'regional World': On Being Caught between Territory and Networks." *Regional Studies*, 1–20. doi:10.1080/00343404.2011.644239.
Harvey, David. 1993. "The Nature of Environment: The Dialectics of Social and Environmental Change." *The Social Register* 29: 1–51.
—. 1998. "What's Green and Makes the World Go Round?" In *The Cultures of Globalization*, edited by Frederic Jameson and Masao Miyoshi, 327–55. Durham; London: Duke University Press.
Holmén, Hans. 1995. "What's New and What's Regional in the 'new Regional Geography'?" *Geografiska Annaler* 77: 47–63.
Holsti, Ole R. 1969. *Content analysis for the social sciences and humanities*, Addison Wesley: Reading, MA.
Jenkins, Richard. 2000. "Categorization: Identity, Social Process and Epistemology." *Current Sociology* 48 (3): 7–25.
Jessop, Bob. 1995. "The Regulation Approach, Governance and Post-Fordism: Alternative Perspectives on Economic and Political Change?" *Economy and Society* 24 (3): 307–33.
Karkkainen, Bradley C. 2004. "Post-Sovereign Environmental Governance." *Global Environmental Politics* 4 (1): 72–96.
Keating, Michael. 2004. *Regions and Regionalism in Europe*. Edward Elgar.
Kohlbacher, Florian. 2006. "The Use of Qualitative Content Analysis in Case Study Research." *Forum Qualitative Sozialforschung / Forum: Qualitative Social Research*, 7, 1 [Art. 21], URL: <http://www.qualitative-research.net/index.php/fqs/article/%20view/75/153January%202006>.
Lalande, François. 2005. *Sauver la montagne*. Genève: Olizane.
Langlois, Simon. 2004. "Identity Movements." In *Encyclopedia of the Social and Behavioral Sciences*, edited by Neil J. Smelser and Paul B. Baltes, 7163–66. Oxford: Elsevier.
Laoureux, Sébastien. 2008. "Du pratique au théorique: La sociologie phénoménologique d'Alfred Schütz et la question de la coupure épistémologique." *Bulletin d'analyse phénoménologique* 4 (3): 169–88.
Leitner, Helga, Claire Pavlik, and Eric Sheppard. 2008. "Networks, Governance, and the Politics of Scale: Inter-Urban Networks and the European Union." In *Geographies of Power: Placing Scale*, edited by Andrew Herod and Melissa W. Wright, 274–303. Oxford: Blackwell Publishers Ltd.

Leitner, Helga, Eric Sheppard, and Kristin M. Sziarto. 2008. "The Spatialities of Contentious Politics." *Transactions of the Institute of British Geographers* 33 (2): 157–72.
Lussault, Michel. 2003. "Constructivisme." In *Dictionnaire de la géographie*, edited by Jacques Lévy, Michel Lussault, 200–203. Paris: Belin.
MacLeod, Gordon. 1998. "In What Sense a Region? Place Hybridity, Symbolic Shape, and Institutional Formation in (post-)modern Scotland." *Political Geography* 17 (7): 833–63.
—. 2001. "New Regionalism Reconsidered: Globalization and the Remaking of Political Economic Space." *International Journal of Urban and Regional Research* 25 (4): 804–29.
MacLeod, Gordon, and Mark Goodwin. 1999. "Space, Scale and State Strategy: Rethinking Urban and Regional Governance." *Progress in Human Geography* 23 (4): 503–27.
Mamadouh, Virginie, and Herman van der Wusten. 2008. "The European Level in EU Governance: Territory, Authority and Trans-Scalar Networks." *GeoJournal* 72: 19–31.
Marston, Sallie A. 2000. "The Social Construction of Scale." *Progress in Human Geography* 24 (2): 219–42.
Marston, Sallie A., John Paul Jones, and Keith Woodward. 2005. "Human Geography without Scale." *Transactions of the Institute of British Geographers* 30 (4): 416–32.
Martin, Denis-Constant. 1994. "Identités et politique. Récit, mythe et idéologie." In *Cartes d'identité. Comment dit-on "nous" en politique?*, edited by Denis-Constant Martin, 304. Paris: Presses de la Fondation nationale des sciences politiques.
Massey, Doreen, and John Allen, eds. 1984. *Geography Matters! A Reader*. Cambridge: Cambridge University Press.
McCann, Eugene, and Kevin Ward. 2010. "Relationality/territoriality: Toward a Conceptualization of Cities in the World." *Geoforum* 41: 175–84.
McMillan, David W., and David M. Chavis. 1986. "Sense of Community: A Definition and Theory." *Journal of Community Psychology* 14 (January): 6–23.
McNeill, Donald. 2004. *New Europe: Imagined Spaces*. London: Hodder Arnold.
Mens, Yann. 2008. "La Méditerranée Est Une Invention Récente." *Alternatives Internationales*, 20–23.
Messerli, Paul. 2010. "Contours and Perspectives for a European Macro-Region 'Alpine Space.'" In *Metropolises and "their" Alps*, edited by Michal Didier, Peter Eggensberger, Silvia Jost, Thomas Scheurer, Dirk Ahner, Christian Ude, Nicolas Evrard, Paul Messerli, and Monika Seynsche, 120–21. Munich: Bayerische Akademie der Wissenschaften. http://epub.oeaw.ac.at/?arp=0x00246f2d.
Michel, Yves, ed. 2007. *Nous les Alpes! Des femmes et des hommes façonnent l'avenir. 3ème rapport sur l'état des Alpes*. CIPRA.
Mintzberg, Henry. 1979. "An Emerging Strategy of 'direct' Research." *Administrative Science Quarterly* 24: 582–89.

Murphy, Alexander B. 1991. "Regions as Social Constructs: The Gap between Theory and Practice." *Progress in Human Geography* 15 (1): 22–35.

Nebe, Tina Maria. 2003. "Le Rôle de l'altérité dans les processus de re-présentation et identification sociale." *Journal International sur les représentations sociales* 1 (1): 1–16.

Nordman, Daniel. 1998. *Frontières de France. De l'espace au territoire XVIe–XIXe siècle*. Paris: Gallimard.

Norer, Roland. 2002. *Die Alpenkonvention. Völkerrechtliches Vertragswerk für den Alpenraum*. Vienne: Institut für Wirtschaft, Politik und Recht. Universität für Bodenkultur Wien. <www.boku.ac.at/wpr/wpr_dp/dp-93.pdf>.

Olsen, Jonathan. 2000. "The Perils of Rootedness: On Bioregionalism and Right Wing Ecology in Germany." *Landscape Journal* 19 (1/2): 73–83.

Oneto, Gilberto. 1996. "Il Po è un drago." *Quaderni Padani* 2 (6): 1–5.

Orlikowski, Wanda J., and Jack J. Baroudi. 1991. "Studying Information Technology in Organizations: Research Approaches and Assumptions." *Information Systems Research* 2 (1): 1–28.

Ozouf-Marignier, Marie-Vic. 1992. *La formation des départements en France. La représentation du territoire français à la fin du 18ème siècle*. Paris: Ecole des Hautes Etudes en Sciences Sociales.

Paasi, Anssi. 1986. "The Institutionalization of Regions: A Theoretical Framework for Understanding the Emergence of Regions and the Constitution of Regional Identity." *Fennia* 164: 105–46.

—. 1991. "Deconstructing Regions: Notes on the Scales of Spatial Life." *Environment and Planning A* 23: 239–56.

—. 2001. "Europe as a Social Process and Discourse. Considerations of Place, Boundaries and Identity." *European Urban and Regional Studies* 8 (1): 7–28.

—. 2002a. "Bounded Spaces in the Mobile World: Deconstructing 'regional Identity.'" *Tijdschrift Voor Economische En Sociale Geografie* 93 (2): 137–48.

—. 2002b. "Place and Region: Regional Worlds and Words." *Progress in Human Geography*, no. 6: 802–11.

—. 2002c. "Regional Transformation in the European Context: Notes on Regions, Boundaries and Identity." *Space & Polity* 6 (2): 197–201.

—. 2003. "Region and Place: Regional Identity in Question." *Progress in Human Geography* 27 (4): 475–85.

—. 2009. "Resurgence of the 'Region' and 'Regional Identity': Theoretical Perspectives and Empirical Observations on Regional Dynamics in Europe." *Review of International Studies* 35 (1): 119–44.

Painter, Joe. 2008. "Cartographic Anxiety and the Search for Regionality." *Environment and Planning A* 40 (2): 342–61.

Perkmann, Markus. 1999. "Building Governance Institutions across European Borders." *Regional Studies* 33 (7): 657–67.

Perruchoud, Michaël. 2010. "Le Derby Tessinois, ou David contre Goliath." *Le Temps*.

Petite, Mathieu. 2009. "Identités en chantiers dans les Alpes: des projets qui mobilisent objets, territoires et réseaux". Université de Genève.

Phillips, Nelson, and Cynthia Hardy. 2002. *Discourse Analysis. Investigating Processes of Social Construction*. Vol. 50. Sage.
Poche, Bernard. 1996. *L'espace fragmenté. Eléments pour une analyse sociologique de la territorialité*. Paris: L'Harmattan.
Pohl, J. 2004. "Regional Identity." In *International Encyclopedia of the Social & Behavioral Sciences*, edited by Neil J. Smelser and Paul B. Baltes, 12917–22. Amsterdam: Elsevier.
Pred, Allan. 1984. "Place as Historically Contingent Process: Structuration and the Time-Geography of Becoming Places." *Annals of the Association of American Geographers* 74: 279–97.
Price, Martin. 2000. "The Alpine Convention: A Model for Other Mountain Regions?" *Mountain Research and Development* 20 (2): 192–94.
Price, Martin F. 1999. "Towards Co-Operation across Mountain Frontiers: The Alpine Convention." *European Environment* 9: 83–89.
Pudup, M.B. 2004. "Geographical Region." In *International Encyclopedia of the Social & Behavioral Sciences*, edited by Neil J. Smelser and Paul B. Baltes, 12905–8. Amsterdam: Elsevier.
Pudup, Mary Beth. 1988. "Arguments within Regional Geography." *Progress in Human Geography* 12: 369–90.
Raagmaa, Garri. 2002. "Regional Identity in Regional Development Planning." *European Planning Studies* 10 (1): 55–76.
Raffestin, Claude, and Ruggero Crivelli. 1992. "Blanche Neige et les Sept Nains ou la Transformation des Alpes en Patrimoine Commun." *Revue de Géographie Alpine* 80 (4): 213–27.
Reichenbach, Hans. 1938. *Experience and Prediction; an Analysis of the Foundations and the Structure of Knowledge*. Chicago: The University of Chicago Press.
Rocha Medeiros, Eduardo José. 2010. "Old vs Recent Cross-Border Cooperation: Portugal-Spain and Norway-Sweden." *Area* 42 (4): 434–43.
Ruby, Christian. 2003. "Réalisme." In *Dictionnaire de la Géographie*, edited by Jacques Lévy and Michel Lussault, 765–66. Belin.
Rudaz, Gilles. 2005. "Porter la voix de la montagne: objectivation et différenciation du territoire par le groupement de la population de montagne du Valais Romand (1945–2004)". Département de géographie, Université de Genève.
Rumford, Chris. 2006. "Theorizing Borders. Introduction." *European Journal of Social Theory* 9 (2): 155–69.
Rumiz, Paolo. 2007. *La Leggenda dei Monti Naviganti*. Feltrinelli.
Sacareau, Isabelle. 2003. *La montagne, une approche géographique*. Paris: Belin.
Sale, Kirkpatrick. 1985. *Dwellers in the Land: The Bioregional Vision*. San Francisco: Sierra Club Books.
Sassen, Saskia. 2003. "Globalization or Denationalization?" *Review of International Political Economy* 10 (1): 1–22.
Sayer, Andrew. 1989. "The 'New' Regional Geography and Problems of Narrative." *Environment and Planning D: Society and Space* 7: 253–76.
Schütz, Alfred. 1954. "Concept and Theory Formation in the Social Sciences." *Journal of Philosophy* 51 (9): 257–73.

Scott, James. 2003. "Cross-Border Governance in the Baltic Sea Region." In *New Borders for a Changing Europe: Cross-Border Cooperation and Governance*, edited by James Anderson, Liam O'Dowd, and Thomas M. Wilson, 135–53. London: F. Cass.

Shields, Rob. 1992. *Places on the Margin: Alternative Geographies of Modernity*. London, New York: Routledge.

Siegrist, Dominik. 2011. "Editorial." *Alpenscène* 95: 2.

Simon, Claire. 2011. "Patient Convention Alpine: Quel Traitement?" *Alpenscène* 95: 17–18.

Simonis, Udo E. 1997. *Bioregionalism. A Pragmatic European Perspective*. Wissenschaftszentrum Berlin. <http://bibliothek.wz-berlin.de/pdf/1997/ii97-407.pdf>.

Snow, David A. 2004. "Collective Identities and Expressive Forms." In *International Encyclopedia of the Social and Behavioural Sciences*, edited by Neil J. Smelser and B. Baltes Paul, 2212–19. Oxford: Elsevier.

Soriga, Flavio. 2008. *Sardinia Blues*. Bompiani.

Strüver, Anke. 2004. "'We Are Only Allowed to Re-Act, Not to Act'. Eurocrat's Strategies and Borderlanders' Tactics in a Dutch-German Cross-Border Region." In *Cross-Border Governance in the European Union*, edited by Olivier Kramsch and Barbara Hooper, 25–40. Routledge.

—. 2005. *Stories of the "Boring Border": The Dutch-German Borderscape in People's Minds*. Lit Verlag.

Swyngedouw, Erik. 2004. "Globalisation or 'Glocalisation'? Networks, Territories and Rescaling." *Cambridge Review of International Affairs* 17 (1): 25–48.

Taylor, P. J. 1991. "A Theory and Practice of Regions: The Case of Europas." *Environment and Planning D: Society and Space* 9: 183–95.

Thrift, Nigel J. 1983. "On the Determination of Social Action in Space and Time." *Environment and Planning D: Society and Space* 1: 23–57.

—. 1990. "For a New Regional Geography 1." *Progress in Human Geography* 14 (3): 272–79.

—. 1991. "For a New Regional Geography 2." *Progress in Human Geography* 15 (4): 456–65.

—. 1998. "Towards a New Regional Geography." *Berichte zur deutschen Landeskunde* 72: 37–46.

Touraine, Alain. 1984. *Le retour de l'acteur*. Fayard.

Tuan, Yi-Fu. 1977. *Space and Place: The Perspective of Experience*. Minneapolis: University of Minnesota Press.

Vertovec, Steven. 2001. "Transnationalism and Identity." *Journal of Ethnic and Migration Studies* 27 (4): 573–82.

Virtanen, Petri. 2004. "Euregios in Changing Europe. Euregio Karelia and Euroregion Pomerania as Examples." In *Cross-Border Governance in the European Union*, edited by Olivier Kramsch and Barbara Hooper, 121–34. New York: Routledge.

Vorum. 2005. "Nutzen der Alpenallianz in Ihrer Gemeinde?" *Vorum, Forum für Gemeindeentwicklung in Vorarlberg* 4: 4.

Wülser, Barbara. 2011. "La convention sert celui qui s'en sert." *Alpenscène* 95: 8–9.

Index

www.ingramcontent.com/pod-product-compliance
Ingram Content Group UK Ltd.
Pitfield, Milton Keynes, MK11 3LW, UK
UKHW021015120926
8556IPUK00040B/1152

9 783034 316309